T0315045

Modular Multilevel Converters

IEEE Press
445 Hoes Lane
Piscataway, NJ 08854

Modular Multilevel Converters

Control, Fault Detection, and Protection

Fujin Deng
Southeast University
Nanjing, China

Chengkai Liu
Southeast University
Nanjing, China

Zhe Chen
Aalborg University
Aalborg, Denmark

Published by John Wiley & Sons, Inc., Hoboken, New Jersey.
Published simultaneously in Canada.

Library of Congress Cataloging-in-Publication Data
Names: Deng, Fujin, author. | Liu, Chengkai, author. | Chen, Zhe, author.
Title: Modular multilevel converters : control, fault detection, and protection / Fujin Deng, South East University, Nanjing, China, Chengkai Liu, South East University, Nanjing, China, Zhe Chen, Aalborg University, Aalborg, DA, Denmark.
Description: Hoboken, New Jersey : Wiley-IEEE Press, [2023] | Series: IEEE Press series on power and energy systems | Includes bibliographical references and index.
Identifiers: LCCN 2022052863 (print) | LCCN 2022052864 (ebook) | ISBN 9781119875604 (cloth) | ISBN 9781119875611 (adobe pdf) | ISBN 9781119875628 (epub)
Subjects: LCSH: Electric current converters. | Modularity (Engineering)
Classification: LCC TK7872.C8 D455 2023 (print) | LCC TK7872.C8 (ebook) | DDC 621.3815/322–dc23/eng/20230103
LC record available at https://lccn.loc.gov/2022052863
LC ebook record available at https://lccn.loc.gov/2022052864

Cover Design: Wiley
Cover Images: © KAY4YK/Shutterstock

Set in 9.5/12.5pt STIXTwoText by Straive, Pondicherry, India

Contents

About the Authors

Fujin Deng received the PhD degree in Energy Technology from the Department of Energy Technology, Aalborg University, Aalborg, Denmark, in 2012. From 2013 to 2015 and from 2015 to 2017, he was a postdoctoral researcher and an assistant professor, respectively, in the Department of Energy Technology, Aalborg University, Aalborg, Denmark. He joined Southeast University in 2017 as a professor in the School of Electrical Engineering, Southeast University, Nanjing, China. He also serves as the Head of the Department of Power Electronics in the School of Electrical Engineering, Southeast University.

Dr. Deng has published more than 110 peer-reviewed journal articles and more than 60 conference papers. And he holds more than 50 issued and pending patents. He is a senior member of the IEEE. His main research interests include modular multilevel converters (MMCs), high-voltage direct current (HVDC) technology, DC/DC solid state transformer, fault diagnosis and tolerant control for power converters, DC grid control, and DFIG/PMSG-based wind turbine/farm modeling and control.

Chengkai Liu received the BEng degree in Electrical Engineering from Chien-Shiung WU College of Southeast University, Nanjing, China, in 2018, where he is currently working toward the PhD degree with the School of Electrical Engineering. He was a guest PhD student in the Department of Energy Technology, Aalborg University, Aalborg, Denmark from 2021 to 2022.

Mr. Liu has published 17 peer-reviewed journal articles and held 6 issued Chinese patents. His main research interests include modular multilevel converters (MMCs), DC grids, fault detection, and DC fault protection. He has participated in several research programs, and his inventions enhance the reliability of the MMC system and reduce the power losses of the converter.

Mr. Liu received a national scholarship for his graduate studies (PhD) from the Chinese Ministry of Education in 2021 and the National First Prize in the National Undergraduate Electronic Design Contest in 2017. He serves as a reviewer for *IEEE Transactions on Power Electronics, IEEE Transactions on Power Delivery, IEEE Journal of Emerging and Selected Topics in Power Electronics, IEEE Transactions on Circuits and Systems II*, and *CSEE Journal of Power and Energy Systems.*

Zhe Chen received his BEng and MSc degrees in Electrical Engineering (power plants and power system automation) from the Northeast China Institute of Electric Power Engineering, Jilin City, China in 1982 and 1986, respectively; MPhil in Power Electronics, from Staffordshire University, England, in 1993; and the PhD degree in Power and Control from the University of Durham, England, in 1997.

Dr. Chen worked as a lecturer and then a senior lecturer in De Montfort University, Leicester, England. He has been a full professor in the Department of Energy Technology, Aalborg University, Denmark, since 2002. He is the leader of Wind Power System Research program at the Department of Energy Technology, Aalborg University. His main research interests are wind power, power electronics, power systems, and multi-energy systems.

In these areas, he has led and participated in many international and national research projects, has supervised many PhDs, postdoctoral researchers, visiting scholars, and visiting PhDs, and has more than 1000 technical publications. He is a panel member or review expert for many international funding organizations. Dr. Chen serves as a member of editorial boards for many international journals, including Associate Editor of the *IEEE Transactions on Power Electronics*, Deputy Editor of *IET Renewable Power Generation* (Wind Turbine Technology and Control), and Editor-in-Chief for the MDPI *Wind*.

Dr. Chen is a Fellow of IEEE, a Fellow of the Institution of Engineering and Technology (IET, London, UK), a Chartered Engineer in the United Kingdom, and a member of the Danish Academy of Technical Sciences.

Preface

With the rapid development of high-power semiconductor devices, power converters have been widely used in electric power conversion systems, converting electric power with high efficiency and economic benefit. The modular multilevel converter (MMC) is considered as one of the most promising converters for medium-/high-voltage and high-power applications because of its superior advantages, such as excellent output power quality, high efficiency, modularity, and scalability.

As increased MMCs are put into practical application, the concerns in terms of reliability and fault protection emerge to front stage. The semiconductors devices and capacitor are two most fragile components in the submodule of the MMC whose failure brings troubles to reliable operation of the MMC. And the MMC is usually required to have an uninterruptable operation ability in case of submodule malfunction. In addition, the AC-side grid fault and DC-side fault also pose challenges to the MMC system.

First, this book provides a brief review of the MMC basic principle and its control method in Chapters 1 and 2. Then, the insulated gate bipolar transistor fault detection methods and capacitor monitoring methods are respectively covered in Chapters 3 and 4. Chapter 5 offers fault-tolerant operation under submodule faults. Finally, Chapter 6 presents the control of MMCs under AC grid faults, and Chapter 7 explores the protection under DC short-circuit fault of the MMC system.

1

Modular Multilevel Converters

1.1 Introduction

Power converters have been widely used in electric power conversion systems that can convert electric power with high conversion efficiency and economic benefits. In the past few decades, several power converters have been developed and commercialized in the industrial community. They can be classified into current source converters (CSCs) and voltage source converters (VSCs).

The modular multilevel converter (MMC), as a kind of VSC, has been considered one of the most promising converters for medium/high-voltage and high-power applications, with the superiority such as excellent output power quality, high efficiency, modularity, and scalability [1–3]. Compared with traditional two-level and multilevel converters, the MMC has prominent advantages such as high efficiency and low power losses. What is more, high modularity enables the MMC to easily expand capacity. Recently, the MMC has become attractive for high-voltage direct-current (HVDC) transmission [4], renewable energy generation [5], electric railway supplies [6], micro power grid [7], etc.

This chapter deals with the fundamental principles of the MMC. The configuration of the MMC is presented in Section 1.2. Section 1.2 gives an introduction of the converter configuration and the submodule (SM) configuration. The operation principles of the SM and the converter are shown in Section 1.3. Section 1.4 demonstrates the modulation schemes, including phase-disposition (PD) pulse width modulation (PWM), phase-shifted (PS) PWM, and nearest level modulation (NLM). Afterward, mathematical models of the SM, the arm, and the MMC are introduced in Section 1.5. Then, Section 1.6 introduces the parameter design methods of power devices, capacitor, and arm inductor in the MMC. An overview of the MMC faults is given in Section 1.7.

Modular Multilevel Converters: Control, Fault Detection, and Protection, First Edition.
Fujin Deng, Chengkai Liu, and Zhe Chen.

Published 2023 by John Wiley & Sons, Inc.

1.2 MMC Configuration

1.2.1 Converter Configuration

A three-phase MMC topology is shown in Figure 1.1, which consists of three phase-legs, each composed of an upper arm and a lower arm. Each arm contains n identical SMs and an arm inductor L_s. The upper and lower arm currents are i_{ju} and i_{jl}, respectively, in phase j ($j = a, b, c$). The AC side of the MMC is equipped with the filter inductor L_f. The MMC links the DC side and the three-phase AC side. Electric power can flow from the DC side to the AC side or from the AC side to the DC side through changing the operation mode of the MMC. The number of SMs can affect the output voltage level and thus power quality. A multilevel voltage would be synthesized at the AC side of the MMC, and output harmonic contents would be reduced with the cascading of SMs [8].

1.2.2 Submodule Configuration

The SM unit is the fundamental component of the MMC. Figure 1.2 shows the typical half-bridge (HB) SM configuration, which is widely used in power

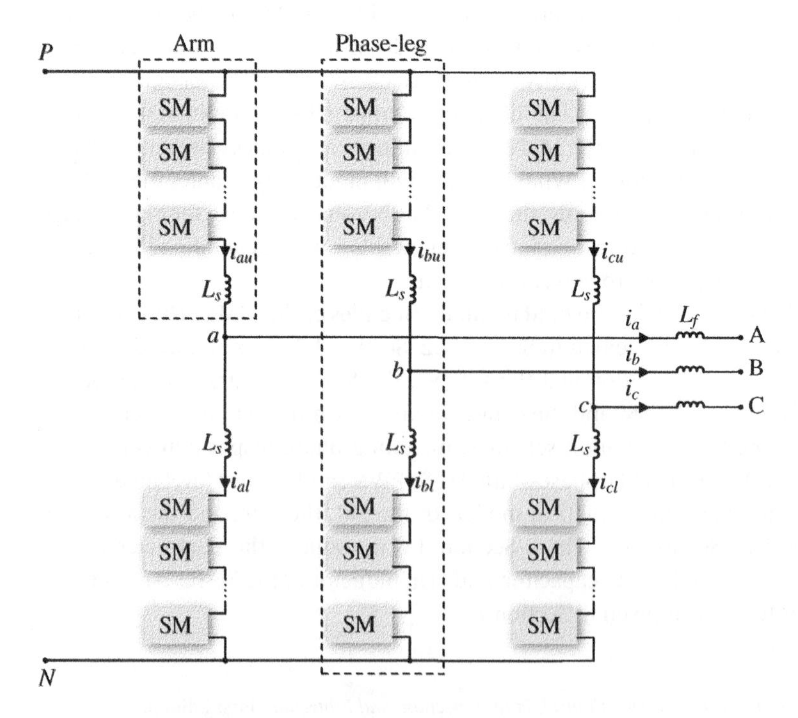

Figure 1.1 Three-phase MMC configuration.

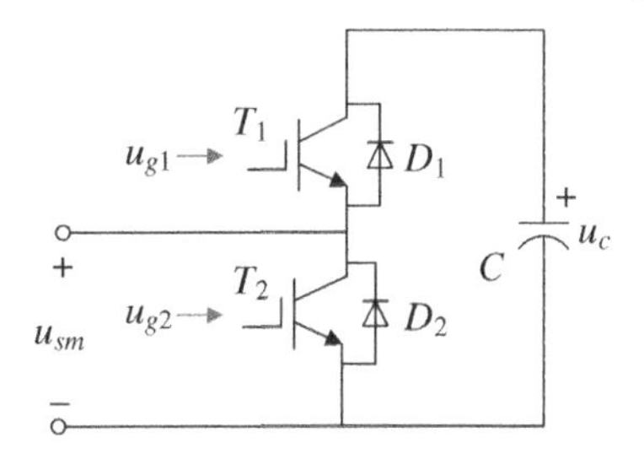

Figure 1.2 Half-bridge SM.

applications. The HB SM consists of two switches/diodes T_1/D_1 and T_2/D_2 and a DC capacitor C. Through the switching of the power switches T_1 and T_2, the HB SM can output two voltage levels, including the capacitor voltage u_c and 0 [8].

1.3 Operation Principles

1.3.1 Submodule Normal Operation

In the normal operation of the MMC, the operation state of the SM is controlled by the switching function S, which is defined as

$$S = \begin{cases} 1, & u_{g1} \text{ is high-level and } u_{g2} \text{ is low-level} \\ 0, & u_{g1} \text{ is low-level and } u_{g2} \text{ is high-level} \end{cases} \tag{1.1}$$

where u_{g1} and u_{g2} are the drive voltages of switches T_1 and T_2, respectively, as shown in Figure 1.2. The switching modes of the HB SM are listed in Table 1.1.

- The T is turned on when the u_g is high-level
- The T is turned off when the u_g is low-level.

The relationship between the SM's output voltage u_{sm} and the SM's capacitor voltage u_c is

$$u_{sm} = S \cdot u_c \tag{1.2}$$

The normal operation of the HB SM has four operation modes, as shown in Table 1.2 and Figure 1.3, which are decided by the switching function S and the direction of the arm current i_{arm}, as follows:

Table 1.1 Switching modes of the HB SM.

S	u_{g1}	u_{g2}	T_1	T_2
1	High-level	Low-level	On	Off
0	Low-level	High-level	Off	On

Table 1.2 Four normal operation modes of the HB SM.

Mode	S	i_{arm}	Circuit state	SM state	T_1	T_2	u_{sm}	C	u_c
1	1	>0	Inserted	On	On	Off	u_c	Charged	Increased
2		<0	Inserted	On	On	Off	u_c	Discharged	Reduced
3	0	>0	Bypassed	Off	Off	On	0	Bypassed	Unchanged
4		<0	Bypassed	Off	Off	On	0	Bypassed	Unchanged

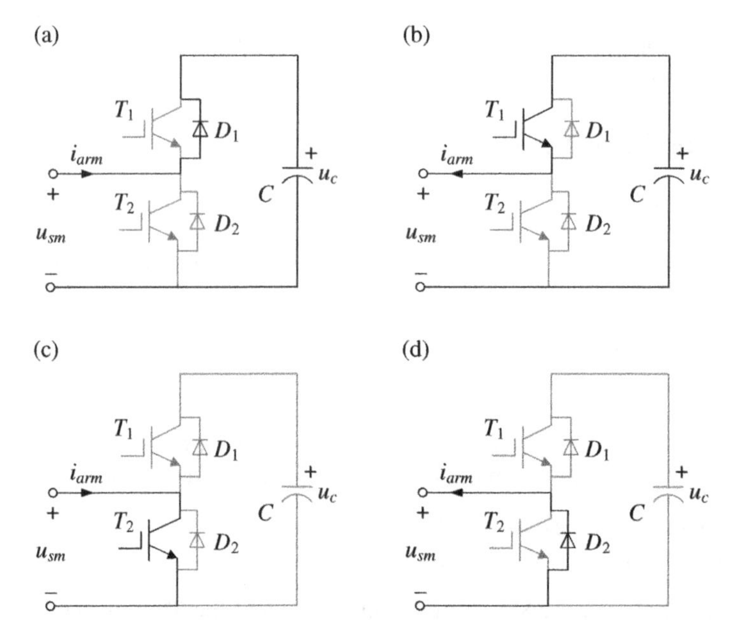

Figure 1.3 Four normal operation modes of the HB SM. (a) Mode 1 (normal operation). (b) Mode 2 (normal operation). (c) Mode 3 (normal operation). (d) Mode 4 (normal operation).

- *Mode 1 (normal operation)*: When $S = 1$ and $i_{arm} > 0$, as shown in Figure 1.3a, the T_1 is turned on, the T_2 is turned off, the SM is inserted into the arm circuit, the SM state is on, and the SM output voltage u_{sm} is equal to the SM capacitor voltage u_c. In this case, the SM capacitor C is charged by i_{arm}, and the capacitor voltage u_c is increased.
- *Mode 2 (normal operation)*: When $S = 1$ and $i_{arm} < 0$, as shown in Figure 1.3b, the T_1 is turned on, the T_2 is turned off, the SM is inserted into the arm circuit, the SM state is on, and the SM output voltage u_{sm} is equal to the SM capacitor

voltage u_c. In this case, the SM capacitor C is discharged by i_{arm}, and the capacitor voltage u_c is reduced.

- *Mode 3 (normal operation)*: When $S = 0$ and $i_{arm} > 0$, as shown in Figure 1.3c, the T_1 is turned off, the T_2 is turned on, the SM is bypassed from the arm circuit, the SM state is off, and the SM output voltage u_{sm} is equal to 0. In this case, the SM capacitor C is bypassed, and the capacitor voltage u_c is unchanged.
- *Mode 4 (normal operation)*: When $S = 0$ and $i_{arm} < 0$, as shown in Figure 1.3d, the T_1 is turned off, the T_2 is turned on, the SM is bypassed from the arm circuit, the SM state is off, and the SM output voltage u_{sm} is equal to 0. In this case, the SM capacitor C is bypassed, and the capacitor voltage u_c is unchanged.

1.3.2 Submodule Blocking Operation

In the blocking operation of the HB SM, the drive voltages u_{g1} and u_{g2} are both low-level and the switches T_1 and T_2 are both turned off, which has two operation modes and is decided by the arm current i_{arm}, as shown in Table 1.3 and Figure 1.4, as follows:

- *Mode 1 (blocking operation)*: When T_1 and T_2 are both turned off and $i_{arm} > 0$, as shown in Figure 1.4a, the SM is inserted into the arm circuit, the SM state is on, and the SM's output voltage u_{sm} is u_c. In this case, the capacitor C is charged by i_{arm}, and the capacitor voltage u_c is increased.

Table 1.3 Two blocking operation modes of the HB SM.

Mode	i_{arm}	Circuit state	SM state	T_1	T_2	u_{sm}	C	u_c
1	>0	Inserted	On	Off	Off	u_c	Charged	Increased
2	<0	Bypassed	Off	Off	Off	0	Bypassed	Unchanged

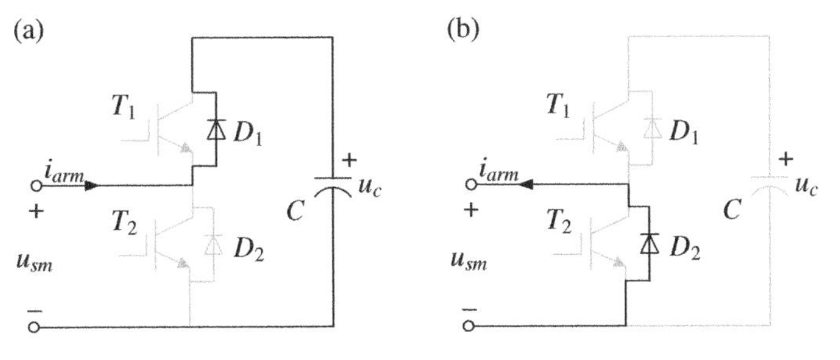

Figure 1.4 Two blocking operation modes of the HB SM. (a) Mode 1 (blocking operation). (b) Mode 2 (blocking operation).

- *Mode 2 (blocking operation)*: When T_1 and T_2 are both turned off and $i_{arm} < 0$, as shown in Figure 1.4b, the SM is bypassed from the arm circuit, the SM state is off, and the SM's output voltage u_{sm} is 0. In this case, the capacitor C is bypassed, and the capacitor voltage u_c is unchanged.

1.3.3 Converter Operation

The MMC generates the multilevel stepped waveform at its AC side by controlling the number of inserted SMs in the arm. To get an intuitive understanding of the operation principle of the MMC, Figure 1.5a shows an example of the upper arm of phase A, where four HB SMs (SM1–SM4) per arm are considered for the MMC. Here, all SMs' capacitor voltages are supposed to be the same as u_c, and the output voltages of the SM1–SM4 are u_{sm_au1}, u_{sm_au2}, u_{sm_au3}, and u_{sm_au4}, respectively. The waveform of the upper arm voltage u_{au} in phase A in one fundamental period is shown in Figure 1.5b, which is the sum of all SMs' output voltages u_{sm_au1}, u_{sm_au2}, u_{sm_au3}, and u_{sm_au4} in the upper arm of phase A. The switching functions of the SM1–SM4 and the corresponding output voltages of the SM1–SM4 are shown in Table 1.4, where the switching functions of

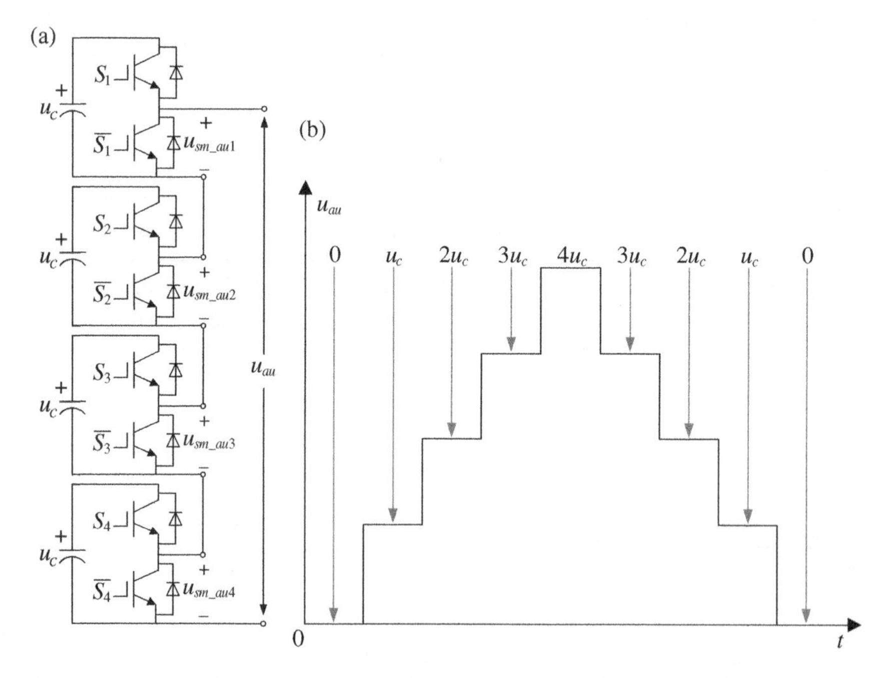

Figure 1.5 Operation principle of the MMC. (a) Upper arm of phase A. (b) Upper arm voltage of phase A.

Table 1.4 Switching functions and output voltages of SMs.

n_{on}	S_{au1}	S_{au2}	S_{au3}	S_{au4}	u_{sm_au1}	u_{sm_au2}	u_{sm_au3}	u_{sm_au4}	u_{au}
0	0	0	0	0	0	0	0	0	0
1	1	0	0	0	u_c	0	0	0	u_c
	0	1	0	0	0	u_c	0	0	
	0	0	1	0	0	0	u_c	0	
	0	0	0	1	0	0	0	u_c	
2	1	1	0	0	u_c	u_c	0	0	$2u_c$
	1	0	1	0	u_c	0	u_c	0	
	1	0	0	1	u_c	0	0	u_c	
	0	1	1	0	0	u_c	u_c	0	
	0	1	0	1	0	u_c	0	u_c	
	0	0	1	1	0	0	u_c	u_c	
3	1	1	1	0	u_c	u_c	u_c	0	$3u_c$
	1	1	0	1	u_c	u_c	0	u_c	
	1	0	1	1	u_c	0	u_c	u_c	
	0	1	1	1	0	u_c	u_c	u_c	
4	1	1	1	1	u_c	u_c	u_c	u_c	$4u_c$

SM1–SM4 are S_{au1}, S_{au2}, S_{au3}, and S_{au4}, respectively. The arm voltage u_{au} has five different levels including 0, u_c, $2u_c$, $3u_c$, and $4u_c$, which is expressed as

$$u_{au} = \sum_{i=1}^{4} u_{sm\_aui} = \sum_{i=1}^{4} \left(S_{aui} \cdot u_c\right) = n_{on} \cdot u_c \tag{1.3}$$

with

$$n_{on} = \sum_{i=1}^{4} S_{aui} \tag{1.4}$$

where n_{on} is the number of inserted SMs in the arm of the MMC. The arm voltage u_{au} depends on different switching combinations corresponding to the SMs in the arm, as shown in Figure 1.5 and Table 1.4.

- *Fifth level ($u_{au} = 4u_c$)*: The highest arm voltage $4u_c$ is generated when all four switching functions S_{au1}, S_{au2}, S_{au3}, and S_{au4} corresponding to the SM1–SM4 are all switched to 1.

- *Forth level ($u_{au}=3u_c$)*: The arm voltage $3u_c$ is generated when three out of four switching functions S_{au1}, S_{au2}, S_{au3}, and S_{au4} corresponding to the SM1–SM4 are switched to 1.
- *Third level ($u_{au}=2u_c$)*: The arm voltage $2u_c$ is generated when two out of four switching functions S_{au1}, S_{au2}, S_{au3}, and S_{au4} corresponding to the SM1–SM4 are switched to 1.
- *Second level ($u_{au}=u_c$)*: The arm voltage u_c is generated when one out of four switching functions S_{au1}, S_{au2}, S_{au3}, and S_{au4} corresponding to the SM1–SM4 is switched to 1.
- *First level ($u_{au}=0$)*: The lowest arm voltage 0 is generated when all four switching functions S_{au1}, S_{au2}, S_{au3}, and S_{au4} corresponding to the SM1–SM4 are all switched to 0.

1.4 Modulation Scheme

Modulation is a technique that produces the desired voltage in the arm or at the AC side of the MMC by controlling the drive voltage of switching devices in the MMC. It affects not only the MMC's external performance, such as AC-side voltage harmonics and AC-side current harmonics, but also the internal characteristics, such as capacitor voltage fluctuation, distribution of energy, and power losses distribution among SMs [9]. Currently, there are three modulation schemes commonly used in the MMC, which are PD-PWM, PS-PWM, and NLM, as shown in Table 1.5. Based on the modulation and the reference y for the arm of the MMC, the number n_{on} of SMs to be inserted into the arm of the MMC can be decided, as shown in Figure 1.6.

Table 1.5 Three modulation schemes [9–16].

Modulation scheme	Characteristics
PD-PWM	• High switching frequency modulation • Unevenly distributed pulses • Suitable for MMCs with not so many SMs per arm
PS-PWM	• High switching frequency modulation • Evenly distributed pulses • Suitable for MMCs with not so many SMs per arm
NLM	• Low switching frequency modulation • Easy for implementation • Suitable for MMCs with large number of SMs per arm

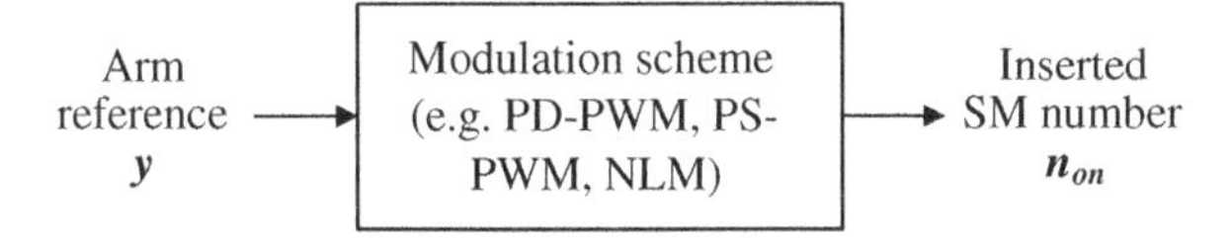

Figure 1.6 Modulation for MMCs.

1.4.1 Phase-Disposition PWM

The PD-PWM is typically suitable for the MMC with not so many SMs per arm [9–11]. For the MMC with n SM per arm, it is realized by applying n number of identical triangular carriers W_1–W_n stacked evenly in the vertical direction between −1 and 1. Through the comparison between the carriers and the reference signal y, the PD-PWM can be produced.

Figure 1.7 illustrates the implementation principle of the PD-PWM ($n = 4$) for the MMC with n SMs per arm. The n carriers W_1–W_n are at the same phase angle. The carrier frequency is f_s. The height Δh of each carrier is equal to $2/n$.

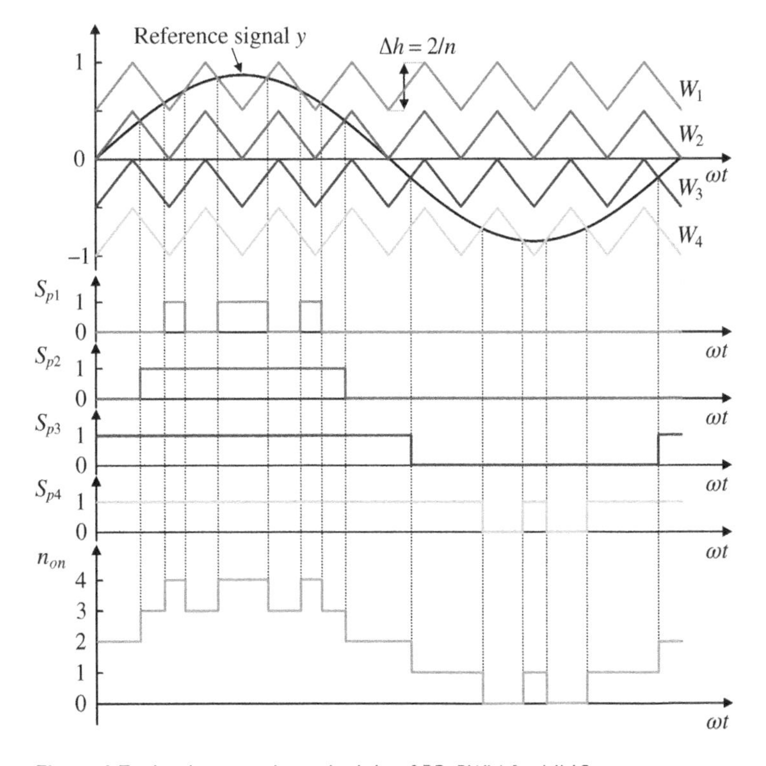

Figure 1.7 Implementation principle of PD-PWM for MMCs.

The reference signal y is compared with the carriers W_1–W_n to generate the pulses S_{p1}–S_{pn}, respectively, as follows:

- The S_{pi} is 1 if the reference y is higher than the carrier W_i ($i = 1, 2, \ldots, n$)
- The S_{pi} is 0 if the reference is lower than the carrier W_i ($i = 1, 2, \ldots, n$)

The total number n_{on} of SMs to be inserted into each arm at each instant can be expressed as the sum of S_{p1}–S_{pn}, as

$$n_{on} = \sum_{i=1}^{n} S_{pi} \tag{1.5}$$

Figure 1.7 shows that the PD-PWM results in an unevenly distributed switching frequency among S_{p1}–S_{pn}. The value of n_{on} varies in the range of 0–n in one fundamental period, which achieves the multilevel synthesized waveform. In addition, it is apparent that the switching actions of S_{p1}–S_{pn} are affected by the amplitude of the reference signal y.

The typical characteristics of the PD-PWM are summarized in Table 1.5.

1.4.2 Phase-Shifted PWM

The PS-PWM is suitable for the MMC with not so many SMs per arm [12, 13]. For the MMC with n SMs per arm, it is realized by applying n number of identical triangular carriers W_1–W_n stacked evenly in horizontal direction. The carrier is between −1 and 1. Through the comparison between the carriers W_1–W_n and the reference signal y, the PS-PWM can be achieved.

Figure 1.8 illustrates the implementation principle of PS-PWM ($n = 4$) for the MMC with n SMs per arm. The carriers W_1–W_n are all with the same carrier frequency f_s. The phase angle between two adjacent carriers of W_1–W_n is denoted as $\Delta\theta$. Generally, the $\Delta\theta$ is

$$\Delta\theta = 2\pi / n \tag{1.6}$$

The reference signal y is compared with the carriers W_1–W_n to generate the pulses S_{p1}–S_{pn}, as follows:

- The S_{pi} is 1 if the reference y is more than the carrier W_i ($i = 1, 2, \ldots, n$)
- The S_{pi} is 0 if the reference is less than the carrier W_i ($i = 1, 2, \ldots, n$)

The total number n_{on} of the SMs to be inserted into each arm at each instant can be expressed as the sum of S_{p1}–S_{pn}, as

$$n_{on} = \sum_{i=1}^{n} S_{pi} \tag{1.7}$$

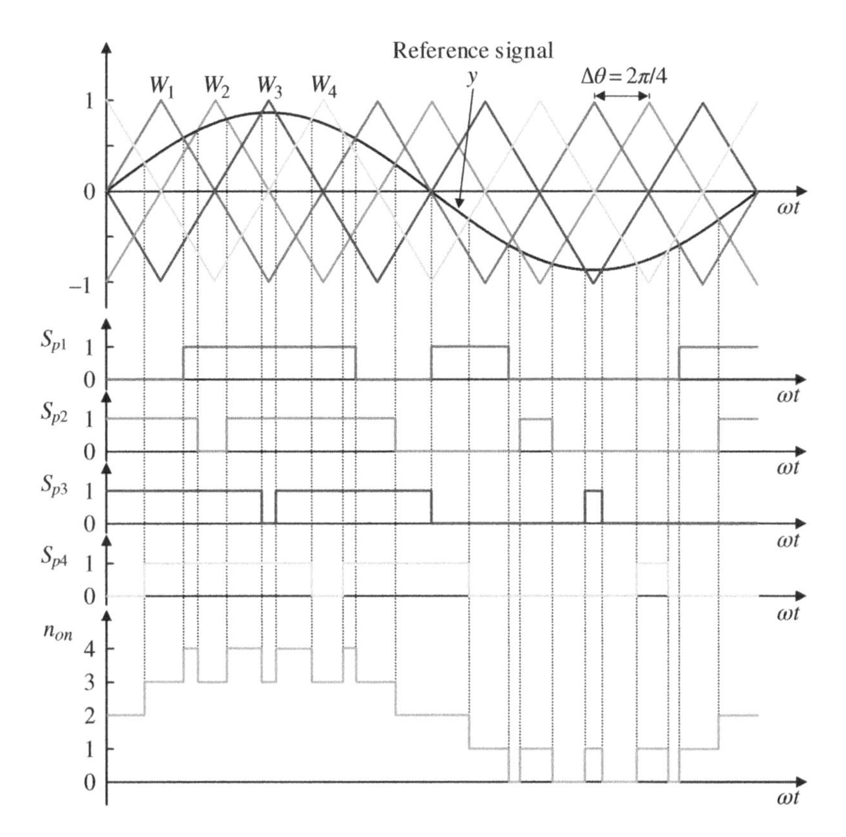

Figure 1.8 Implementation principle of PS-PWM for MMCs.

In Figure 1.8, the reference signal y is always modulated by all carriers W_1–W_n to generate the PWM pulses. The pulses and the switching frequency are distributed evenly among S_{p1}–S_{pn}. The value of n_{on} varies in the range of 0–n in one fundamental period, which achieves the multilevel synthesized waveform. In addition, the switching actions of S_{p1}–S_{pn} remain constant regardless of the modulation index of the reference signal.

The typical characteristics of the PS-PWM are summarized in Table 1.5.

1.4.3 Nearest Level Modulation

The NLM is suitable for the MMC applications with large number of SMs, e.g. high voltage direct current (HVDC) transmission [9, 14–16]. The NLM uses the staircase waveform nearest to the desired reference signal. For the MMC with n SMs per arm, the NLM can be produced with the simple rounding function for the reference y.

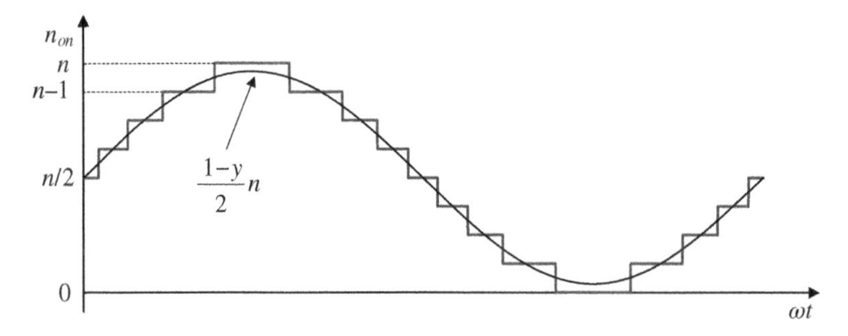

Figure 1.9 Implementation principle of NLM for MMCs.

Figure 1.9 illustrates the implementation principle of the NLM for the upper arm of the MMC ($n = 8$) with n SMs per arm. Based on the reference signal y for the phase unit, the total number n_{on} of SMs to be inserted into the upper arm at each instant can be obtained as

$$n_{on} = \mathrm{round}\left(\frac{1-y}{2}n\right) \tag{1.8}$$

The NLM directly generates the number of inserted SMs in the arm to yield the multilevel waveforms. The waveform of n_{on} is a staircase with step height of 1. The value of n_{on} varies in the range of 0–n. Therefore, the maximum level number of n_{on} is $n+1$. The harmonic spectrum of the NLM is dependent on n. If n is small, the staircase number is small and the harmonics will be a little high. If n is large, the staircase number is large and the multilevel waveforms will be very close to the reference signal and have very little harmonics.

The typical characteristics of the NLM are summarized in Table 1.5.

1.5 Mathematical Model

1.5.1 Submodule Mathematical Model

Figure 1.10 shows the equivalent circuit diagram of the i-th SM in the upper arm of phase A. Figure 1.10a shows the equivalent circuit diagram of the SM when its switching function S_{aui} is 1, and Figure 1.10b shows the equivalent circuit diagram of the SM when its switching function S_{aui} is 0. According to the equivalent circuit diagrams, the switching-function based model and the reference-based model of the HB SM can be derived as discussed in the following sections.

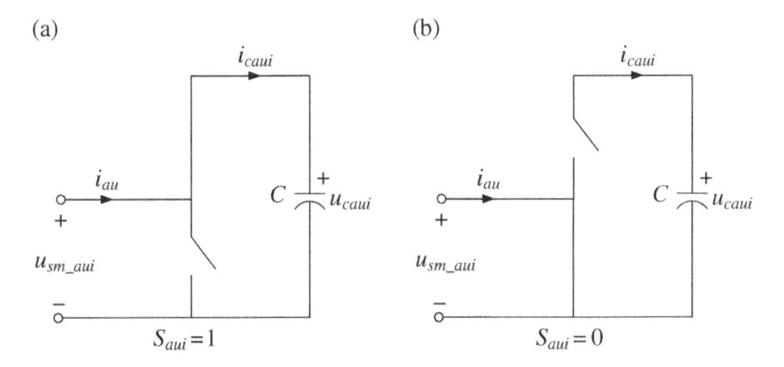

Figure 1.10 HB SM equivalent circuit. (a) S_{aui} = 1. (b) S_{aui} = 0.

1.5.1.1 Switching-Function Based Model

The relationship between the i-th SM's capacitor current i_{caui} and arm current i_{au} can be expressed as

$$i_{caui} = S_{aui} \cdot i_{au} \tag{1.9}$$

The relationship between SM's output voltage u_{sm_aui} and SM's capacitor voltage u_{caui} can be expressed as

$$u_{sm\_aui} = S_{aui} \cdot u_{caui} \tag{1.10}$$

According to the Volt–Ampere characteristic of the capacitor [17], the voltage u_{caui} imposed on the SM's capacitor C can be expressed as

$$u_{caui} = \frac{1}{C}\int i_{caui} dt = \frac{1}{C}\int \left(S_{aui} \cdot i_{au}\right) dt \tag{1.11}$$

Combining equations (1.10) and (1.11), the SM's output voltage u_{sm_uai} can be rewritten as

$$u_{sm\_aui} = \frac{S_{aui}}{C}\int \left(S_{aui} \cdot i_{au}\right) dt \tag{1.12}$$

1.5.1.2 Reference-Based Model

With the Fourier transform, the SM's switching function S_{aui} can be expressed using Fourier series as

$$S_{aui} = \frac{1}{2} + \frac{1}{2} m \sin\left(\omega t + \phi\right) + \sum_{h=2}^{\infty} A_h \sin\left(h\omega t + \phi_h\right) \tag{1.13}$$

where m is the modulation index, ω is the fundamental angular frequency, ϕ is the phase angle of the fundamental component, and A_h and θ_h are the amplitude and phase angle for the h-th harmonic, respectively.

Neglecting the harmonic components and supposing that the SMs in the arm are the same, the SM's switching function can be simplified as

$$S_{aui} \approx \frac{1}{2} + \frac{1}{2} y_{au} \tag{1.14}$$

with

$$y_{au} = m \sin(\omega t + \phi) \tag{1.15}$$

where y_{au} is the reference signal for upper arm of phase A.

Combining equations (1.9) and (1.14), the SM's capacitor current i_{caui} can be expressed as

$$i_{caui} = \frac{1 + y_{au}}{2} \cdot i_{au} \tag{1.16}$$

Based on equations (1.11) and (1.16), the SM's capacitor voltage u_{caui} can be expressed as

$$u_{caui} = \frac{1}{C} \int \left(\frac{1 + y_{au}}{2} \cdot i_{au} \right) dt \tag{1.17}$$

Combining equations (1.10), (1.14), and (1.17), the SM's output voltage u_{sm_uai} can be expressed as

$$u_{sm\_aui} = \frac{1 + y_{au}}{2} \cdot u_{caui} = \frac{1 + y_{au}}{2C} \int \left(\frac{1 + y_{au}}{2} \cdot i_{au} \right) dt \tag{1.18}$$

1.5.2 Arm Mathematical Model

Figure 1.11 shows the configurations of the upper and lower arms of phase j ($j = a$, b, c). Figure 1.11a shows the configuration of the upper arm of phase j, and Figure 1.11b shows the configuration of the lower arm of phase j. According to the arm configurations, the switching-function based model and the reference-based model of upper and lower arms can be derived as discussed in the following sections.

1.5.2.1 Switching-Function Based Model

In Figure 1.11, the upper arm voltage u_{ju} and lower arm voltage u_{jl} of phase j are equal to the sum of SM's output voltages [18]. The upper arm voltage u_{ju} and lower arm voltage u_{jl} can be expressed as

$$\begin{cases} u_{ju} = \sum_{i=1}^{n} u_{sm\_jui} \\ u_{jl} = \sum_{i=1}^{n} u_{sm\_jli} \end{cases} \tag{1.19}$$

where u_{sm_jui} is the output voltage of the i-th SM in the upper arm of phase j and u_{sm_jli} is the output voltage of the i-th SM in the lower arm of phase j.

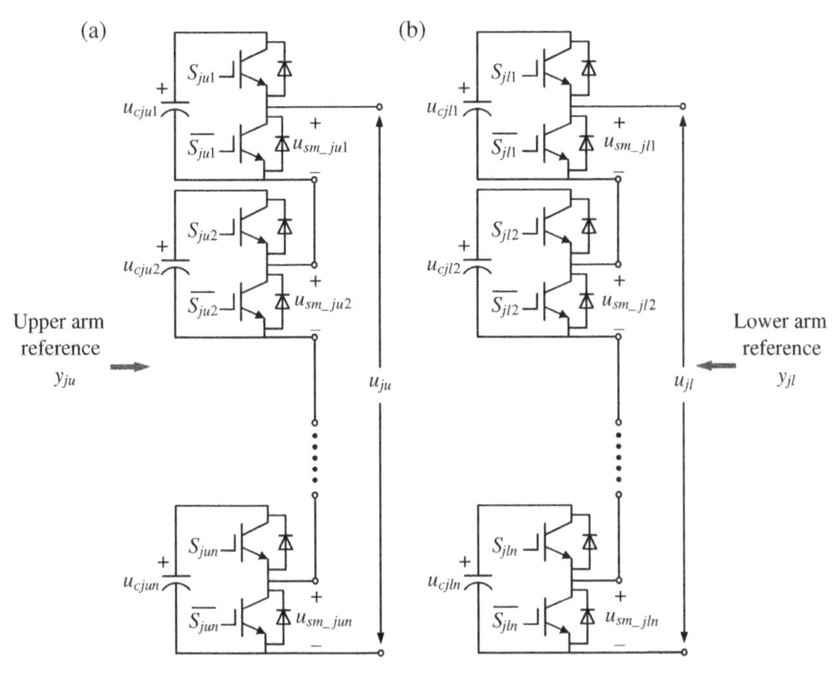

Figure 1.11 Arm configurations. (a) Upper arm of phase j. (b) Lower arm of phase j.

Combining equations (1.10) and (1.19), the upper arm voltage u_{ju} and lower arm voltage u_{jl} can be expressed as

$$\begin{cases} u_{ju} = \sum_{i=1}^{n} \left(S_{jui} \cdot u_{cjui} \right) \\ u_{jl} = \sum_{i=1}^{n} \left(S_{jli} \cdot u_{cjli} \right) \end{cases} \tag{1.20}$$

where S_{jui} is the switching function of the i-th SM in the upper arm of phase j, S_{jli} is the switching function of the i-th SM in the lower arm of phase j, u_{cjui} is the capacitor voltage of the i-th SM in the upper arm of phase j, and u_{cjli} is the capacitor voltage of the i-th SM in the lower arm of phase j.

1.5.2.2 Reference-Based Model

Combining equations (1.14) and (1.20), the upper arm voltage u_{ju} and lower arm voltage u_{jl} can be expressed as

$$\begin{cases} u_{ju} = \dfrac{1 + y_{ju}}{2} \cdot \sum_{i=1}^{n} u_{cjui} \\ u_{jl} = \dfrac{1 + y_{jl}}{2} \cdot \sum_{i=1}^{n} u_{cjli} \end{cases} \tag{1.21}$$

where y_{ju} and y_{jl} are the reference signals for the upper arm and lower arm of phase j, respectively, as shown in Figure 1.11.

1.5.3 Three-Phase MMC Mathematical Model

The configuration of the three-phase MMC is shown in Figure 1.12, and its mathematical model can be derived, as follows.

According to the Kirchhoff's voltage law, the upper arm voltage u_{ju} and the lower arm voltage u_{jl} can be expressed as

$$\begin{cases} u_{ju} = \dfrac{V_{dc}}{2} - u_j - L_s \dfrac{di_{ju}}{dt} \\ u_{jl} = \dfrac{V_{dc}}{2} + u_j - L_s \dfrac{di_{jl}}{dt} \end{cases} \tag{1.22}$$

where u_j is the AC-side voltage of phase j, i_{ju} is the upper arm current of phase j, and i_{jl} is the lower arm current of phase j.

According to the Kirchhoff's current law, the AC-side current i_j can be expressed as

$$i_j = i_{ju} - i_{jl} \tag{1.23}$$

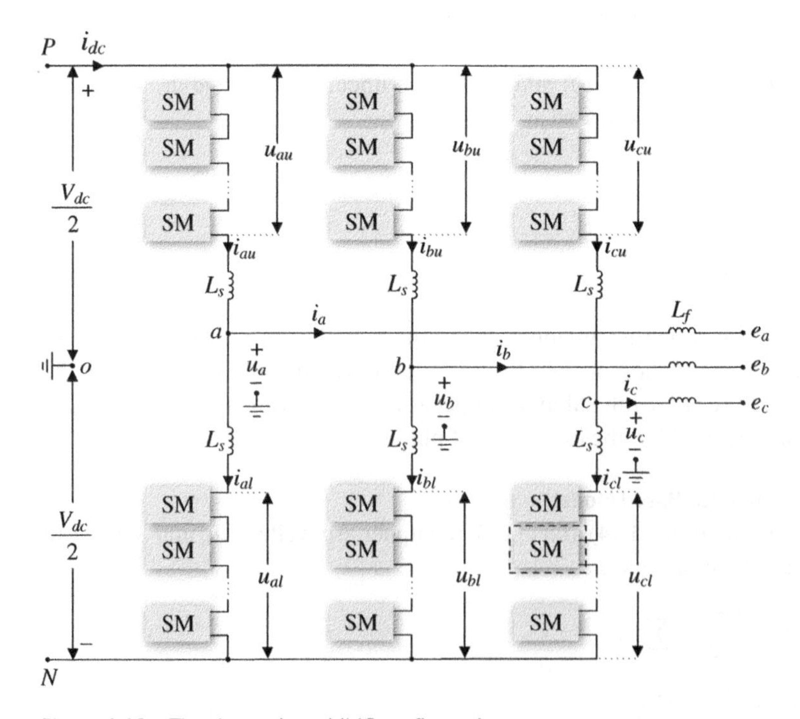

Figure 1.12 The three-phase MMC configuration.

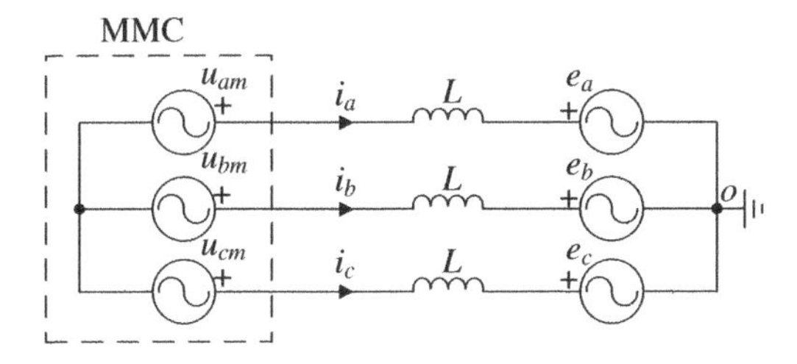

Figure 1.13 The AC-side equivalent circuit of the MMC.

1.5.3.1 AC-Side Mathematical Model

Figure 1.13 shows the AC-side equivalent circuit of the MMC. Combining equation (1.22) and Figure 1.13, the AC-side dynamics of the MMC can be expressed as

$$e_j = u_{jm} - L\frac{di_j}{dt} \tag{1.24}$$

where e_j is the AC-side voltage in phase j. u_{jm} is the internal converter voltage of the MMC and is expressed as

$$u_{jm} = \frac{u_{jl} - u_{ju}}{2} \tag{1.25}$$

L is the equivalent AC-side inductance and is expressed as

$$L = \frac{L_s}{2} + L_f \tag{1.26}$$

1.5.3.2 DC-Side Mathematical Model

Figure 1.14 shows the DC-side equivalent circuit of the MMC. According to equation (1.22) and Figure 1.14, the DC-side voltage V_{dc} of the MMC can be expressed as

$$V_{dc} = u_{ju} + u_{jl} + 2L_s\frac{di_{diff\_j}}{dt} \tag{1.27}$$

with

$$i_{diff\_j} = \frac{i_{ju} + i_{jl}}{2} \tag{1.28}$$

where i_{diff_j} ($j = a, b, c$) is the circulating current in phase j.

The i_{diff_j} mainly contains DC component and the second-order harmonic component [18], as

$$i_{diff\_j} = \frac{i_{dc}}{3} + i_{2f\_j} \tag{1.29}$$

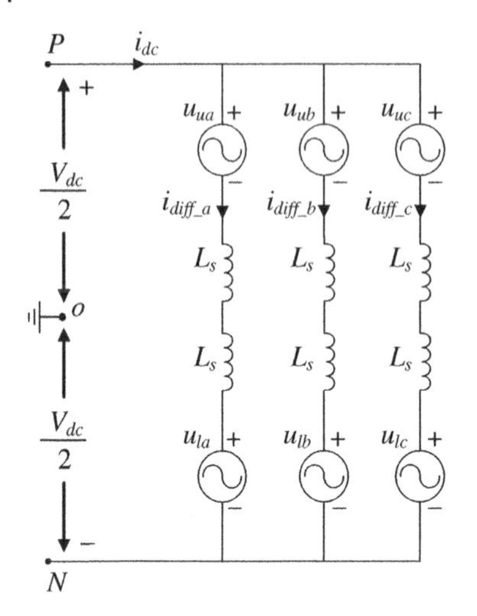

Figure 1.14 The DC-side equivalent circuit of the MMC.

where i_{dc} is the DC-side current. i_{2f_j} is the second-order circulating current in phase j. The upper arm current and the lower arm current in phase j of the MMC can be expressed as

$$\begin{cases} i_{ju} = \dfrac{i_j}{2} + i_{diff\_j} \\ i_{jl} = -\dfrac{i_j}{2} + i_{diff\_j} \end{cases} \tag{1.30}$$

1.6 Design Constraints

Parameter designs of the MMC components are important because reasonable parameters can improve the stability and dynamic performance of the system [19–23]. This section focuses on the parameter design methods of power devices, capacitor, and arm inductor in the MMC.

1.6.1 Power Device Design

The configuration of the HB SM is shown in Figure 1.15, which has four power devices, including switches T_1, T_2 and diodes D_1, D_2. According to Section 1.3, the HB SM has four operation modes under normal operation and two operation modes under blocking operation.

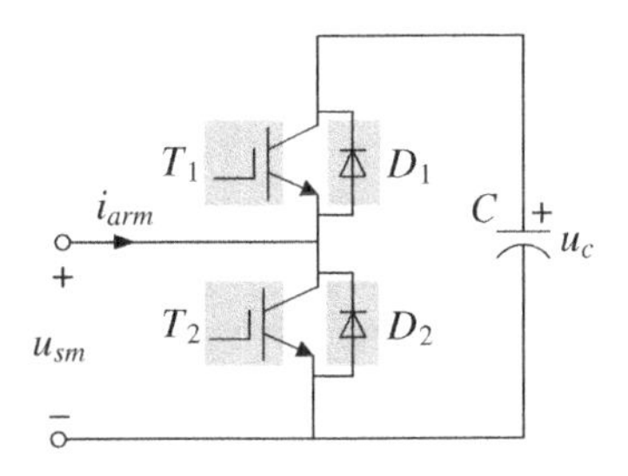

Figure 1.15 HB SM configuration.

Table 1.6 Voltage stresses of power devices in the HB SM under normal operation.

Mode		S	i_{arm}	T_1	T_2	Power device voltage stress			
						u_{T1}	u_{D1}	u_{T2}	u_{D2}
Normal	1	1	>0	On	Off	0	0	u_c	u_c
	2		<0	On	Off	0	0	u_c	u_c
	3	0	>0	Off	On	u_c	u_c	0	0
	4		<0	Off	On	u_c	u_c	0	0

1.6.1.1 Rated Voltage of Power Devices

The voltages u_{T1}, u_{D1}, u_{T2}, u_{D2} of power devices T_1, D_1, T_2, D_2, respectively, in the MMC under normal operation are shown in Table 1.6. According to Table 1.6, the maximum voltage stress of power devices in four operation modes is the capacitor voltage u_c. Since the rated voltage parameters of power devices need to be selected with a margin, the rated voltage parameters of power devices can be chosen as 1.5–2 times the capacitor voltage u_c.

1.6.1.2 Rated Current of Power Devices

The current i_{T1}, i_{D1}, i_{T2}, i_{D2} flowing through power devices T_1, D_1, T_2, D_2, respectively, in the MMC under normal operation are shown in Table 1.7, as follows:

- The arm current i_{arm} flows through T_1 when $i_{arm} < 0$ and $S = 1$.
- The arm current i_{arm} flows through T_2 when $i_{arm} > 0$ and $S = 0$.
- The arm current i_{arm} flows through D_1 when $i_{arm} > 0$ and $S = 1$.
- The arm current i_{arm} flows through D_2 when $i_{arm} < 0$ and $S = 0$.

Taking the upper arm of phase A as an example, Figure 1.16 shows the arm current i_{au} in one fundamental cycle. Suppose that t_1, t_2, and t_3 are the times when $i_{au} = 0$, as shown in Figure 1.16. Then, the root mean square (RMS) values I_{rms_T1},

Table 1.7 Current through power devices in the MMC under normal operation.

Mode		S	i_{arm}	T_1	T_2	Power device current			
						i_{T1}	i_{D1}	i_{T2}	i_{D2}
Normal	1	1	>0	On	Off	0	i_{arm}	0	0
	2		<0	On	Off	i_{arm}	0	0	0
	3	0	>0	Off	On	0	0	i_{arm}	0
	4		<0	Off	On	0	0	0	i_{arm}

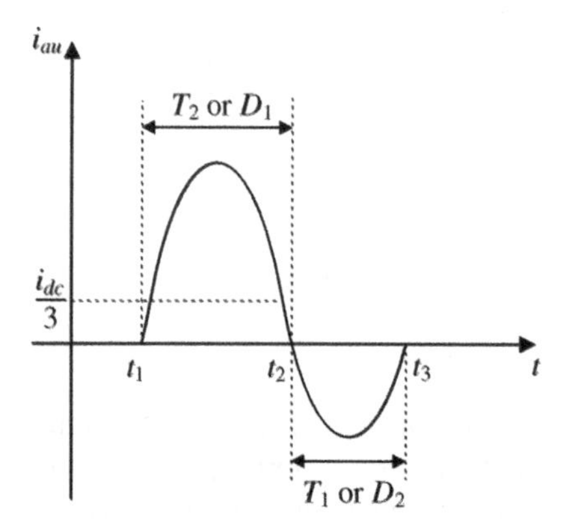

Figure 1.16 Upper arm current of phase A.

I_{rms_T2}, I_{rms_D1}, I_{rms_D2} of the current flowing through power devices T_1, T_2, D_1, D_2, respectively, in the i-th SM of the upper arm of phase A can be obtained as

$$\begin{cases} I_{rms\_T_1} = \sqrt{\dfrac{1}{T_f}\displaystyle\int_{t_2}^{t_3} S_{aui}^{\,2} i_{au}^{\,2} dt} \\ I_{rms\_T_2} = \sqrt{\dfrac{1}{T_f}\displaystyle\int_{t_1}^{t_2} (1-S_{aui})^2 i_{au}^{\,2} dt} \\ I_{rms\_D_1} = \sqrt{\dfrac{1}{T_f}\displaystyle\int_{t_1}^{t_2} S_{aui}^{\,2} i_{au}^{\,2} dt} \\ I_{rms\_D_2} = \sqrt{\dfrac{1}{T_f}\displaystyle\int_{t_2}^{t_3} (1-S_{aui})^2 i_{au}^{\,2} dt} \end{cases} \tag{1.31}$$

where T_f is the fundamental period of upper arm current i_{au}.

Since the rated current parameters of power devices need to be selected with a margin, the rated current parameters of power devices can be chosen as 1.5–2 times of the RMS value of the current flowing through power devices.

1.6.2 Capacitor Design

The SM's capacitor voltage u_c contains the DC component and the fluctuation component under normal operation of the MMC. The SM's capacitor voltage u_c can be expressed as

$$u_c = U_c + \Delta u_c \tag{1.32}$$

where U_c is the DC component of the SM's capacitor voltage with $U_c = V_{dc}/n$ and Δu_c is the fluctuation component of the SM's capacitor voltage.

The capacitor voltage fluctuation ratio ε is

$$\varepsilon = \frac{\max\left|\Delta u_c\right|}{U_c} \tag{1.33}$$

To simplify the analysis while neglecting the voltage of the arm inductance and the second-order circulating current, the voltage u_{au} and the current i_{au} of the upper arm in phase A of the MMC can be expressed as

$$\begin{cases} u_{au} = \dfrac{V_{dc}}{2} - u_a \\ i_{au} = \dfrac{i_a}{2} + \dfrac{i_{dc}}{3} \end{cases} \tag{1.34}$$

Suppose that the voltage u_a and current i_a of phase A are

$$\begin{cases} u_a = U_m \sin(\omega t) \\ i_a = I_m \sin(\omega t + \theta) \end{cases} \tag{1.35}$$

where U_m and I_m are the amplitudes of u_a and i_a, respectively. θ is the phase angle.

Combining equations (1.34) and (1.35), the instantaneous power p_{au} of the upper arm in phase A is

$$p_{au} = u_{au} \cdot i_{au} = \frac{U_m I_m}{4}\left[\frac{I_m}{i_{dc}}\cos\theta - 2\sin(\omega t)\right]\cdot\left[-\sin(\omega t + \theta) - \frac{I_m}{i_{dc}}\right] \tag{1.36}$$

The instantaneous power p_{au} has two zero-crossing points in one fundamental period T_f, $T_f = 2\pi/\omega$, defined as t_1 and t_2, respectively, as shown in Figure 1.17, which can be expressed as

$$\begin{cases} t_1 = \dfrac{1}{\omega}\cdot\left[-\arcsin\left(\dfrac{i_{dc}}{I_m}\right) + \theta\right] \\ t_2 = \dfrac{1}{\omega}\cdot\left[\pi + \arcsin\left(\dfrac{i_{dc}}{I_m}\right) - \theta\right] \end{cases} \tag{1.37}$$

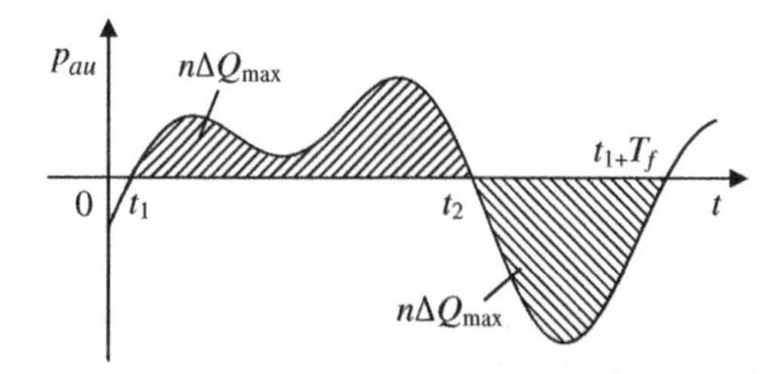

Figure 1.17 Instantaneous power of upper arm in phase A.

In Figure 1.17, the energy stored in the upper arm of phase A would be increased during the period when $p_{au} > 0$, while the arm energy would be reduced during the period when $p_{au} < 0$. In addition, the energy absorbed by the arm when $p_{au} > 0$ equals the energy discharged when $p_{au} < 0$ during one fundamental period T_f, so as to ensure the balance of the arm energy. As a result, the arm energy reaches its minimum and maximum at t_1 and t_2, respectively. Assuming that the arm energy is equally distributed among ***n*** SM's capacitors, the largest change of the capacitor energy ΔQ_{max} during one fundamental cycle T_f can be obtained as [22]

$$\Delta Q_{max} = Q_{max} - Q_{min} = \frac{1}{n} \cdot \int_{t_1}^{t_2} |p_{au}| dt = \frac{2}{3} \frac{S_N}{mn\omega} \left[1 - \left(\frac{m\cos\theta}{2} \right)^2 \right]^{3/2} \tag{1.38}$$

where Q_{max} and Q_{min} are the maximum and minimum capacitor energy, respectively. S_N is the rated apparent power of the MMC, and ***m*** is the modulation index.

In each SM of upper arm of phase A, the SM maximum capacitor energy Q_{max} and the SM minimum capacitor energy Q_{min} are

$$\begin{cases} Q_{max} = \frac{1}{2} C \left[U_c \cdot (1 + \varepsilon) \right]^2 \\ Q_{min} = \frac{1}{2} C \left[U_c \cdot (1 - \varepsilon) \right]^2 \end{cases} \tag{1.39}$$

Substituting equation (1.39) into (1.38), there is

$$\varepsilon = \frac{1}{3} \frac{S_N}{mn\omega C U_c^2} \left[1 - \left(\frac{m\cos\theta}{2} \right)^2 \right]^{3/2} \tag{1.40}$$

The basic consideration for selecting the SM's capacitance value is to ensure that the capacitor fluctuation ratio ε is less than the maximum allowed voltage fluctuation ε_{max} under any operation conditions of the MMC, which is usually selected within 5% to 10%. According to (1.40), when $\cos\theta = 1$, ***m*** = 1, the capacitor voltage fluctuation ratio ε reaches its maximum [23], as

$$\varepsilon = \frac{1}{3} \cdot \frac{S_N}{n\omega C U_c^2} \tag{1.41}$$

The SM's capacitor can be designed as

$$C \geq \frac{1}{3} \cdot \frac{S_N}{n\omega U_c^2 \varepsilon_{\max}} \tag{1.42}$$

1.6.3 Arm Inductor Design

Limiting the DC-side short-circuit fault current rise rate is the distinctive function of the arm inductor in the MMC [24]. When a short circuit fault occurs at the DC side of the MMC, the duration of the transient process is so short that the voltages of the SM capacitors are supposed to be constant. Suppose that the sum of the capacitor voltages of the SMs inserted in phase j is equal to V_{dc} after the fault

$$u_{jl} + u_{ju} = V_{dc} \tag{1.43}$$

In addition, a huge fault current would be caused under the DC-side short-circuit fault, which will flow to the fault point. According to (1.27), the DC-side mathematical model of the MMC can be expressed as

$$u_{jl} + u_{ju} + \left(L_s \frac{di_{ju}}{dt} + L_s \frac{di_{jl}}{dt} \right) = 0 \tag{1.44}$$

where $|di_{ju}/dt|$ is the fault current rise rate in the upper arm in phase j, and $|di_{jl}/dt|$ is the fault current rise rate in the lower arm in phase j. In a short period after the fault, the majority of the arm current is the transient component. Thus, the upper and lower arm currents are supposed to be equal as

$$i_{ju} = i_{jl} \tag{1.45}$$

Therefore, the fault current rise rate $|di_{ju}/dt|$ in the upper arm and the fault current rise rate $|di_{jl}/dt|$ in the lower arm of phase j are supposed to be equal as

$$\left| \frac{di_{ju}}{dt} \right| = \left| \frac{di_{jl}}{dt} \right| = \lambda \tag{1.46}$$

where λ is the fault current rise rate.

Combining equations (1.43), (1.44), and (1.46), there is

$$\lambda = \frac{V_{dc}}{2L_s} \tag{1.47}$$

To limit the DC-side short-circuit fault current rise rate, the arm inductor L_s can be designed as

$$L_s \geq \frac{V_{dc}}{2\lambda_{\max}} \tag{1.48}$$

where $\lambda_{\max}$ is the maximum allowed fault current rise rate.

1.7 Faults Overview of MMCs

This section introduces the internal and external faults of the MMC. The internal faults are caused by component failures in the MMC, while the external faults are induced by grid faults at the AC or DC side of the MMC, as shown in Figure 1.18.

1.7.1 Internal Faults of MMCs

The internal faults of the MMC can be mainly classified into two categories, namely, the IGBT faults and the capacitor faults.

The IGBT faults can be further classified into two categories based on the fault characteristics, namely the IGBT short-circuit fault and the IGBT open-circuit fault [25]. The IGBT short-circuit fault is mainly caused by incorrect driving signals or device over-stress, and it will lead to an abnormal overcurrent and cause serious damage to other components in a short time. The IGBT open-circuit fault is mainly caused by the failure of the switch gate driver and the failure of the bond wire, and it will cause an unbalanced state in the circuit, which produces distorted waveforms and voltage/current stresses and obviously deteriorates the system performance. Therefore, it is necessary to detect the IGBT faults, which are introduced in detail in Chapter 3.

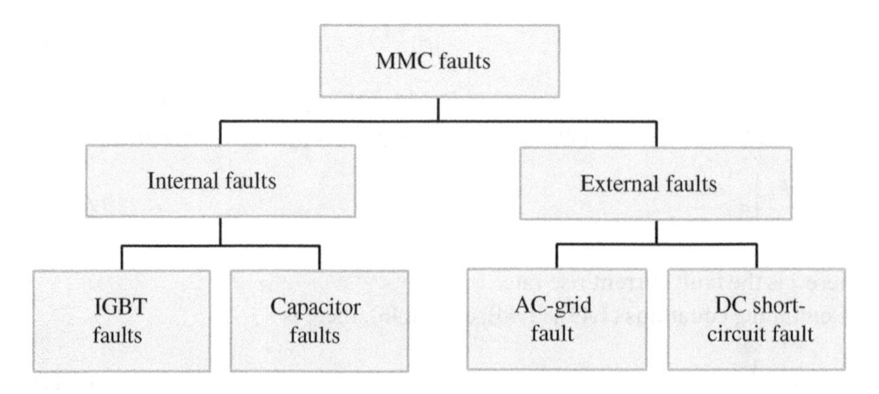

Figure 1.18 MMC faults overview.

The capacitor faults are caused by combined effects of thermal, electrical, mechanical, and environmental stress and can result in a decrease in capacitance and an increase in equivalent series resistance [26]. Once one of the capacitors fails, the normal operation of the MMC would be disrupted. Therefore, it is also necessary to provide an effective capacitor condition monitoring method for reliable, efficient, and safe operation of the MMC, which is introduced in detail in Chapter 4.

Fault tolerant operation is one of the most important challenges for the MMC, where it is desired that the MMC can continue operating without any interruption, despite some of the SMs malfunctioning. Therefore, in addition to the method of fault detection, fault tolerant control of MMCs under SM faults is introduced in detail in Chapter 5.

1.7.2 External Faults of MMCs

The external faults of the MMC can be mainly classified into two categories, namely the AC-grid fault and the DC-grid short-circuit fault.

When the AC-grid fault occurs, the AC-side current may consist of a negative-sequence component and the zero-sequence current may also exist in transformerless applications of the MMC [27]. The unbalanced AC-side current can cause overcurrent and the oscillation of active and reactive power, which is harmful to the whole MMC system. Therefore, the operation and control of the MMC system in the presence of grid imbalances is very important, which is introduced in detail in Chapter 6.

When the DC grid short-circuit fault occurs [28], the MMC will enter blocking mode and the AC grid will feed the fault on the DC side through the anti-parallel diodes of the HB SMs. In such an uncontrolled operation mode, the converter has to withstand large fault current from the AC grid until the DC fault is interrupted and the large fault current can damage the MMC due to the limited current capability of power devices. Therefore, it is also important to adopt DC-grid fault protection methods, which are introduced in detail in Chapter 7.

1.8 Summary

In this chapter, a comprehensive overview of MMCs, including MMC configuration and operation principles, are presented. The MMC can be realized by using a number of SMs. The most popular and widely used SM configuration with HB SM configuration is discussed. The mathematical relationships of the MMC, including the mathematical relationship in the SM, the mathematical relationship in the arm, and the mathematical relationship in the three-phase MMC, are presented.

PWM schemes are commonly employed to control the MMC. A comprehensive overview of the modulation schemes for the MMC is presented, including

PD-PWM, PS-PWM, and NLM. Considering the computational complexity, the PD-PWM and PS-PWM are suitable for the MMC with not so many SM per arm, while the NLM is suitable for the MMC with a large number of SMs. The power devices, capacitor, and arm inductor are the important components in the MMC. The design constraints of the power devices, capacitor, and arm inductor are presented based on their voltage and current constraints.

The internal faults of the MMC, including the power device faults such as the switch open-circuit fault, switch short-circuit fault, and capacitor faults, are harmful to the MMC, and fault detection and fault tolerant control are essential for the MMC, which are further presented in Chapters 3–5. The external faults of the MMC, including the AC-grid faults and the DC-grid faults, would harm the MMC, and the corresponding control and protection are also essential for the MMC, which are further presented in Chapters 6 and 7.

References

1 J. Lyu, X. Cai, and M. Molinas, "Optimal design of controller parameters for improving the stability of MMC-HVDC for wind farm integration," *IEEE Trans. Emerg. Sel. Top. Power Electron.*, vol. 6, no. 1, pp. 40–53, Mar. 2018.

2 S. Debnath, J. Qin, B. Bahrani, M. Saeedifard, and P. Barbosa, "Operation, control, and applications of the modular multilevel converter: a review," *IEEE Trans. Power Electron.*, vol. 30, no. 1, pp. 37–53, Mar. 2014.

3 M. A. Perez, S. Bernet, J. Rodriguez, S. Kouro, and R. Lizana, "Circuit topologies, modeling, control schemes, and applications of modular multilevel converters," *IEEE Trans. Power Electron.*, vol. 30, no. 1, pp. 4–17, Jan. 2015.

4 N. Flourentzou, V. G. Agelidis, and G. D. Demetriades, "VSC-based HVDC power transmission systems: an overview," *IEEE Trans. Power Electron.*, vol. 24, no. 3, pp. 592–602, Mar. 2009.

5 M. Vasiladiotis, N. Cherix, and A. Rufer, "Impact of grid asymmetries on the operation and capacitive energy storage design of modular multilevel converters," *IEEE Trans. Ind. Electron.*, vol. 62, no. 11, pp. 6697–6707, Nov. 2015.

6 F. Ma, Z. He, Q. Xu, A. Luo, L. Zhou, and M. Li, "Multilevel power conditioner and its model predictive control for railway traction system," *IEEE Trans. Ind. Electron.*, vol. 63, no. 11, pp. 7275–7285, Nov. 2016.

7 P. Wu, W. Huang, and N. Tai, "Advanced design of microgrid interface for multiple microgrids based on MMC and energy storage unit," *J. Eng. Technol.*, vol. 2017, no. 13, pp. 2231–2235, Oct. 2017.

8 Q. Tu, Z. Xu, and L. Xu, "Reduced switching-frequency modulation and circulating current suppression for modular multilevel converters," *IEEE Trans. Power Del.*, vol. 26, no. 3, pp. 2009–2017, Jul. 2011.

9 F. Deng, Y. Lü, C. Liu, Q. Heng, Q. Yu, and J. Zhao, "Overview on submodule topologies, modeling, modulation, control schemes, fault diagnosis, and tolerant control strategies of modular multilevel converters," *Chin. J. Elect. Eng.*, vol. 6, no. 1, pp. 1–21, Mar. 2020.

10 Q. Yu, F. Deng, C. Liu, J. Zhao, F. Blaabjerg, and S. Abulanwar, "DC-link high-frequency current ripple elimination strategy for MMCs using phase-shifted double-group multicarrier-based phase-disposition PWM," *IEEE Trans. Power Electron.*, vol. 36, no. 8, pp. 8872–8886, Aug. 2021.

11 B. P. McGrath, C. A. Teixeira, and D. G. Holmes, "Optimized phase disposition (PD) modulation of a modular multilevel converter," *IEEE Trans. Ind. Appl.*, vol. 53, no. 5, pp. 4624–4633, Oct. 2017.

12 J. Wang and Y. Tang, "A fault-tolerant operation method for medium voltage modular multilevel converters with phase-shifted carrier modulation," *IEEE Trans. Power Electron.*, vol. 34, no. 10, pp. 9459–9470, Oct. 2019.

13 B. Li, R. Yang, D. Xu, G. Wang, W. Wang, and D. Xu, "Analysis of the phase-shifted carrier modulation for modular multilevel converters," *IEEE Trans. Power Electron.*, vol. 30, no. 1, pp. 297–310, Jan. 2015.

14 Q. Tu and Z. Xu, "Impact of sampling frequency on harmonic distortion for modular multilevel converter," *IEEE Trans. Power Del.*, vol. 26, no. 1, pp. 298–306, Jan. 2011.

15 P. Hu and D. Jiang, "A level-increased nearest level modulation method for modular multilevel converters," *IEEE Trans. Power Electron.*, vol. 30, no. 4, pp. 1836–1842, Apr. 2015.

16 L. Lin, Y. Lin, Z. He, Y. Chen, J. Hu, and W. Li, "Improved nearest-level modulation for a modular multilevel converter with a lower submodule number," *IEEE Trans. Power Electron.*, vol. 31, no. 8, pp. 5369–5377, Aug. 2016.

17 A. Antonopoulos, L. Angquist, and H. Nee, "On dynamics and voltage control of the modular multilevel converter," in *13th European Conference on Power Electronics and Applications*, 2009, pp. 1–10.

18 M. Guan and Z. Xu, "Modeling and control of a modular multilevel converter-based HVDC system under unbalanced grid conditions," *IEEE Trans. Power Electron.*, vol. 27, no. 12, pp. 4858–4867, Dec. 2012.

19 S. Yang, A. Bryant, P. Mawby, D. Xiang, L. Ran, and P. Tavner, "An industry-based survey of reliability in power electronic converters," *IEEE Trans. Ind. Appl.*, vol. 47, no. 3, pp. 1441–1451, Aug. 2011.

20 H. Qiu, J. Wang, P. Tu, and Y. Tang, "Device-level loss balancing control for modular multilevel converters," *IEEE Trans. Power Electron.*, vol. 36, no. 4, pp. 4778–4790, Apr. 2021.

21 T. Mutou and H. Shirahama, "Multiphases PWM and capacitor voltage fluctuation of modular multilevel converter," in *2018 21st International Conference on Electrical Machines and Systems (ICEMS)*, 2018, pp. 2189–2193.

22 A. Lesnicar, "*Neuartiger Modularer Mehrpunktumrichter M2C für Netzkupplungsanwendungen*," Shaker Verlag, München, Germany, 2008.

23 Z. Xu, H. Xiao, and Z. Zhang, "Selection methods of main circuit parameters for modular multilevel converters," *IET Renew. Power Gener.*, vol. 10, no. 6, pp. 788–797, Jul. 2016.

24 Q. Tu, Z. Xu, H. Huang, and J. Zhang, "Parameter design principle of the arm inductor in modular multilevel converter based HVDC," in *2010 International Conference on Power System Technology*, 2010, pp. 1–6.

25 K. B. Pedersen and K. Pedersen, "Bond wire lift-off in IGBT modules due to thermomechanical induced stress," in *2012 3rd IEEE International Symposium on Power Electronics for Distributed Generation Systems (PEDG)*, 2012, pp. 519–526.

26 S. Ochi and N. Zommer, "Driving and protecting the latest high voltage and current power MOS and IGBTs," in *Conference Record of the 1992 IEEE Industry Applications Society Annual Meeting*, 1992, pp. 1196–1203.

27 J. He, Q. Yang, and Z. Wang, "On-line fault diagnosis and fault-tolerant operation of modular multilevel converters — a comprehensive review," *CES Trans. Electr. Mach. Syst.*, vol. 4, no. 4, pp. 360–372, Dec. 2020.

28 C. Liu, F. Deng, Q. Heng, X. Cai, R. Zhu, and M. Liserre, "Crossing thyristor branches based hybrid modular multilevel converters for DC line faults," *IEEE Trans. Ind. Electron.*, vol. 68, no. 10, pp. 9719–9730, 2021.

2

Control of MMCs

2.1 Introduction

The control is important for the modular multilevel converter (MMC), which regulates multiple external and internal control objectives. The external control objectives include the MMC's AC-side current, AC-side voltage, DC-side voltage, active power, and reactive power, which are regulated by the output control. The internal control objectives include the capacitor voltage of each submodule (SM) and circulating current among three phases, which are regulated by the capacitor voltage balancing control (VBC) and circulating current control (CCC), respectively.

The output control affects the external performance of the MMC. Since the MMC can be applied in various scenarios, the objectives of the output control are determined by the specific MMC and grid structures. Based on different control objectives, the output control are classified into current control, DC-link voltage and power control, and the grid forming control.

The MMC contains a large number of floating capacitors, which can buffer the power imbalance between the DC side and the AC side of the MMC. During the steady operation of the MMC, the energy stored in the MMC should be fixed and evenly distributed in the capacitor of each SM, which means the capacitor voltages of each SM should be balanced. Therefore, an effective capacitor VBC is essential for reliable and safe operation of the MMC.

The circulating current among the three phases is a typical problem for the MMC. The circulating current has no influence on the AC side but increases the arm current, which potentially leads to extra power losses of the switches. Therefore, an effective CCC is necessary for the efficient operation of the MMC.

Modular Multilevel Converters: Control, Fault Detection, and Protection, First Edition.
Fujin Deng, Chengkai Liu, and Zhe Chen.

Published 2023 by John Wiley & Sons, Inc.

This chapter deals with the recent advancements in the control of the MMC. The overall control of the MMC is introduced in Section 2.2. The output control of the MMC is introduced in Section 2.3. The centralized capacitor VBC of the MMC is introduced in Section 2.4. The individual capacitor VBC of the MMC is introduced in Section 2.5. The CCC is introduced in Section 2.6 to suppress the circulating current of the MMC. Finally, the summary of this chapter is discussed in Section 2.7.

2.2 Overall Control of MMCs

The overall control of phase j ($j = a, b, c$) of the MMC is shown in Figure 2.1, including the output control, CCC, and VBC [1], as follows.

- Output control
  The output control derives the reference voltage u_{jm_ref}, which can regulate the DC-side voltage, active power, reactive power, or AC-side voltage of phase j.
- CCC
  The CCC derives the reference voltage $u_{diff\_j\_ref}$, which can regulate the circulating current of phase j.
- VBC
  The VBC reasonably distributes the switching signals in each arm of the MMC to achieve the SM capacitor voltage balancing in each arm of the MMC.

With the output control and CCC, the upper arm reference voltage u_{ju_ref} and the lower arm reference voltage u_{jl_ref} can be expressed as

$$\begin{cases} u_{ju\_ref} = -u_{jm\_ref} + u_{diff\_j\_ref} \\ u_{jl\_ref} = u_{jm\_ref} + u_{diff\_j\_ref} \end{cases} \tag{2.1}$$

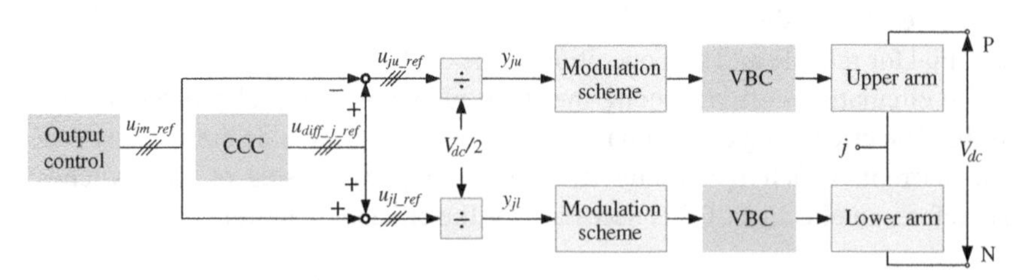

Figure 2.1 Overall control of the MMC.

The upper arm reference y_{ju} and the lower arm reference y_{jl} can be expressed as

$$\begin{cases} y_{ju} = \dfrac{u_{ju\_ref}}{V_{dc}/2} \\ y_{jl} = \dfrac{u_{jl\_ref}}{V_{dc}/2} \end{cases} \tag{2.2}$$

Afterwards, based on the modulation and VBC, the switching signal of each SM in the upper arm and the lower arm of phase j is obtained, which can ensure the SM capacitor voltage balancing in the upper arm and the lower arm of phase j.

2.3 Output Control of MMCs

This section describes the output control of MMCs. The current control regulates the AC-side current, which is usually adopted as the inner loop control for output control of the MMC. Based on the current controller, the control objectives of the outer loop control are determined by the structure of the MMC and grid. When the AC side of the MMC is connected to the AC power grid, the DC-link voltage, active power, and reactive power are controlled to regulate the power exchange between the MMC and grid. When the AC side of the MMC is connected to passive loads, the amplitude and frequency of the MMC's AC voltage is controlled to form the AC grid.

2.3.1 Current Control

The current control [2, 3] is the basic inner control loop for the MMC, which regulates the AC-side currents i_a, i_b, and i_c, as shown in Figure 2.2. Referring to (1.24), the mathematical model of the MMC in the stationary-abc reference frame can be expressed with the MMC's internal converter voltage u_{am}, u_{bm}, u_{cm} and grid voltage e_a, e_b, e_c as

$$\begin{cases} L\dfrac{di_a}{dt} = u_{am} - e_a \\ L\dfrac{di_b}{dt} = u_{bm} - e_b \\ L\dfrac{di_c}{dt} = u_{cm} - e_c \end{cases} \tag{2.3}$$

The voltages and currents of the MMC are time-varying variables, which are complex to control. To simplify the analysis, the time-varying variables in

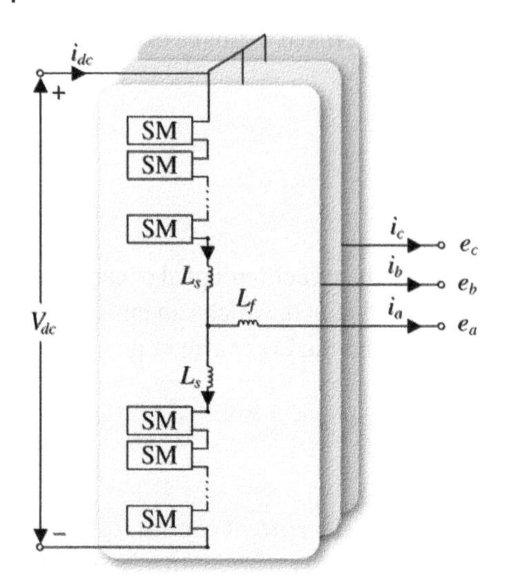

Figure 2.2 Structure of the three-phase MMC.

stationary-*abc* reference frame is transformed to DC variables in synchronous-*dq* reference frame, as

$$\begin{bmatrix} i_d \\ i_q \end{bmatrix} = T_{abc/dq} \cdot \begin{bmatrix} i_a \\ i_b \\ i_c \end{bmatrix} \tag{2.4}$$

$$\begin{bmatrix} e_d \\ e_q \end{bmatrix} = T_{abc/dq} \cdot \begin{bmatrix} e_a \\ e_b \\ e_c \end{bmatrix} \tag{2.5}$$

$$\begin{bmatrix} u_{dm} \\ u_{qm} \end{bmatrix} = T_{abc/dq} \cdot \begin{bmatrix} u_{am} \\ u_{bm} \\ u_{cm} \end{bmatrix} \tag{2.6}$$

with

$$T_{abc/dq} = \frac{2}{3} \begin{bmatrix} \cos\theta_e & \cos\left(\theta_e - \frac{2\pi}{3}\right) & \cos\left(\theta_e + \frac{2\pi}{3}\right) \\ -\sin\theta_e & -\sin\left(\theta_e - \frac{2\pi}{3}\right) & -\sin\left(\theta_e + \frac{2\pi}{3}\right) \end{bmatrix} \tag{2.7}$$

where $T_{abc/dq}$ is the *dq* transformation matrix, and θ_e is the phase angle of AC-grid voltage. i_d and i_q are the *d*-axis component and the *q*-axis component of the current i_a, i_b, i_c, respectively. e_d and e_q are the *d*-axis component and the *q*-axis component of the voltage e_a, e_b, e_c, respectively. u_{dm} and u_{qm} are the d-axis component and the q-axis component of the voltage u_{am}, u_{bm}, u_{cm}, respectively.

According to equations (2.3)–(2.7), the mathematical model of the MMC in the synchronous-dq reference frame in time-domain is derived as

$$L\frac{d}{dt}\begin{bmatrix} i_d \\ i_q \end{bmatrix} = \begin{bmatrix} u_{dm} \\ u_{qm} \end{bmatrix} - \begin{bmatrix} e_d \\ e_q \end{bmatrix} + \begin{bmatrix} 0 & \omega L \\ -\omega L & 0 \end{bmatrix}\begin{bmatrix} i_d \\ i_q \end{bmatrix} \tag{2.8}$$

With the Laplace transformation, the model in the Laplace-domain is derived in equation (2.9) and Figure. 2.3. According to Figure 2.3, the model has coupling between d-axis and q-axis.

$$sL\begin{bmatrix} i_d \\ i_q \end{bmatrix} = \begin{bmatrix} u_{dm} \\ u_{qm} \end{bmatrix} - \begin{bmatrix} e_d \\ e_q \end{bmatrix} + \begin{bmatrix} 0 & \omega L \\ -\omega L & 0 \end{bmatrix}\begin{bmatrix} i_d \\ i_q \end{bmatrix} \tag{2.9}$$

To simplify the control of i_d and i_q, the coupling should be avoided in the current control. The decoupling can be achieved through the feedforward of i_d and i_q, as shown in Figure 2.4, where the feedforward components counteract the coupling components in Figure 2.3. The decoupling ensures the independent control of i_d and i_q. According to the current references i_{d_ref} and i_{q_ref}, the current control produces the d-axis and q-axis reference voltage u_{dm_ref} and u_{qm_ref} with the proportional-integral (PI) controller. Then the three-phase reference voltage u_{am_ref}, u_{bm_ref}, and u_{cm_ref} can be calculated as

$$\begin{bmatrix} u_{am\_ref} \\ u_{bm\_ref} \\ u_{cm\_ref} \end{bmatrix} = T_{dq/abc} \cdot \begin{bmatrix} u_{dm\_ref} \\ u_{qm\_ref} \end{bmatrix} \tag{2.10}$$

with

$$T_{dq/abc} = \begin{bmatrix} \cos\theta_e & -\sin\theta_e \\ \cos\left(\theta_e - \frac{2\pi}{3}\right) & -\sin\left(\theta_e - \frac{2\pi}{3}\right) \\ \cos\left(\theta_e + \frac{2\pi}{3}\right) & -\sin\left(\theta_e + \frac{2\pi}{3}\right) \end{bmatrix} \tag{2.11}$$

where $T_{dq/abc}$ is the inverse dq transformation matrix.

2.3.2 Power and DC-Link Voltage Control

When the MMC is connected to the AC power grid, the DC-link voltage V_{dc}, active power P, or reactive power Q of the MMC are controlled to regulate the power exchange between the MMC and the AC power grid, as shown in Figure 2.5.

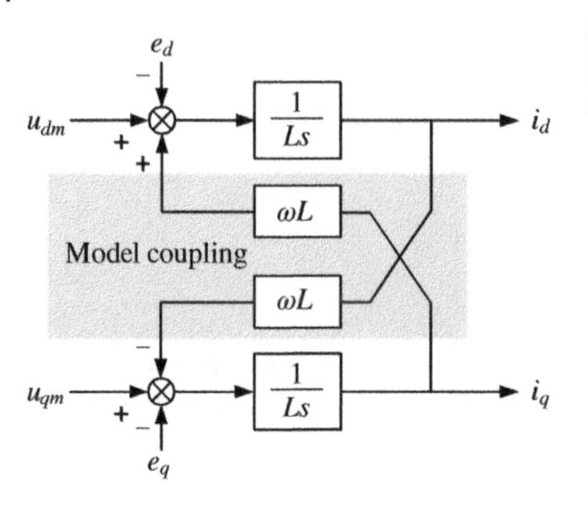

Figure 2.3 Mathematical model of the MMC in the synchronous-*dq* reference frame.

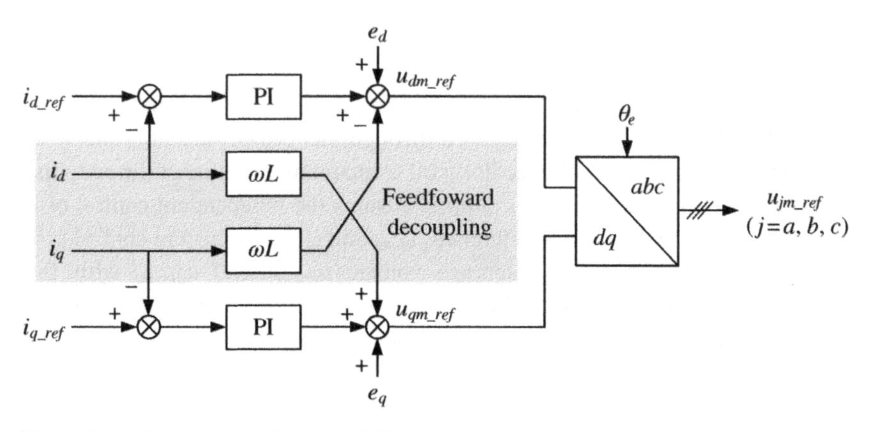

Figure 2.4 Current control of the MMC.

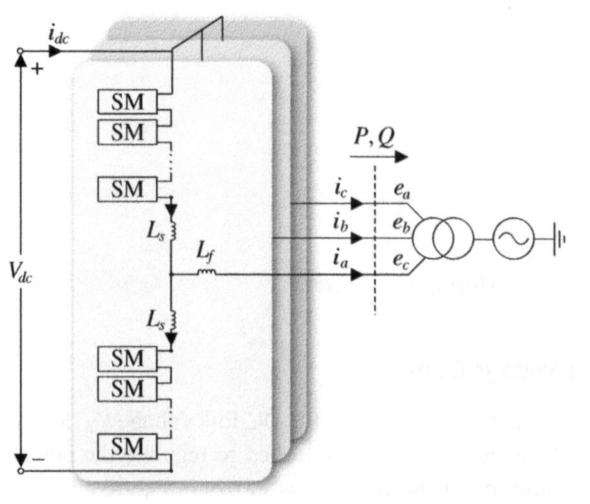

Figure 2.5 Structure of the three-phase MMC connected to the AC grid.

The active power P and reactive power Q in the synchronous-dq reference frame [4] can be expressed as

$$\begin{bmatrix} P \\ Q \end{bmatrix} = \frac{3}{2} \begin{bmatrix} e_d & e_q \\ e_q & -e_d \end{bmatrix} \begin{bmatrix} i_d \\ i_q \end{bmatrix} \tag{2.12}$$

Under balanced grid, if the grid voltage vector is aligned to the d-axis of synchronous reference frame, the d-axis grid voltage e_d is equal to the phase voltage amplitude E_m, and the q-axis grid voltage e_q is equal to 0, as

$$\begin{cases} e_d = E_m \\ e_q = 0 \end{cases} \tag{2.13}$$

As a result, the P and Q can be expressed as

$$\begin{cases} P = \dfrac{3}{2} e_d i_d = \dfrac{3}{2} E_m i_d \\ Q = -\dfrac{3}{2} e_d i_q = -\dfrac{3}{2} E_m i_q \end{cases} \tag{2.14}$$

The active power P is related to i_d, and the reactive power Q is related to i_q. Therefore, i_d and i_q can be controlled to regulate the P and Q exchange between the MMC and the AC grid, respectively.

If the DC-side voltage V_{dc} is supported by external voltage sources, the active power P and reactive power Q are controlled. The structures of active power and reactive power control of the MMC are shown in Figure 2.6a,b, respectively. According to the active power reference P_{ref} and the active power P, the d-axis current reference i_{d_ref} is derived by the PI control. According to the reactive power reference Q_{ref} and the reactive power Q, the q-axis current reference i_{q_ref} is derived by the PI control. Based on the current references i_{d_ref} and i_{q_ref}, the P and Q of the MMC can be regulated by the current control in Figure 2.4.

If the DC-side voltage V_{dc} is not supported by external voltage sources, the DC-link voltage V_{dc} and reactive power Q are controlled. V_{dc} is regulated by controlling i_d, because the V_{dc} is related to the stored energy of the MMC, which can be adjusted by the active power P. The structures of the DC-link voltage and reactive

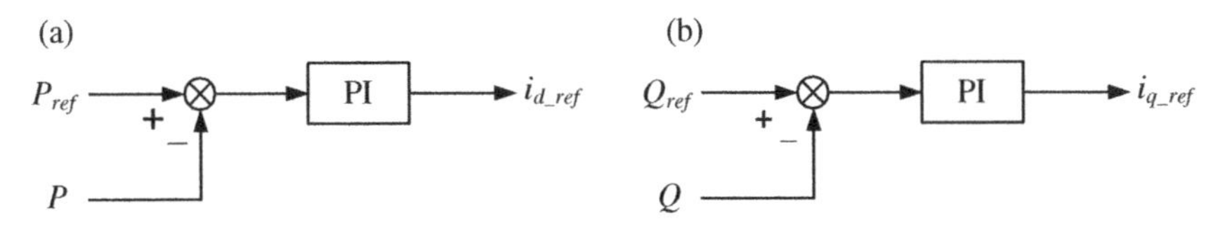

Figure 2.6 (a) Active power control of the MMC. (b) Reactive power control of the MMC.

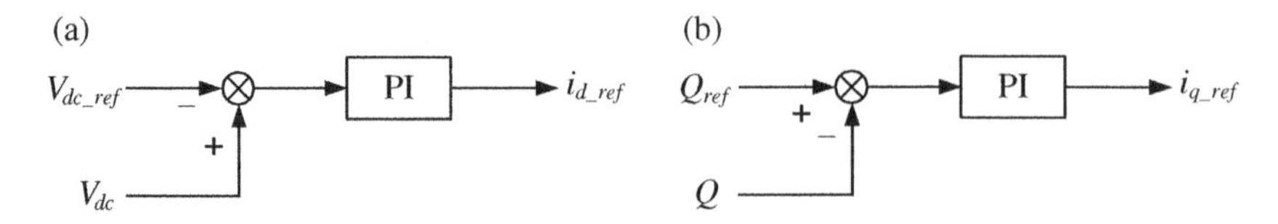

Figure 2.7 (a) DC-link voltage control of the MMC. (b) Reactive power control of the MMC.

power control of the MMC are shown in Figure 2.7a,b, respectively. According to the DC-link voltage reference V_{dc_ref} and DC-link voltage V_{dc}, the d-axis current reference i_{d_ref} is derived by the PI control. According to the reactive power reference Q_{ref} and the reactive power Q, the q-axis current reference i_{q_ref} is derived by the PI control. Based on the current references i_{d_ref} and i_{q_ref}, the V_{dc} and Q of the MMC can be regulated by the current control in Figure 2.4.

2.3.3 Grid Forming Control

The MMC operates in the grid forming mode for offshore and microgrid applications [5–7], where the amplitude E_m and angular frequency ω of AC grid voltages are regulated by the MMC. As shown in Figure 2.8, the output AC side of the MMC is connected to passive loads. The C_{ac} is considered in grid forming applications, which is the equivalent capacitance of added filter capacitors and network cables [7, 8].

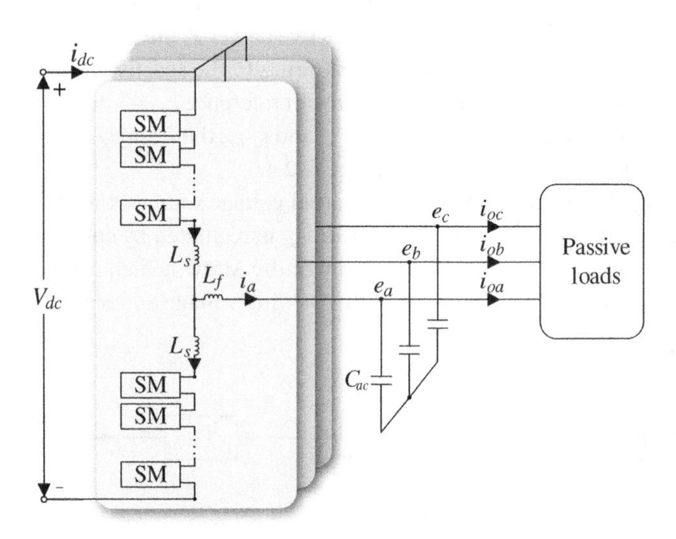

Figure 2.8 Structure of the three-phase MMC connected to passive loads.

In the grid forming mode, the mathematical model of the MMC in the stationary-*abc* reference frame is expressed as

$$\begin{cases} L\dfrac{di_a}{dt} = u_{am} - e_a \\ L\dfrac{di_b}{dt} = u_{bm} - e_b \\ L\dfrac{di_c}{dt} = u_{cm} - e_c \end{cases} \tag{2.15}$$

$$\begin{cases} C_{ac}\dfrac{de_a}{dt} = i_a - i_{oa} \\ C_{ac}\dfrac{de_b}{dt} = i_b - i_{ob} \\ C_{ac}\dfrac{de_c}{dt} = i_c - i_{oc} \end{cases} \tag{2.16}$$

where i_{oa}, i_{ob}, i_{oc} are the output current of the MMC. According to equations (2.7), (2.15), and (2.16), the mathematical model in synchronous-*dq* reference frame in the time-domain is expressed as

$$L\frac{d}{dt}\begin{bmatrix} i_d \\ i_q \end{bmatrix} = \begin{bmatrix} u_{dm} \\ u_{qm} \end{bmatrix} - \begin{bmatrix} e_d \\ e_q \end{bmatrix} + \begin{bmatrix} 0 & \omega L \\ -\omega L & 0 \end{bmatrix}\begin{bmatrix} i_d \\ i_q \end{bmatrix} \tag{2.17}$$

$$C_{ac}\frac{d}{dt}\begin{bmatrix} e_d \\ e_q \end{bmatrix} = \begin{bmatrix} i_d \\ i_q \end{bmatrix} - \begin{bmatrix} i_{od} \\ i_{oq} \end{bmatrix} + \begin{bmatrix} 0 & \omega C_{ac} \\ -\omega C_{ac} & 0 \end{bmatrix}\begin{bmatrix} e_d \\ e_q \end{bmatrix} \tag{2.18}$$

where i_{od} and i_{oq} are the *d*-axis and *q*-axis components of the MMC's output currents i_{oa}, i_{ob}, and i_{oc}, respectively.

With the Laplace transformation, the model in the Laplace-domain is derived in equations (2.19) and (2.20) and Figure. 2.9.

$$sL\begin{bmatrix} i_d \\ i_q \end{bmatrix} = \begin{bmatrix} u_{dm} \\ u_{qm} \end{bmatrix} - \begin{bmatrix} e_d \\ e_q \end{bmatrix} + \begin{bmatrix} 0 & \omega L \\ -\omega L & 0 \end{bmatrix}\begin{bmatrix} i_d \\ i_q \end{bmatrix} \tag{2.19}$$

$$sC_{ac}\begin{bmatrix} e_d \\ e_q \end{bmatrix} = \begin{bmatrix} i_d \\ i_q \end{bmatrix} - \begin{bmatrix} i_{od} \\ i_{oq} \end{bmatrix} + \begin{bmatrix} 0 & \omega C_{ac} \\ -\omega C_{ac} & 0 \end{bmatrix}\begin{bmatrix} e_d \\ e_q \end{bmatrix} \tag{2.20}$$

Figure 2.10 shows the structure of grid forming control for the MMC, and the coupling components between the *d*-axis and *q*-axis in Figure 2.9 are counteracted by the feedforward components in Figure 2.10. The grid forming control includes

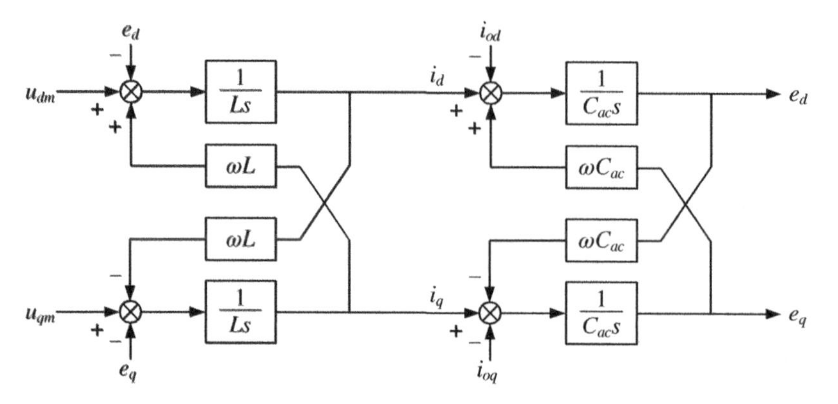

Figure 2.9 Model of the MMC in grid forming mode in the synchronous-*dq* reference frame.

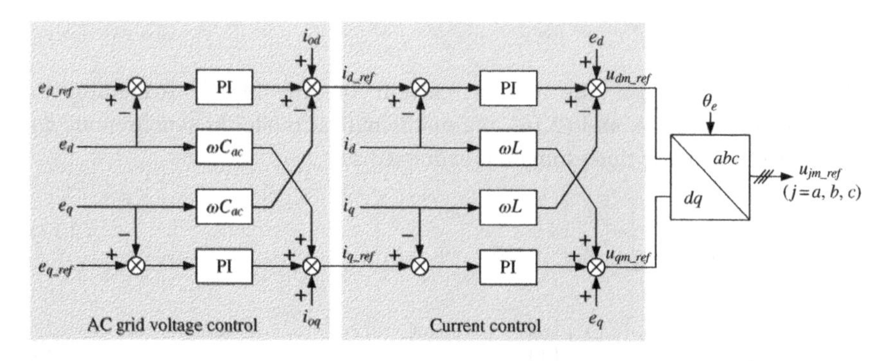

Figure 2.10 Grid forming control of the three-phase MMC.

the current control in Figure 2.4 and the AC grid voltage control. The phase angle θ_e of the AC grid voltage is generated by the controller. e_{d_ref} and e_{q_ref} are the d-axis and q-axis AC grid voltage references, where $e_{d\_ref} = E_m$ and $e_{q\_ref} = 0$. According to these references, the d-axis current reference i_{d_ref} and q-axis current reference i_{q_ref} are calculated by the AC grid voltage control. Then the current control produces the d-axis reference voltage u_{dm_ref} and q-axis reference voltage u_{qm_ref}. Finally, the three-phase reference voltage can be calculated by the inverse transformation $T_{dq/abc}$ from u_{dm_ref} and u_{qm_ref} to u_{am_ref}, u_{bm_ref}, and u_{cm_ref}.

2.4 Centralized Capacitor Voltage Balancing Control

The safe operation of MMCs relies on SM capacitor voltage balancing. This section mainly introduces several sorting-based centralized capacitor VBC.

2.4.1 On-State SMs Number Based VBC

The centralized on-state SMs number-based capacitor VBC [9] is based on the SM capacitor voltages in the arm, arm current, and the modulation scheme. Figure 2.11 shows the centralized on-state SMs number based VBC for the upper arm of phase A. According to the modulation scheme in Section 1.4, the number n_{on} of the SMs to be switched on in the upper arm of phase A is determined. The capacitor voltages u_{cau1}–u_{caun} of the n SMs in the upper arm of phase A are sorted in ascending order. The n_{on} SMs to be switched on in the upper arm of phase A are determined as follows.

- If $i_{au} > 0$, the n_{on} SMs with the lowest capacitor voltages in the upper arm are switched on, and the other $n - n_{on}$ SMs in the upper arm are switched off.
- If $i_{au} < 0$, the n_{on} SMs with the highest capacitor voltages in the upper arm are switched on, and the other $n - n_{on}$ SMs in the upper arm are switched off.

The on-state SMs number-based capacitor VBC can achieve excellent capacitor voltage balancing in the MMC. However, its calculation burden for SMs sorting would be large. Besides, additional unnecessary switching states are produced, which increases the total switching losses of the MMC.

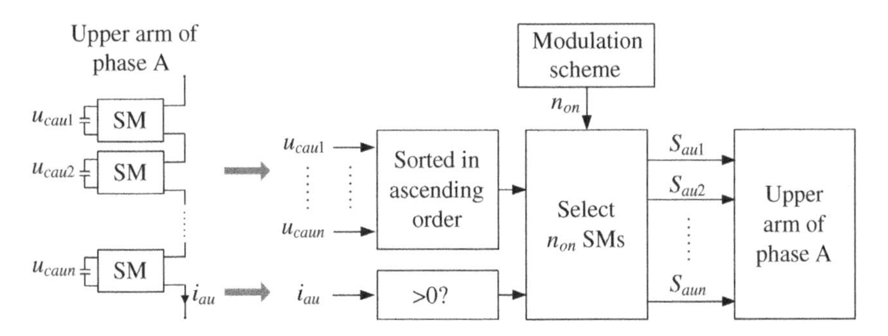

Figure 2.11 On-state SMs number based capacitor VBC.

2.4.2 Balancing Adjusting Number Based VBC

Compared with the on-state SMs number-based capacitor VBC, the balancing adjusting number-based capacitor VBC has the advantage of low switching frequency [10]. This control is quite suitable for MMCs where the switching frequency must be kept low to reduce switching losses. Besides, under different balancing adjusting number (BAN), the switching frequency can be calculated, which is beneficial to the design of cooling system.

2.4.2.1 Capacitor VBC

Figure 2.12 shows the BAN-based capacitor VBC for the MMC in the upper arm of phase A [10]. The number of on-state SMs n_{on} in the arm can be obtained by the reference signal y_{au} and the modulation strategy within each control period T_c. The incremental number ΔN_{incre} of the on-state SMs in the arm can be calculated as the difference of the on-state SMs number n_{on_new} at current control period and the on-state SMs number n_{on_old} at previous control period, as

$$\Delta N_{incre} = n_{on\_new} - n_{on\_old} \tag{2.21}$$

The adjusted number N_{adj} of the SMs to be switched on or switched off within this sampling cycle is

$$N_{adj} = |\Delta N_{incre}| + N_{ban} \tag{2.22}$$

where the N_{ban} is the balancing adjusting number.

The actual switching number of the SMs to be switched on among the $n-n_{on\_old}$ off-state SMs and the actual switching number of the SMs to be switched off among the n_{on_old} on-state SMs are decided based on the direction of the arm current i_{au} and the incremental number ΔN_{incre}, as follows.

- $i_{au} > 0$ *and* $\Delta N_{incre} = 0$: Switch on N_{ban} SMs with the lowest voltages among the $n-n_{on\_old}$ off-state SMs. In addition, switch off N_{ban} SMs with the highest voltages among the n_{on_old} on-state SMs.
- $i_{au} > 0$ *and* $\Delta N_{incre} > 0$: Switch on N_{adj} SMs with the lowest voltages among the $n-n_{on\_old}$ off-state SMs. In addition, switch off N_{ban} SMs with the highest voltages among the n_{on_old} on-state SMs.
- $i_{au} > 0$ *and* $\Delta N_{incre} < 0$: Switch on N_{ban} SMs with the lowest voltages among the $n-n_{on\_old}$ off-state SMs. In addition, switch off N_{adj} SMs with the highest voltages among the n_{on_old} on-state SMs.
- $i_{au} < 0$ *and* $\Delta N_{incre} = 0$: Switch on N_{ban} SMs with the highest voltages among the $n-n_{on\_old}$ off-state SMs. In addition, switch off N_{ban} SMs with the lowest voltages among the n_{on_old} on-state SMs.
- $i_{au} < 0$ *and* $\Delta N_{incre} > 0$: Switch on N_{adj} SMs with the highest voltages among the $n-n_{on\_old}$ off-state SMs. In addition, switch off N_{ban} SMs with the lowest voltages among the n_{on_old} on-state SMs.
- $i_{au} < 0$ *and* $\Delta N_{incre} < 0$: Switch on N_{ban} SMs with the highest voltages among the $n-n_{on\_old}$ off-state SMs. In addition, switch off N_{adj} SMs with the lowest voltages among the n_{on_old} on-state SMs.

2.4.2.2 SM Switching Frequency

In the BAN based capacitor VBC, when ΔN_{incre} is 0, N_{ban} "off" ("on") state SMs will be switched to on (off) state within each control period. Here, the switching times of all SMs in one arm is N_{ban} within the control period. When ΔN_{incre} is not equal

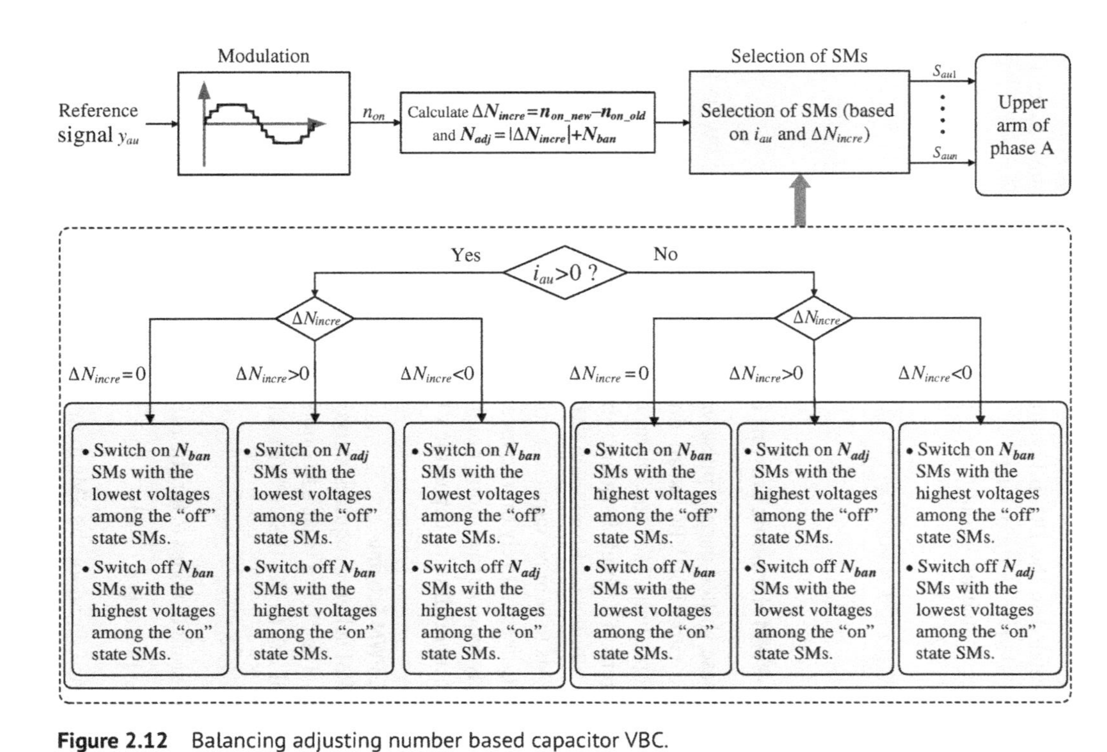

Figure 2.12 Balancing adjusting number based capacitor VBC.

to 0, the switching times will be more than N_{ban} within the control period. Suppose that the fundamental period is T_f, the control period is T_c, the total switching times N_{total} of all SMs in one arm within one fundamental period can be expressed as [10]

$$N_{total} = N_{ban} \times T_f / T_c + N_e \tag{2.23}$$

where N_e is the extra switching times introduced by the modulation algorithm. The N_e within each fundamental period T_f is about

$$N_e = \frac{(0.5V_{dc} + 0.5mV_{dc}) - (0.5V_{dc} - 0.5mV_{dc})}{u_c} = mn \tag{2.24}$$

The N_e is related to the DC-link voltage V_{dc}, capacitor voltage u_c, modulation index m, and SM number n per arm.

The average switching frequency f_{sw} of each SM can then be calculated based on (2.23) and (2.24), as

$$f_{sw} = \frac{N_{ban} T_f / T_c + mn}{n \cdot T_f} = \frac{N_{ban}}{n \cdot T_c} + \frac{m}{T_f} \tag{2.25}$$

2.4.3 IPSC-PWM Harmonic Current Based VBC

Improved phase-shifted carrier-pulse width modulation (IPSC-PWM) harmonic current based capacitor VBC is an interesting control, where the IPSC-PWM scheme is adopted and used to control the high-frequency current component in the arm current. The high-frequency energy based on the high-frequency arm current is distributed to the suitable SMs through assigning these PWM pulses to the suitable SMs to achieve capacitor voltage balancing in the arm. The presented capacitor VBC does not rely on the measurement of the arm currents, which not only effectively reduces the number of the sensors and decreases the costs but also simplifies the algorithm for capacitor VBC.

2.4.3.1 IPSC-PWM Scheme

The IPSC-PWM scheme requires n triangular carrier waves for the MMC with n SMs per arm. The carrier wave's frequency is f_s. Each carrier wave is phase-shifted by an angle of $\Delta\theta$. In the IPSC-PWM scheme, the $\Delta\theta$ is

$$0 \le \Delta\theta < 2\pi / n \tag{2.26}$$

The phase-shifted angles θ_i for the i-th carrier wave W_{ari} is

$$\theta_i = (i-1) \cdot \Delta\theta \ (1 \le i \le n) \tag{2.27}$$

The reference signal for the SMs in the upper arm is $-y_j$ and the reference signal for the SMs in the lower arm is y_j, ($j = a, b, c$). The n pulses S_{p_1}–S_{p_n} for the SMs in each arm can be generated by comparing the corresponding reference signal with the n phase-shifted carrier waves $W_{ar1}-W_{arn}$.

Figure 2.13 shows the waveforms for phase A ($n = 3$) under the IPSC-PWM scheme with the assumption that the frequency f_s of the carrier wave is far higher than that of the reference signal, and the capacitor voltages in each arm are the same. The three triangular carrier waves W_{ar1}, W_{ar2}, and W_{ar3} are considered for each arm, and each is shifted by an angle $\Delta\theta$. $\omega_s = 2\pi f_s$ is the angular frequency of the carrier wave. The switching functions are produced as follows.

- Through the comparison between the upper arm reference $-y_a$ and the carriers W_{ar1}, W_{ar2}, and W_{ar3}, the pulse S_{p_au1}, S_{p_au2}, S_{p_au3} for the upper arm are generated, whose summation is $S_{p\_au\_sum}$.
- Through the comparison between the lower arm reference y_a and the carriers W_{ar1}, W_{ar2}, and W_{ar3}, the pulse S_{p_al1}, S_{p_al2}, S_{p_al3} for the lower arm are generated, whose summation is $S_{p\_al\_sum}$.

The $S_{p\_a\_sum}$ can be produced as the sum of $S_{p\_au\_sum}$ and $S_{p\_al\_sum}$.

2.4.3.2 High-Frequency Arm Current

To analyze the high-frequency component at the frequency of f_s in each arm current, some assumptions are given as follows.

- The carrier wave frequency f_s is far higher than that of the reference wave. The widths of the n pulses for the upper arm SMs are assumed to be the same in each period of 2π, and the phase-shifted angles between these pulses are the same as $\Delta\theta$, as shown in Figure 2.13. Moreover, the widths of the n pulses for the lower arm SMs are assumed to be the same in each period of 2π, and the phase-shifted angles between these pulses are also the same as $\Delta\theta$, as shown in Figure 2.13.
- The capacitor voltages in each arm are the same.

The output voltage $u_{sm\_ui}(t)$ of the SM in the upper arm with the i-th pulse S_{p_aui} can be analyzed in each period of 2π from $(n-1)\Delta\theta/2+2(j-1)\pi$ to $(n-1)\Delta\theta/2+j2\pi$ ($j = 1, 2\ldots$), which can be expressed as a Fourier series expansion as

$$u_{sm\_ui}(t) = \frac{u_c \Delta\theta_u}{2\pi} + \frac{2u_c}{\pi}\sum_{m=1}^{\infty}(-1)^m \frac{1}{m}\sin\left(\frac{m\Delta\theta_u}{2}\right)\cos\left[m\left(\omega_s t-(i-1)\Delta\theta\right)\right] \quad (2.28)$$

where $\Delta\theta_u$ is the pulse width in the upper arm, as shown in Figure 2.13. The high-frequency voltage component at the frequency of f_s in $u_{sm\_ui}(t)$ is

$$u_{sm\_fs\_ui}(t) = -\frac{2u_c}{\pi}\sin\left(\frac{\Delta\theta_u}{2}\right)\cos\left[\omega_s t-(i-1)\Delta\theta\right] \quad (2.29)$$

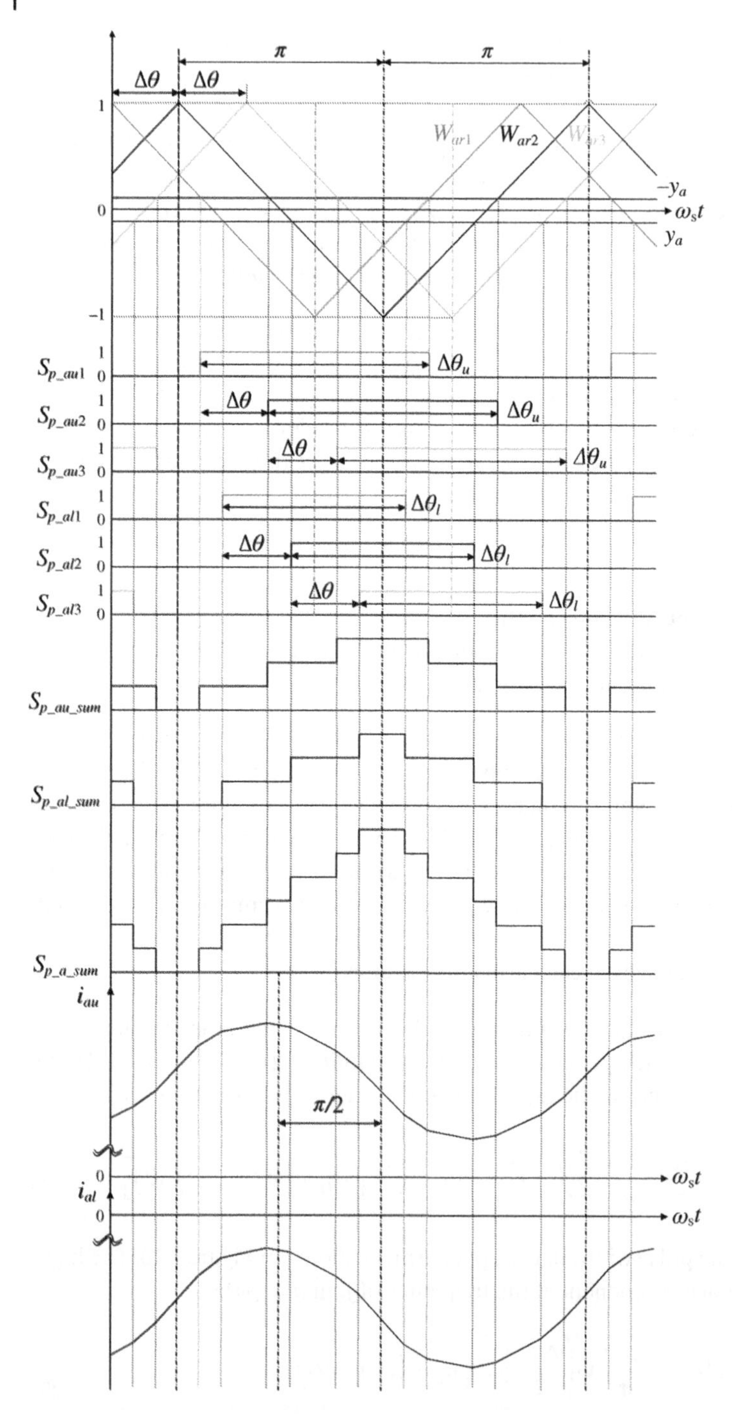

Figure 2.13 PSC-PWM scheme for phase A.

The high-frequency voltage components $u_{smu\_fs\_sum}(t)$ at the frequency of f_s in the upper arm can be expressed as

$$u_{smu\_fs\_sum}(t)=\sum_{i=1}^{n}u_{sm\_fs\_ui}(t)=-\frac{2u_c}{\pi}\sin\frac{\Delta\theta_u}{2}\frac{\sin\frac{n\Delta\theta}{2}}{\sin\frac{\Delta\theta}{2}}\cos\left(\omega_s t-\frac{n-1}{2}\Delta\theta\right) \tag{2.30}$$

With the same method, the high-frequency voltage components $u_{sml\_fs\_sum}(t)$ at the frequency of f_s in the lower arm can be expressed as

$$u_{sml\_fs\_sum}(t)=\sum_{i=1}^{n}u_{sm\_fs\_li}(t)=-\frac{2u_c}{\pi}\sin\frac{\Delta\theta_l}{2}\frac{\sin\frac{n\Delta\theta}{2}}{\sin\frac{\Delta\theta}{2}}\cos\left(\omega_s t-\frac{n-1}{2}\Delta\theta\right) \tag{2.31}$$

where $\Delta\theta_l$ is the pulse width in the lower arm, as shown in Figure 2.13. The voltage $u_{fs\_sum}(t)$ with the frequency of f_s can be obtained with the addition of $u_{smu\_fs\_sum}(t)$ and $u_{sml\_fs\_sum}(t)$.

$$\begin{aligned}u_{fs\_sum}(t)&=u_{smu\_fs\_sum}(t)+u_{sml\_fs\_sum}(t)\\&=-\frac{4u_c}{\pi}\frac{\sin\frac{n\Delta\theta}{2}}{\sin\frac{\Delta\theta}{2}}\cos\left(\frac{\Delta\theta_u-\Delta\theta_l}{4}\right)\cos\left(\omega_s t-\frac{n-1}{2}\Delta\theta\right)\end{aligned} \tag{2.32}$$

The voltage $u_{fs\_sum}(t)$ may be caused in the MMC under the IPSC-PWM scheme. The peak value of the $u_{fs\_sum}(t)$ appears at the phase angle of $(n-1)\Delta\theta/2+(2j-1)\pi$ ($j = 1, 2...$), which is the middle point of each period, as shown in Figure 2.13. Owing to the voltage $u_{fs\_sum}(t)$, a high-frequency current $i_{fs\_sum}(t)$ with a frequency of f_s may also be caused, which can be expressed as

$$\begin{aligned}i_{fs\_sum}(t)&=\frac{u_{fs\_sum}(t)}{2\omega_s L_s}\\&=\frac{2u_c}{\pi\omega_s L_s}\frac{\sin\frac{n\Delta\theta}{2}}{\sin\frac{\Delta\theta}{2}}\cos\left(\frac{\Delta\theta_u-\Delta\theta_l}{4}\right)\cos\left(\omega_s t-\frac{n-1}{2}\Delta\theta-\frac{\pi}{2}\right)\end{aligned} \tag{2.33}$$

The high-frequency current $i_{fs\_sum}(t)$ leads the $u_{fs\_sum}(t)$ by $\pi/2$, as shown in Figure 2.13. This means that the zero-crossing point of the caused high-frequency current with the frequency of f_s occurs at the phase angle of $(n-1)\Delta\theta/2+(2j-1)\pi$ ($j = 1, 2...$), which is the middle point of each period. The peak value of the high-frequency current with the frequency of f_s appears at approximately $\pi/2$ in each period, as shown in Figure 2.13.

Table 2.1 Capacitor state.

Arm current (i_{au} or i_{al})	SM state	Capacitor state	Capacitor voltage u_c
Positive	On	Charge	Increased
	Off	Bypass	Unchanged
Negative	On	Discharge	Decreased
	Off	Bypass	Unchanged

2.4.3.3 Arm Capacitor Voltage Analysis

The capacitor voltage in each SM is related to its switching state and the direction of the arm current, as shown in Table 2.1. In principle, if the arm current i_{au} (or i_{al}) is positive, as shown in Figure 1.3, and the SM's state is on, the corresponding capacitor would be charged and its voltage increased. On the contrary, the capacitor would be discharged and its voltage reduced if the arm current i_{au} (or i_{al}) is negative and the SM's state is on. In contrast, if the SM's state is off, which means the capacitor is bypassed, the capacitor voltage remains unchanged.

Under the IPSC-PWM scheme, the capacitor state including charge and discharge can be analyzed within each period of 2π, as shown in Figure 2.13. Owing to the aforementioned assumption that the widths of the n pulses for the SMs in the upper arm are the same in each period, the capacitor voltage will be decided by the arm current during the different on-times of the pulse, as follows.

- *Positive arm current ($i_{au} > 0$)*
  The capacitors in the upper arm may absorb power and be charged in this case. According to the above analysis, the pulse with its middle point close to $\pi/2$ may make the SM absorb more power and cause the capacitor voltage to be higher than the pulse with its middle point far from $\pi/2$.
- *Negative arm current ($i_{au} < 0$)*
  The capacitors in the upper arm may emit power and be discharged in this case. According to the above analysis, the pulse with its middle point close to $\pi/2$ may make the SM emit less power and cause the capacitor voltage to be higher than the pulse with its middle point far from $\pi/2$.

Regardless of the direction of the arm current, the SM with the pulse whose middle point is close to $\pi/2$ may absorb more power or emit less power, and its corresponding voltage tends to be higher in comparison with the SM with the pulse whose middle point is far away from $\pi/2$. The analysis for the lower arm SM capacitor voltage is the same as that for the upper arm SM capacitor voltage.

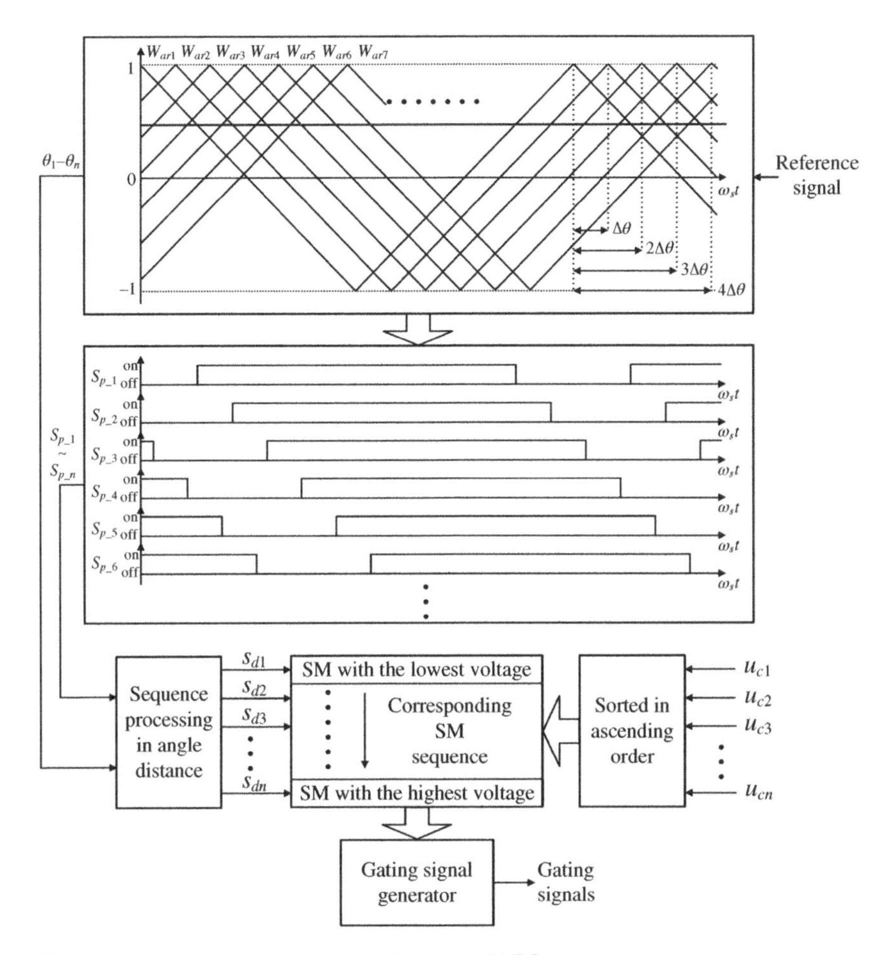

Figure 2.14 IPSC-PWM harmonic current based VBC.

2.4.3.4 Voltage Balancing Control

Figure 2.14 shows the IPSC-PWM harmonic current based VBC for the capacitor voltage balancing of the SMs in each arm, which is implemented in each arm. With the IPSC-PWM scheme, the n carrier waves W_{ar1}–W_{arn} with the phase-shifted angle $\Delta\theta$ are compared with the reference signal and used to generate n phase-shifted pulses S_{p_1}–S_{p_n}.

To achieve the capacitor voltage balancing of the SMs in each arm, the capacitor voltage u_{c1}–u_{cn} in each arm is monitored in real time and sampled at the control frequency f_s and then sorted in ascending order. The VBC is implemented in each period of 2π from $(n-1)\Delta\theta/2+(j-1)2\pi$ to $(n-1)\Delta\theta/2+j2\pi$ ($j = 1, 2...$), as shown in Figure 2.13.

If the capacitor voltage is low, the pulse with its middle point close to $\pi/2$ may be assigned to the SM. Consequently, the corresponding SM capacitor will absorb more power when the arm current is positive and the capacitor voltage increases more or emits less power when the arm current is negative and the capacitor voltage decreases less. In contrast, if the capacitor voltage is high, the pulse with its middle point far from $\pi/2$ may be assigned to the SM. Consequently, the corresponding SM capacitor will absorb less power when the arm current is positive and the capacitor voltage increases less or emits more power when the arm current is negative and the capacitor voltage decreases more. As a consequence, the pulses S_{p_1}–S_{p_n} are sequenced in ascending order according to the distances between the middle point of the pulses and $\pi/2$. Actually, the middle point of the pulse is also the time point of the minimum value of the carrier wave, as shown in Figure 2.13. Finally, the pulses S_{d1}–S_{dn} are obtained as shown in Figure 2.14, in which the distance between the middle point of S_{d1} and $\pi/2$ is minimum and the distance between the middle point of S_{dn} and $\pi/2$ is maximum. The SMs in the arm are also ordered according to the capacitor voltage in ascending order and driven with the pulses in a sequence from S_{d1} to S_{dn}, which can effectively ensure capacitor voltage balancing in each arm.

Case Study 2.1 Analysis of the IPS-PWM Harmonic Current Based VBC

Objective: To verify the IPSC-PWM harmonic current based VBC, a three-phase MMC system is modeled with the professional tool power systems computer aided design/electromagnetic transients including DC (PSCAD/EMTDC), as shown in Figure 2.15. The three-phase MMC system is linked to a three-phase AC grid and works in the inverter mode. The system parameters are shown in Table 2.2.

Simulation results and analysis:

Figure 2.16 shows the performance of the three-phase MMC under the presented voltage balancing control, in which the phase-shifted angle $\Delta\theta$ is set as $2\pi/10.5$. The three-phase voltage and current waveforms of the MMC are

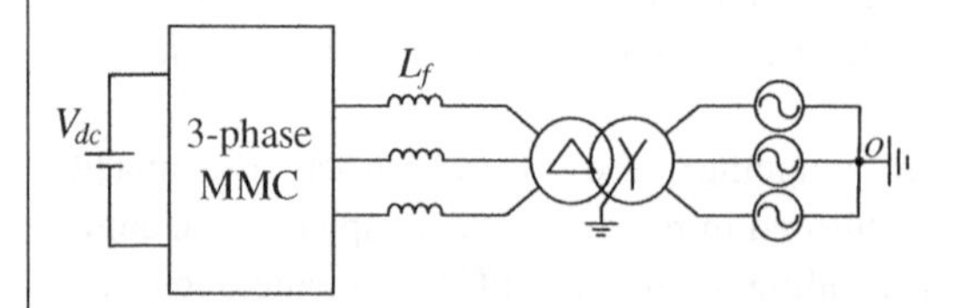

Figure 2.15 Block diagram of the simulation system.

Table 2.2 Simulation system parameters.

Parameters	Value
Active power P (MW)	20
DC-link voltage V_{dc} (kV)	20
Grid line-to-line voltage (kV)	35
Grid frequency (Hz)	50
Transformer voltage rating (kV)	10/35
Transformer leakage reactance	5%
Number of SMs per arm n	10
SM capacitance C (mF)	15
Arm inductance L_s (mH)	6
Inductance L_f (mH)	4
Carrier frequency f_s (kHz)	2.5

shown in Figure 2.16a,b. Here, the system active power is 20 MW, and the reactive power is 0. The upper arm current i_{au}, the lower arm current i_{al}, and the $(i_{au} + i_{al})/2$ are shown in Figure 2.16c, in which the total harmonic distortion of the arm current is approximately 0.55% and the ratio of the high-frequency current with the frequency of f_s to the 50 Hz fundamental current is 0.45%. The circulating current suppression method introduced in [1] is used in the MMC. Hence, the $(i_{au} + i_{al})/2$ only contains a DC component. Figure 2.16d shows the upper arm capacitor voltage $u_{cau1} - u_{cau10}$ and the lower arm capacitor voltage $u_{cal1} - u_{cal10}$. The capacitor voltages of the MMC are kept balanced.

The high-frequency component of the arm current and the stability of the capacitor voltage is investigated with different phase-shifted angles. Figures 2.17–2.19 show the upper arm current i_{au} and the capacitor voltage in phase A under the phase-shifted angles $\Delta\theta$ as $2\pi/15$, $2\pi/12$, and $2\pi/10$, respectively. The high-frequency current with the frequency of f_s appears in the i_{au} when $\Delta\theta$ is $2\pi/15$ and $2\pi/12$, which accounts for 3.85% and 1.79% in the 50 Hz fundamental current, respectively. Under the voltage balancing control, the capacitor voltage balancing can be effectively maintained when $\Delta\theta$ is $2\pi/15$ and $2\pi/12$, as shown in Figures 2.17b and 2.18b. When $\Delta\theta$ is $2\pi/10$, the high-frequency current with the frequency of f_s accounts for 0.01% in the 50 Hz fundamental current i_{au}, which is quite small and can be nearly neglected, as shown in Figure 2.19a. As a consequence, the capacitor voltage balancing cannot be maintained, as shown in Figure 2.19b.

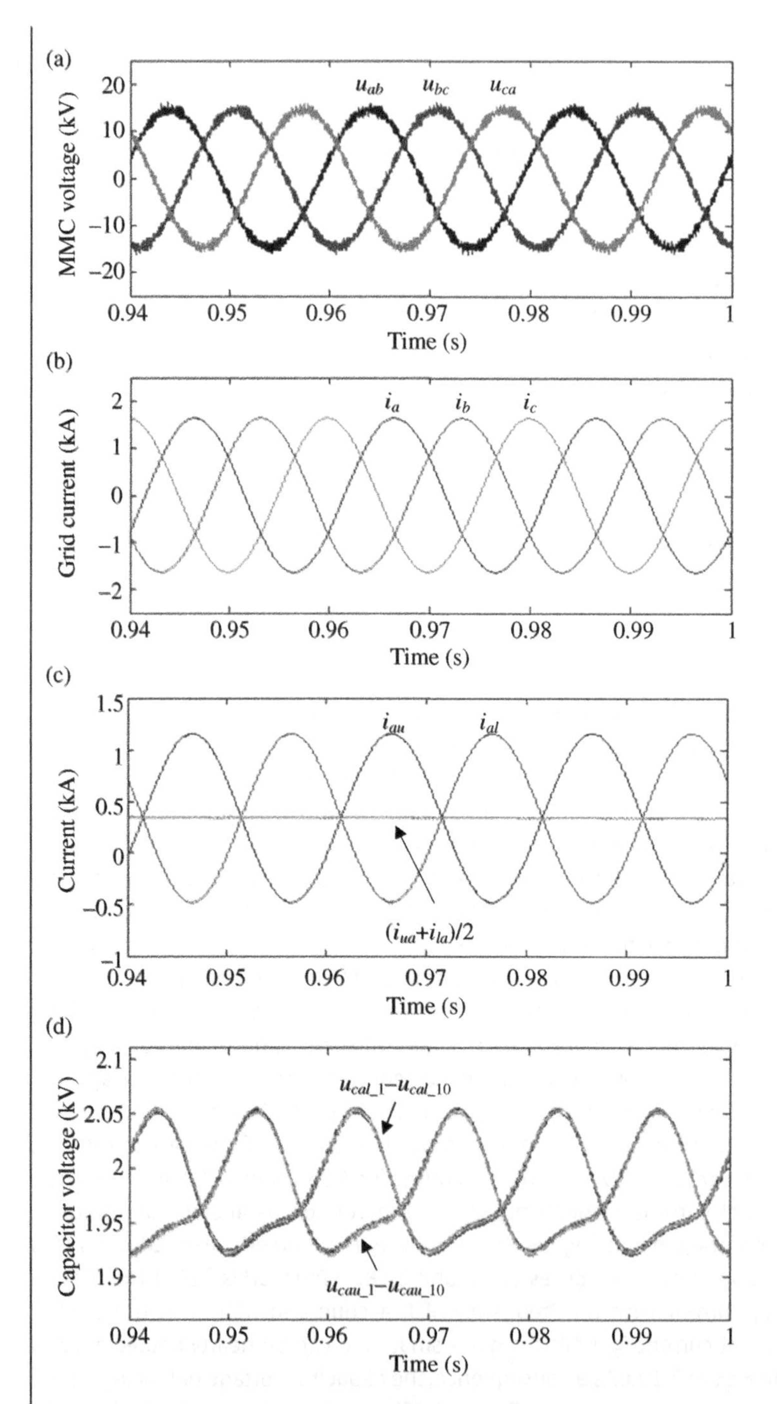

Figure 2.16 Simulated waveforms of the MMC with the voltage balancing control. (a) Voltage u_{ab}, u_{bc}, and u_{ca}. (b) Grid current i_a, i_b, and i_c. (c) Phase A arm current i_{au}, i_{al}, and $(i_{au}+i_{al})/2$. (d) Capacitor voltage of phase A.

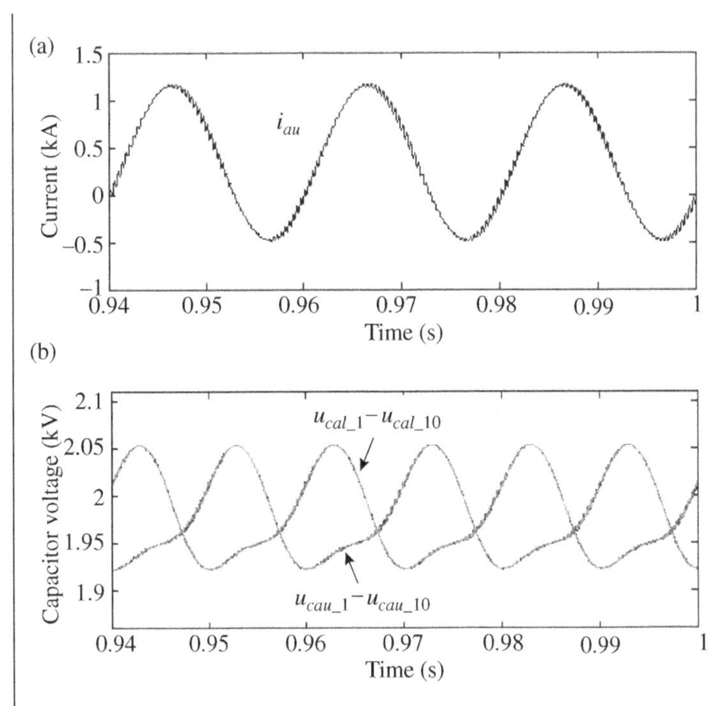

Figure 2.17 (a) Arm current i_{au} under a phase-shifted angle of $2\pi/15$. (b) Capacitor voltage in phase A under a phase-shifted angle of $2\pi/15$.

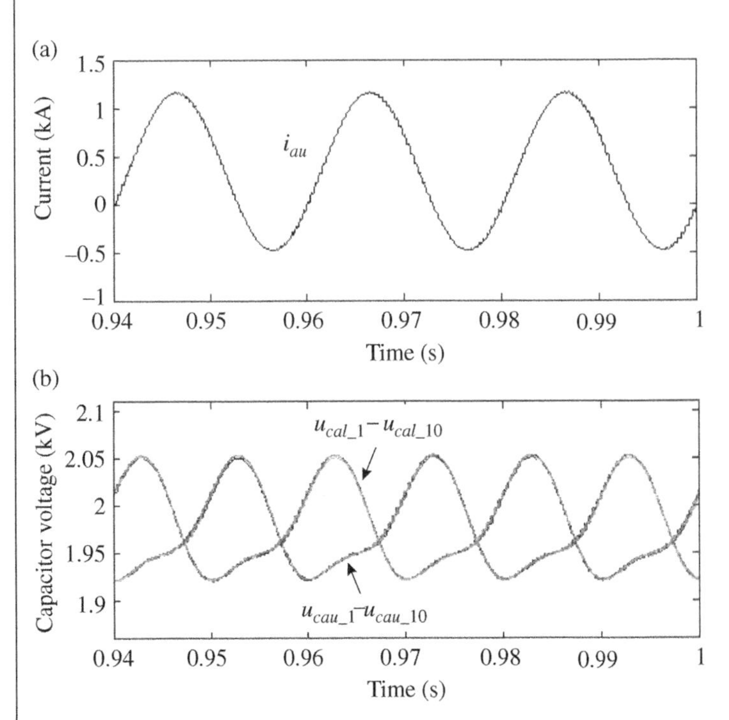

Figure 2.18 (a) Arm current i_{au} under a phase-shifted angle of $2\pi/12$. (b) Capacitor voltage in phase A under a phase-shifted angle of $2\pi/12$.

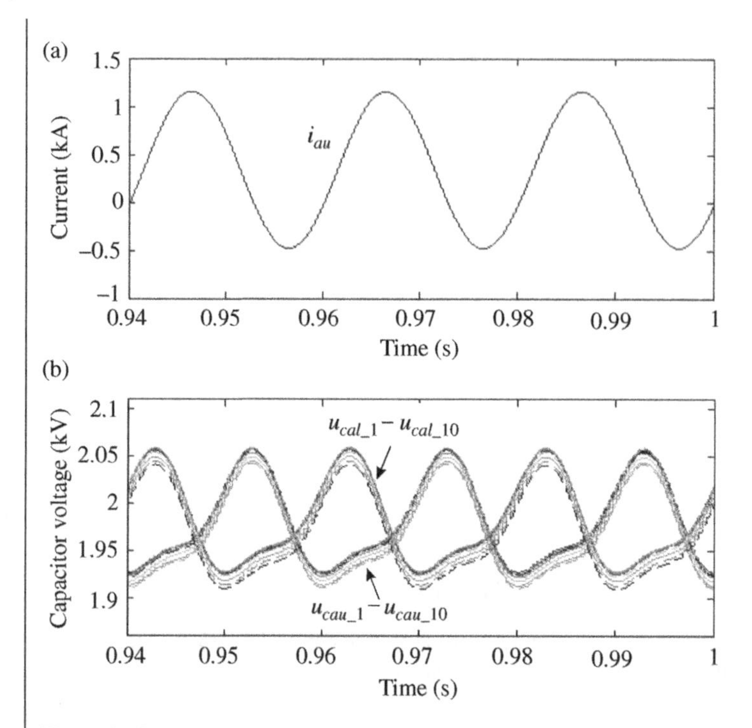

Figure 2.19 (a) Arm current i_{au} under a phase-shifted angle of $2\pi/10$. (b) Capacitor voltage in phase A under a phase-shifted angle of $2\pi/10$.

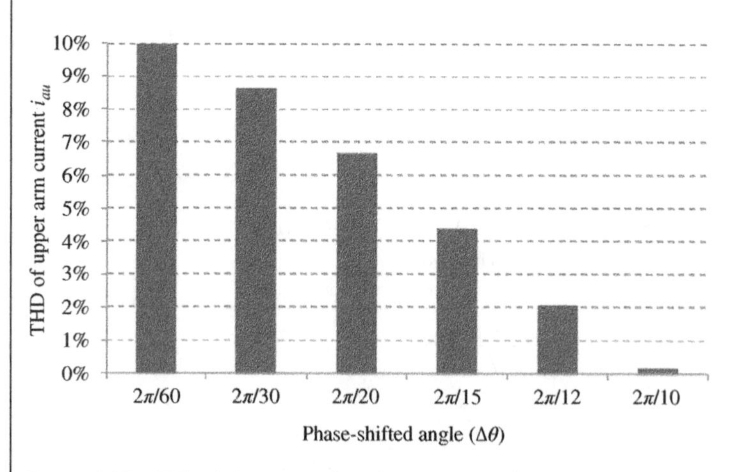

Figure 2.20 THD of the upper arm current i_{au}.

Figure 2.20 shows the total harmonic distortion (THD) of the arm current under the different phase-shifted angles, including $2\pi/60, 2\pi/30, 2\pi/20, 2\pi/15$, $2\pi/12$, and $2\pi/10$, where the larger the phase-shifted angle, the smaller the

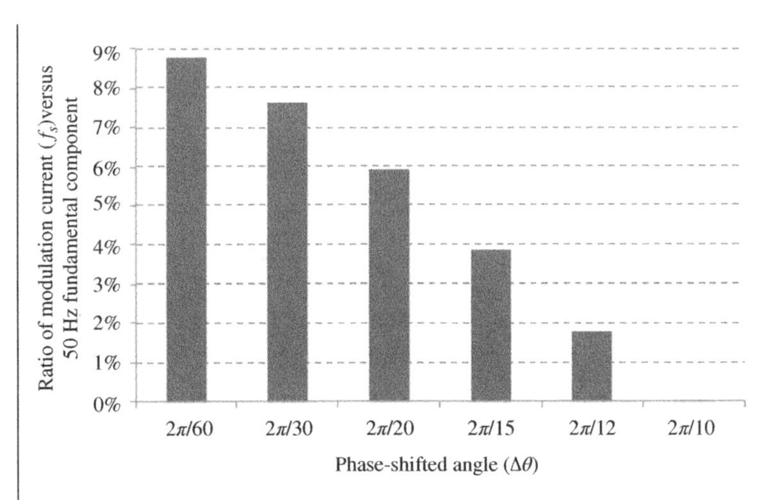

Figure 2.21 The ratio of the high-frequency current with the frequency of f_s to the 50 Hz fundamental current in the upper arm of phase A.

THD of the arm current. Figure 2.21 shows the ratio of the high-frequency current with the frequency of f_s to the 50 Hz fundamental current. Along with the increase of the phase-shifted angle, the current component with the frequency of f_s in the arm current will be decreased and will be nearly zero when the phase-shifted angle is $2\pi/10$.

2.4.4 SHE-PWM Pulse Energy Sorting Based VBC

Selected harmonic elimination (SHE)-PWM pulse energy sorting based control is with the same switching frequency as grid frequency, where the drive pulses have different pulse widths but the same phase angle in each switching period. The SHE-PWM pulse energy sorting based VBC is achieved by sorting and assigning the optimal low-frequency pulse width to the corresponding SM and does not rely on the measurement of the arm currents, which potentially contributes to the improvement of the reliability of the system by reducing the number of current sensors that may potentially fail.

2.4.4.1 MMCs Analysis with Grid-Frequency Pulses

The MMC has n number of SMs in each arm, where n is even. Each SM in the arm is driven with a grid-frequency pulse, as shown in Figure 2.22 ($n = 4$). $\omega = 2\pi f_g$, f_g is grid frequency. In Figure 2.22, S_{p_l1}–S_{p_ln} and S_{p_u1}–S_{p_un} are the pulses for the lower and upper arms, respectively.

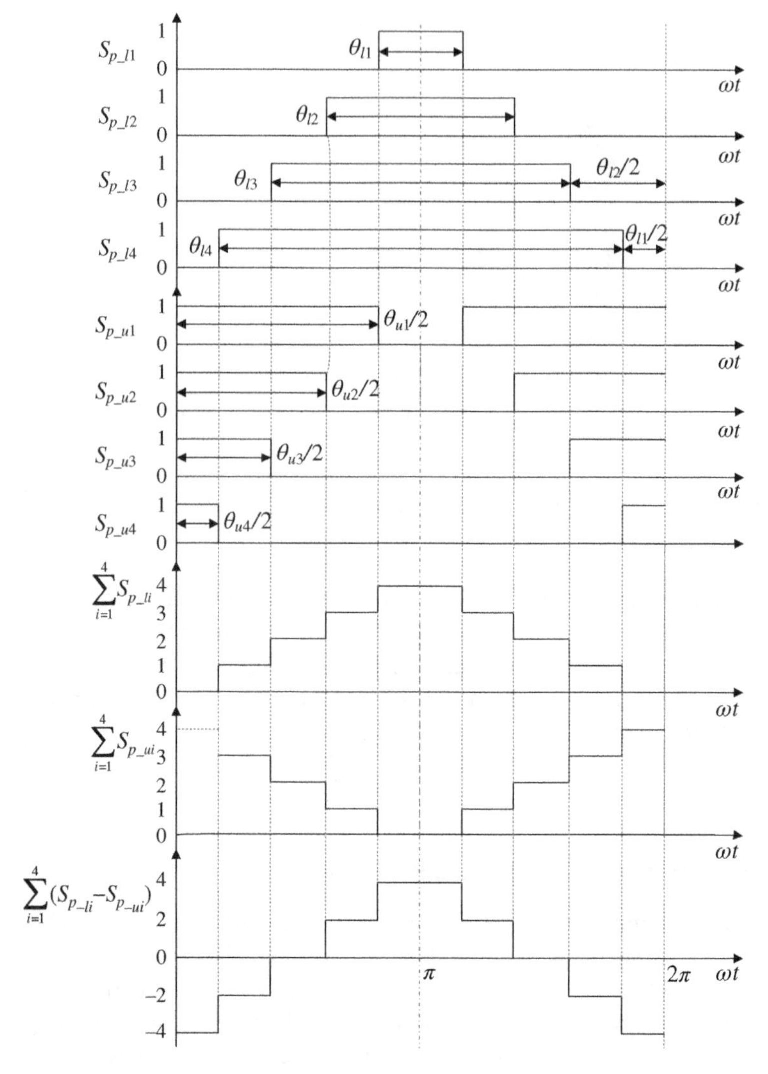

Figure 2.22 Upper and lower arm pulses for the MMC.

The pulse width θ_{l1}–θ_{ln} ($0 \leq \theta_{l1} \leq \theta_{l2} \leq \ldots \leq \theta_{ln} \leq 2\pi$) corresponding to the n pulses S_{p_l1}–S_{p_ln} in the lower arm has the relationship as

$$\theta_{li} = 2\pi - \theta_{l(n+1-i)} \left(1 \leq i \leq n\right) \tag{2.34}$$

The n pulses S_{p_u1}–S_{p_un} for the SMs in the upper arm are complementary to those for the lower arm SMs as $S_{p\_ui}=/S_{p\_li}$ ($1 \leq i \leq n$). Hence, the upper arm pulse width θ_{u1}–θ_{un} can be obtained as

$$\theta_{ui} = 2\pi - \theta_{li}\left(1 \le i \le n\right) \tag{2.35}$$

Suppose the capacitor voltage in each arm of the MMC is kept the same, the i-th SM output voltage u_{sm_aui} and u_{sm_ali} with the pulse S_{p_ui} and S_{p_li} in the upper and lower arms of phase A can be expressed as

$$\begin{cases} u_{sm\_aui} = \dfrac{V_{dc}}{n} S_{p\_ui} \\ u_{sm\_ali} = \dfrac{V_{dc}}{n} S_{p\_li} \end{cases} \tag{2.36}$$

The total output voltage u_{au} and u_{al} of the series-connected SMs in the upper and lower arms of phase A can be expressed as

$$\begin{cases} u_{au} = \sum_{i=1}^{n} u_{sm\_aui} = \dfrac{V_{dc}}{n} \sum_{i=1}^{n} S_{p\_ui} \\ u_{al} = \sum_{i=1}^{n} u_{sm\_ali} = \dfrac{V_{dc}}{n} \sum_{i=1}^{n} S_{p\_li} \end{cases} \tag{2.37}$$

The converter output voltage u_{am} of phase A can be expressed as

$$u_{am} = \frac{u_{al} - u_{au}}{2} = \frac{V_{dc}}{2n} \sum_{i=1}^{n} \left(S_{p\_li} - S_{p\_ui}\right) \tag{2.38}$$

Combining equations (2.36), (2.37), and (2.38) and Figure 2.22, the n lower arm pulses S_{p_l1}–S_{p_ln} and the n upper arm pulses S_{p_u1}–S_{p_un} synthesize a symmetric $(n+1)$-level output voltage u_{am} for the MMC. The Fourier series expansion of the voltage u_{am} can be expressed as

$$u_{am} = \frac{4V_{dc}}{n\pi} \sum_{h=1,3,5}^{\infty} \left[\frac{1}{h} \sum_{i=1}^{n/2} \sin\left(\frac{h\theta_{li}}{2} \right) \right] \cdot \cos\left(h\omega t\right) \tag{2.39}$$

where h is the order of the harmonic. Equation (2.39) shows that the voltage u_{am} only consists of the fundamental component and the odd harmonic components. Normally, the triplen harmonics can be excluded from the converter in a balanced three-phase system by a Y/Δ-connected three-phase transformer, with the Δ-connection on the converter side. In addition, some non-triplen odd harmonics can also be eliminated with the pulse width regulation to reduce the total harmonic distortion [11]. According to (2.39), the fundamental amplitude of u_{am} can be calculated as

$$U_{m} = \frac{4V_{dc}}{n\pi} \sum_{i=1}^{n/2} \sin\left(\frac{\theta_{li}}{2} \right) \tag{2.40}$$

2.4.4.2 Charge Transfer of Capacitors in Lower Arm

Suppose the circulating current is suppressed with the method introduced in [12], the arm current i_{au} and i_{al} can be described as

$$\begin{cases} i_{au} = \dfrac{i_{dc}}{3} + \dfrac{I_m}{2}\sin(\omega t + \theta) \\ i_{al} = \dfrac{i_{dc}}{3} - \dfrac{I_m}{2}\sin(\omega t + \theta) \end{cases} \tag{2.41}$$

Figure 2.23a shows the low-frequency pulses and the arm current in one period. During the inserted state of the SM, its capacitor voltage would be changed due to the charge transfer under the variable arm current. The charge Q_{al} transferred to the lower arm capacitor during one period can be obtained by the integral of the lower arm current i_{al} over the on-time of the lower arm pulse S_{p_l} as

$$Q_{al} = \int_0^{2\pi} S_{p\_l} \cdot i_{al} d(\omega t) = -I_m \sin(\theta)\sin\left(\frac{\theta_l}{2}\right) + \frac{i_{dc}}{3}\theta_l \tag{2.42}$$

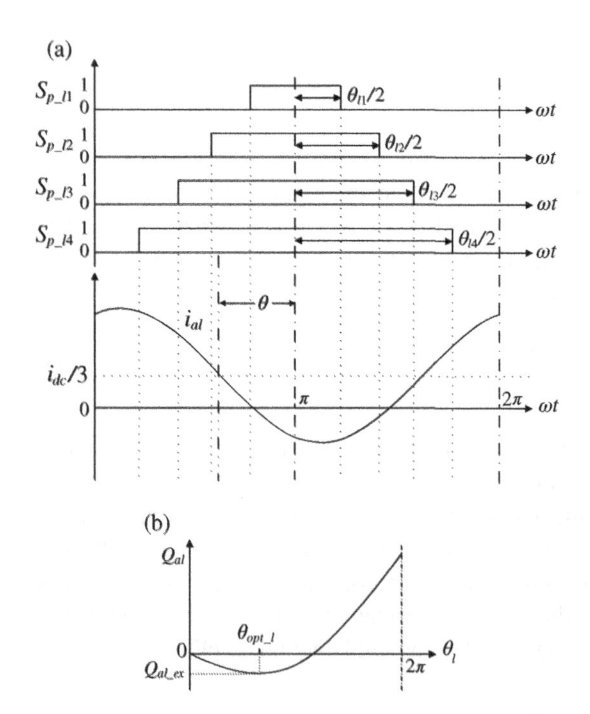

Figure 2.23 (a) Lower arm pulses and current. (b) Charge transfer characteristic of the capacitor under various pulse widths in the lower arm.

The Q_{al} is a function of the width θ_l of the pulse S_{p_l} in the lower arm. Figure 2.23b shows the characteristic of the variable Q_{al} according to the value of the θ_l, where the Q_{al} may be positive or negative.

- If the Q_{al} is positive, the capacitor would be charged and its voltage is increased.
- If the Q_{al} is negative, the capacitor would be discharged and its voltage is decreased.

The capacitor voltage ripple is related to the Q_{al} in one period. In Figure 2.23, the optimal pulse width for the charge transfer extreme Q_{al_ex} is

$$\theta_{opt\_l} = 2\cos^{-1}\left[\frac{2i_{dc}}{3I_m \sin(\theta)}\right] \tag{2.43}$$

Neglecting the losses and combining equations (2.40) and (2.41) and Figure 2.23, the power balancing relationship between the DC side and the AC side of the three-phase MMC can be expressed as

$$V_{dc} \cdot i_{dc} = \frac{3U_m I_m}{2} \cdot \sin(\theta) \tag{2.44}$$

Substituting equations (2.40) and (2.44) into equation (2.43), there will be

$$\theta_{opt\_l} = 2\cos^{-1}\left[\frac{4}{n\pi}\sum_{i=1}^{n/2} \sin\left(\frac{\theta_{li}}{2}\right)\right] \tag{2.45}$$

The value θ_{opt_l} can be calculated with the pulse widths θ_{l1}–θ_{ln}. According to Figure 2.23, the capacitor in the lower arm SM can be charged or discharged to different extents by the pulses with different widths. If the pulse width is close to θ_{opt_l}, the pulse will result in more charge transferred away from the capacitor under $Q_{al} < 0$, or the pulse will cause less charge transferred to the capacitor under $Q_{al} > 0$, in comparison with the pulse whose width is far away from θ_{opt_l}. Consequently, the pulse with its width close to θ_{opt_l} can result in lower capacitor voltage than the pulse with its width far away from θ_{opt_l} in the lower arm. The analysis for phases B and C is the same to phase A, which is not repeated here.

2.4.4.3 Charge Transfer of Capacitors in Upper Arm

With the same method for the lower arm, the charge transfer Q_{au} of the upper arm capacitor in one period can be obtained as equation (2.46) by the integral of the upper arm current i_{au} over the upper arm pulse S_{p_u} as shown in Figure 2.24a.

$$Q_{au} = \int_0^{2\pi} S_{p\_u} \cdot i_{au} d(\omega t) = -I_m \sin(\theta)\sin\left(\frac{\theta_u}{2}\right) + \frac{i_{dc}}{3}\theta_u \tag{2.46}$$

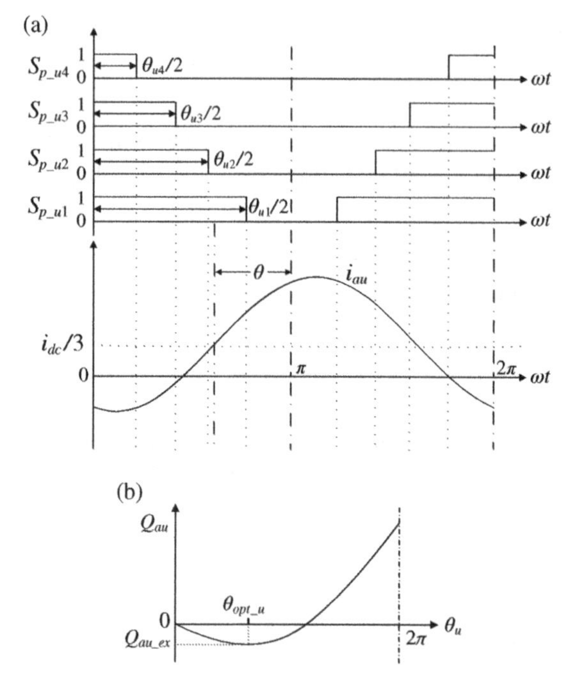

Figure 2.24 (a) Upper arm pulses and current. (b) Charge transfer characteristic of the capacitor under various pulse widths in the upper arm.

The capacitor charge transfer Q_{au} is related to the width θ_u of the pulse S_{p_u} in the upper arm. Figure 2.24b depicts the relationship between the Q_{au} and the θ_u. Based on equation (2.46), the optimal pulse width for the charge transfer extreme Q_{au_ex} in Figure 2.24 can be calculated as

$$\theta_{opt\_u} = 2\cos^{-1}\left[\frac{2i_{dc}}{3I_m \sin(\theta)}\right] \tag{2.47}$$

Substituting equations (2.40) and (2.44) into equation (2.47), the optimal pulse width θ_{opt_u} for the charge transfer extreme can be obtained as

$$\theta_{opt\_u} = 2\cos^{-1}\left[\frac{4}{n\pi}\sum_{i=1}^{n/2}\sin\left(\frac{\theta_{ui}}{2}\right)\right] \tag{2.48}$$

Owing to $\theta_{ui} + \theta_{li} = 2\pi$, there will be

$$\theta_{opt\_u} = \theta_{opt\_l} \tag{2.49}$$

If the width of a pulse is close to θ_{opt_u} in the upper arm, the pulse will result in more charge transferred away from the capacitor under $Q_{au} < 0$, or the pulse will cause less charge transferred to the capacitor under $Q_{au} > 0$, in comparison with the pulse whose width being far away from θ_{opt_u}. Consequently, one pulse with its width close to θ_{opt_u} can result in the lower capacitor voltage than the other one with its width far away from θ_{opt_u} in the upper arm. The analysis for phases B and C is the same to phase A, which is not repeated here.

2.4.4.4 Voltage Balancing Control

Figure 2.25a shows the control structure of the MMC with the SHE-PWM pulse energy sorting based VBC. The phase angle θ_e of the grid voltage is obtained with the phase locked loop (PLL). According to the control objective such as active power, reactive power, and DC-link voltage control, the current references i_{d_ref}

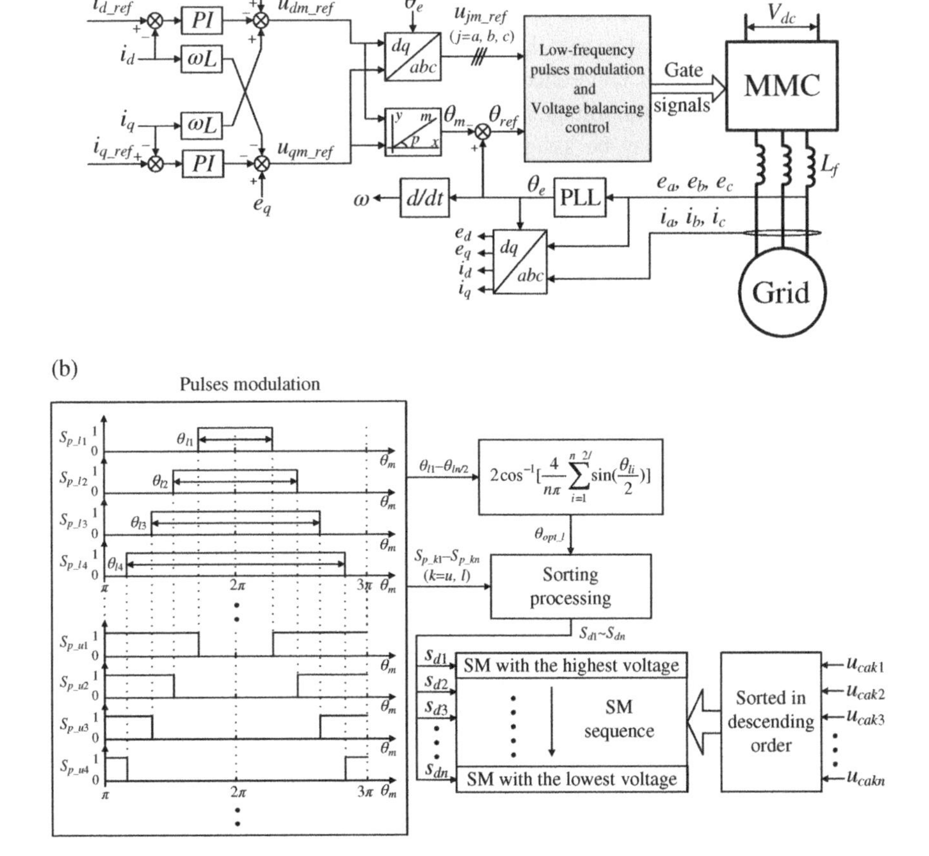

Figure 2.25 Block diagram of (a) SHE-PWM pulse energy sorting based VBC for MMCs. (b) SHE-PWM pulse energy sorting based VBC for phase A.

and i_{q_ref} can be obtained [1, 3]. The vector control method is used here for grid current control [1, 3] and produce the dq-axis reference voltage u_{dm_ref} and u_{qm_ref}. Finally, the three-phase reference voltage can be calculated by the inverse transformation $T_{dq/abc}$ from u_{dm_ref}, u_{qm_ref} to u_{am_ref}, u_{bm_ref}, u_{cm_ref}, which are used to produce grid-frequency pulses with low-frequency pulse modulation methods (e.g. nearest level control [NLC]). In Figure 2.25a, a polar/rectangular coordinate converter is used and the phase angle of the three-phase reference voltage u_{am_ref}, u_{bm_ref}, and u_{cm_ref} can be calculated as $\theta_{ref} = \theta_e - \theta_m$. Referring to Figure 2.22, the pulse modulation and VBC for phases A, B, and C can be implemented cycle by cycle and started from π, $\pi/3$, and $5\pi/3$ of the phase angle θ_{ref}, respectively. The implementation cycle is 2π.

Figure 2.25b shows the VBC for phase A of MMCs switched at grid frequency, where the algorithm is started from $\theta_{ref} = \pi$ and implemented cycle by cycle. With the low-frequency pulse modulation method, the grid-frequency pulses for each period can be produced and the pulse width can be obtained. And then the optimal pulse width θ_{opt_l} can be calculated based on (2.45) or (2.48) for this period of 2π, which will be used as a reference value to assign the produced pulses, as shown in Table 2.3. If a capacitor voltage is high, a pulse in S_{p_k1}–S_{p_kn} ($k = u, l$) with its width close to θ_{opt_l} may be assigned to the SM. Consequently, more charge is transferred away from the capacitor if the charge transferred to the capacitor is negative and the capacitor voltage decreases more, or less charge is transferred to the capacitor if the charge transferred to the capacitor is positive and the capacitor voltage increases less. In contrast, if the capacitor voltage is low, a pulse in S_{p_k1}–S_{p_kn} with its width far from θ_{opt_l} may be assigned to the SM. Consequently, less charge is transferred away from the capacitor if the charge transferred to the capacitor is negative and the capacitor voltage decreases less, or more charge is transferred to the capacitor if the charge transferred to the capacitor is positive and the capacitor voltage increases more. As a consequence, the pulses S_{p_k1}–S_{p_kn} are sorted in ascending order according to the width of the pulses S_{p_k1}–S_{p_kn} close to θ_{opt_l}. Finally, the pulses S_{d1}–S_{dn} are obtained as shown in Figure 2.25, in which the width of S_{d1} is most close to θ_{opt_l} and the width of S_{dn} is most far from θ_{opt_l}.

Table 2.3 SM capacitor voltage control.

SM capacitor voltage	Pulse assignment	Charge transfer to capacitor under the assigned pulse	SM capacitor voltage trend
High	Pulse with width close to θ_{opt_l}	Negative ($Q<0$)	Decreased more
		Positive ($Q>0$)	Increased less
Low	Pulse with width far away from θ_{opt_l}	Negative ($Q<0$)	Decreased less
		Positive ($Q>0$)	Increased more

To achieve the capacitor voltage balancing task, the capacitor voltages u_{cak1}–u_{cakn} ($k = u, l$) in each arm are monitored in real time and sampled when θ_{ref} is π, $\pi/3$, and $5\pi/3$ for phases A, B, and C, respectively, in each period. And then, the SMs in the arm are sorted according to the capacitor voltage in descending order and driven with the pulses in a sequence from S_{d1} to S_{dn}, which can effectively ensure capacitor voltage balancing in each arm. The control for phases B and C is the same to that for phase A, which is not repeated here.

Case Study 2.2 Analysis of SHE-PWM Pulse Energy Sorting Based VBC

Objective: To verify the SHE-PWM pulse energy sorting-based VBC, a three-phase MMC system is modeled with the time-domain simulation tool PSCAD/EMTDC, as shown in Figure 2.26. The system parameters are shown in Table 2.4.

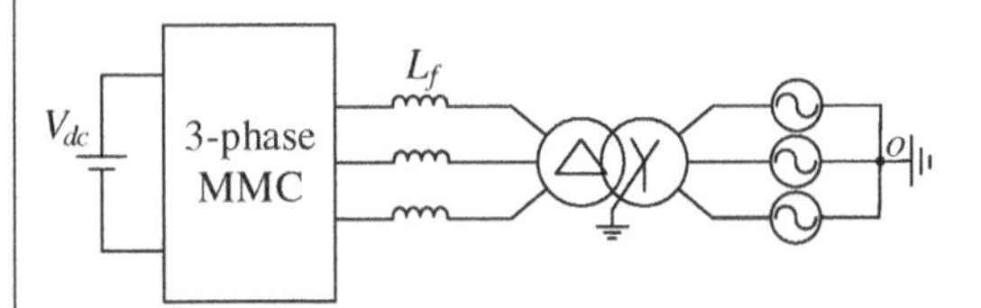

Figure 2.26 Block diagram of the simulation system.

Table 2.4 Simulation system parameters.

Parameter	Value
Rated active power P (MW)	65
Rated reactive power Q (Mvar)	21
Grid line-to-line voltage (kV)	32.5
Line frequency (Hz)	60
Transformer voltage rating (kV)	32.5/110
Transformer leakage reactance	5%
DC bus voltage V_{dc} (kV)	60
Number of SMs per arm n	32
SM capacitance C_{sm} (mF)	7.5 (35 kJ/MVA)
Arm inductance L_s (mH)	8
Inductance L_f (mH)	4
Switching frequency (Hz)	60

Simulation results and analysis:
Figure 2.27 shows the performance of the MMC. The active power P and reactive power Q of the three-phase MMC system is initially controlled as 30 MW and −21 Mvar, respectively, as shown in Figure 2.27a,b, with the power control strategy [1]. The NLC is used here to produce pulses at grid frequency [11]. The circulating current suppressing refers to [12]. In the simulation, at 3.5 seconds, a 65 MW active power command and a 0 reactive power

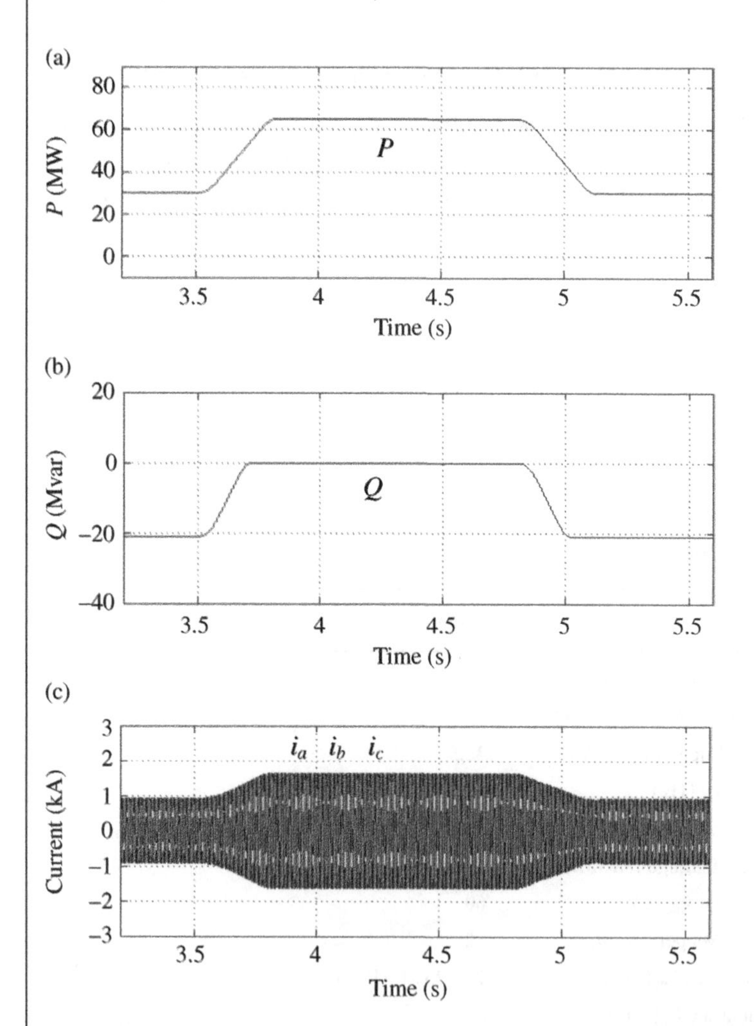

Figure 2.27 Simulated waveforms of MMCs. (a) Active power P. (b) Reactive power Q. (c) Three-phase current i_a, i_b, and i_c. (d) Arm current i_{au} and i_{al}. (e) Upper arm capacitor voltage of phase A. (f) Lower arm capacitor voltage of phase A. (g) DC-link voltage V_{dc}.

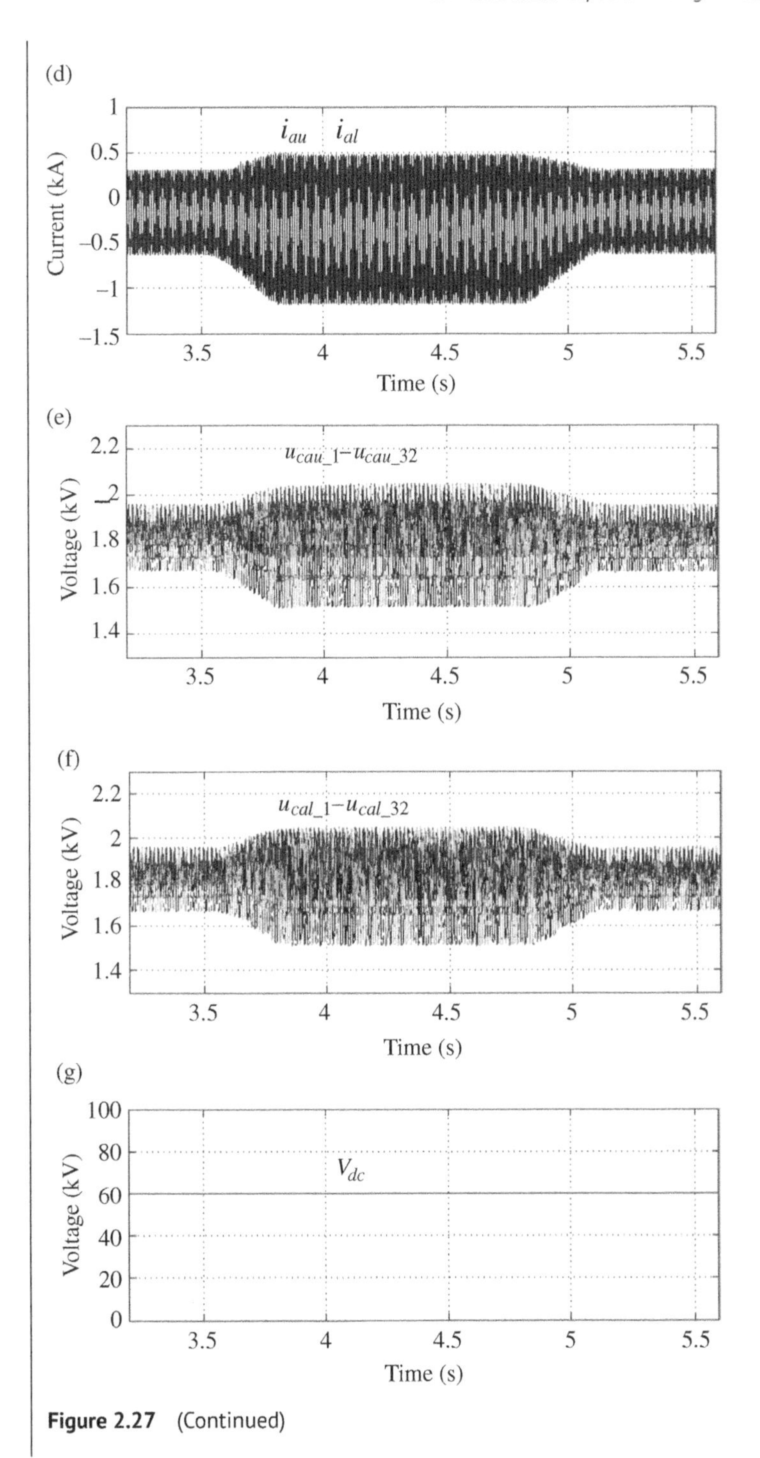

Figure 2.27 (Continued)

command are ramped up, respectively. At 4.8 seconds, the active power is ramped down from 65 to 30 MW again and the reactive power is ramped down from 0 to −21 Mvar again. Figure 2.27c shows the line current i_a, i_b, and i_c. Figure 2.27d shows the arm current i_{au} and i_{al} of phase A. The upper arm capacitor voltage u_{cau1}–u_{cau32} and the lower arm capacitor voltage u_{cal1}–u_{cal32} are shown in Figure 2.27e,f, respectively, which are kept balanced. The maximum peak-to-peak voltage ripple is about 29%, which is similar to [13] in rectifier mode and higher than [3, 14–18] with a higher switching frequency and a bigger capacitance. Figure 2.27g shows the DC-link voltage.

Figure 2.28 shows the performance of the MMC in a short time scale of Figure 2.27, where the active power and reactive power is 65 MW and 0, respectively. Figure 2.28a shows the line-to-line voltage u_{ab}, u_{bc}, and u_{ca}. The line current i_a, i_b, and i_c is shown in Figure 2.28b. The arm current i_{au} and i_{al} of phase A is shown in Figure 2.28c. Figure 2.28d,e shows the upper arm capacitor voltage u_{cau1}–u_{cau32} and the lower arm capacitor voltage u_{cal1}–u_{cal32}, respectively, which are kept balanced.

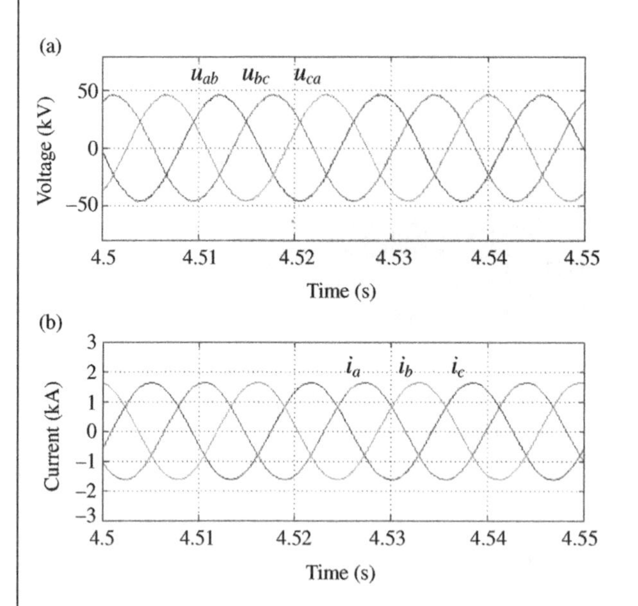

Figure 2.28 Simulated waveforms of MMCs in a short time scale of Figure 2.27. (a) Voltage u_{ab}, u_{bc}, and u_{ca}. (b) Three-phase current i_a, i_b, and i_c. (c) Arm current i_{au} and i_{al}. (d) Upper arm capacitor voltage of phase A. (e) Lower arm capacitor voltage of phase A.

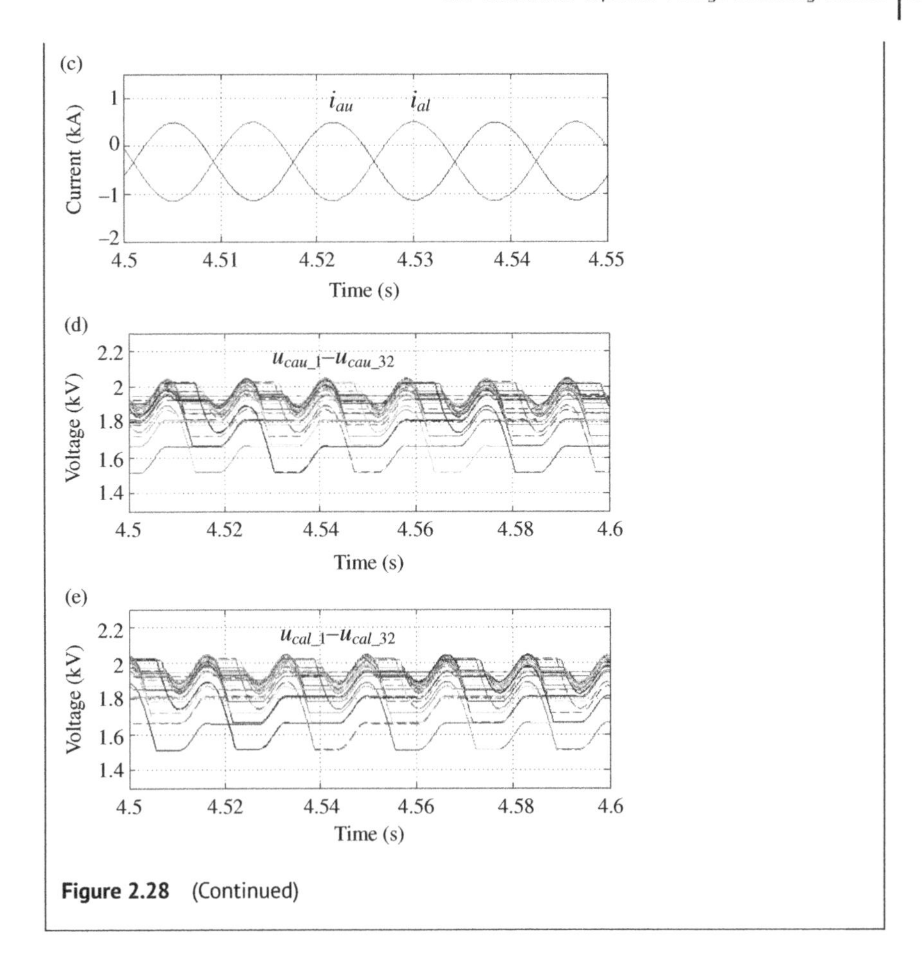

Figure 2.28 (Continued)

2.4.5 PSC-PWM Pulse Energy Sorting Based VBC

PSC-PWM pulse energy sorting based VBC uses linearization method for pulse sorting. In this control, the voltage balancing algorithm is implemented in each carrier wave period and the switching frequency is the same as the carrier wave frequency. The PSC-PWM pulse energy sorting based VBC does not rely on the arm current measurement.

2.4.5.1 MMC with PSC-PWM

Suppose that the MMC has n SMs in each arm, where n is even. The carrier waves are n isosceles triangles with the frequency of f_s for the n SMs in each arm of the

MMC. In order to eliminate the high-frequency harmonic components in the arm current, the phase-shifted angle of each carrier wave is considered as [17]

$$\Delta\theta = 2\pi / n \tag{2.50}$$

Figure 2.29 shows an example of the PSC-PWM scheme for phase A in one carrier wave period of 2π ($n = 4$). The n switching signals S_{p_au1}–S_{p_aun} in upper arm can be generated by comparing the carrier waves W_{ar1}–W_{arn} with the upper arm

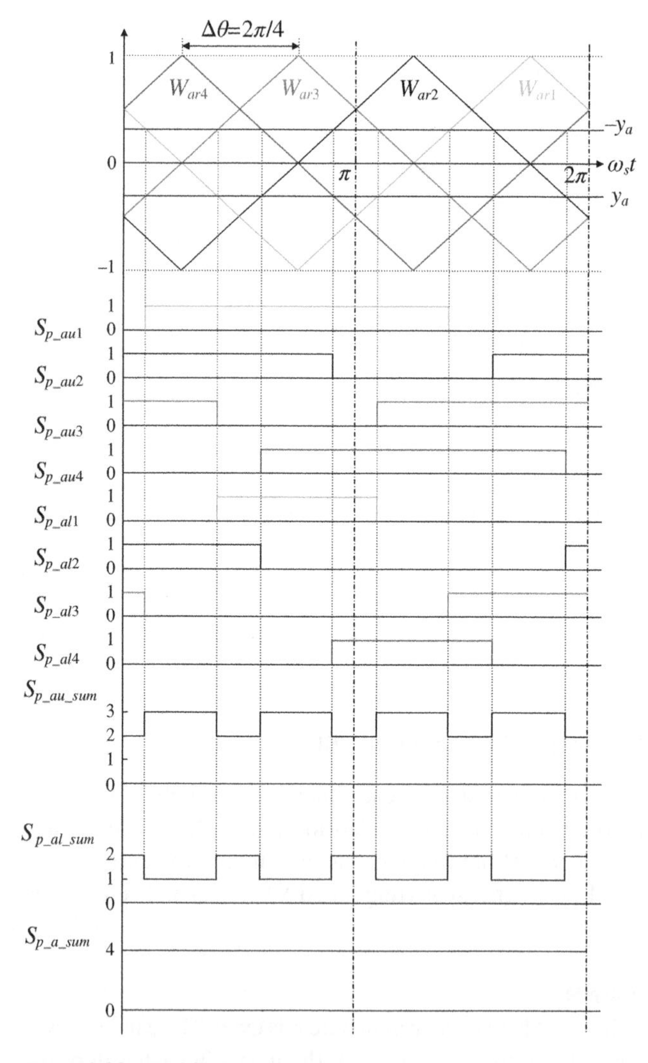

Figure 2.29 PSC-PWM for the MMC.

reference value $-y_a$. The n switching signals S_{p_al1}–S_{p_aln} in lower arm can be generated by comparing the carrier waves W_{ar1}–W_{arn} with the lower arm reference value y_a [17], as shown in Figure 2.29.

Suppose the carrier wave frequency f_s is far higher than that of the reference signal, the widths of the generated upper arm pulses are the same in each carrier wave period and the widths of the generated lower arm pulses are the same in each carrier wave period, as shown in Figure 2.29. Owing to n is even and $\Delta\theta$ is $2\pi/n$, the carrier wave W_{ari} ($i = 1, 2, \ldots, n/2$) is phase-shifted by an angle of π with the carrier wave $W_{ar(i+n/2)}$, which can derive

$$\begin{cases} S_{p\_aui} = /S_{p\_al(i+n/2)} \\ S_{p\_au(i+n/2)} = /S_{p\_ali} \end{cases}, \quad (i = 1,2,\ldots,n/2) \tag{2.51}$$

The summations $S_{p\_au\_sum}$ and $S_{p\_al\_sum}$ of the upper and lower arm switching signals, respectively, as shown in Figure 2.29, has the relationship as

$$S_{p\_au\_sum} + S_{p\_al\_sum} = S_{p\_a\_sum} = n \tag{2.52}$$

There always are n SMs switched on at any time in one phase of the MMC, as shown in Figure 2.29. Suppose the capacitor voltage is kept the same as V_{dc}/n, the total output voltage of the series-connected SMs in the phase is always equal to the DC-link voltage V_{dc} of the MMC, which may effectively eliminate the harmonics and improve the arm current in the MMC [17].

2.4.5.2 Capacitor Charge Transfer Under Linearization Method

Figure 2.30 shows the i-th carrier wave W_{ari} of phase A in one carrier wave period of 2π, whose peak value appears at $2\pi-(i-0.5)\Delta\theta$ and initial value d_i is

$$d_i = \begin{cases} 1-(2i-1)\dfrac{\Delta\theta}{\pi}, (n/2 \geq i > 0) \\ 1-\left[2(n+1-i)-1\right]\dfrac{\Delta\theta}{\pi}, (n \geq i > n/2) \end{cases} \tag{2.53}$$

The generated lower arm switching signal S_{p_ali} by W_{ari} may have two modes in terms of the reference value y_a, as shown in Figure 2.30.

1) *Mode-I*: $y_a \geq d_i$. The S_{p_ali} consists of two separated on-time states in each carrier wave period, as shown in Figure 2.30a. Then the θ_I and $\Delta\theta_I$ in Figure 2.30a can be calculated as

$$\begin{cases} \theta_I = \dfrac{\pi}{2}\left(3 + y_a - \dfrac{2i-1}{\pi}\Delta\theta\right) \\ \Delta\theta_I = \pi(1 - y_a) \end{cases} \tag{2.54}$$

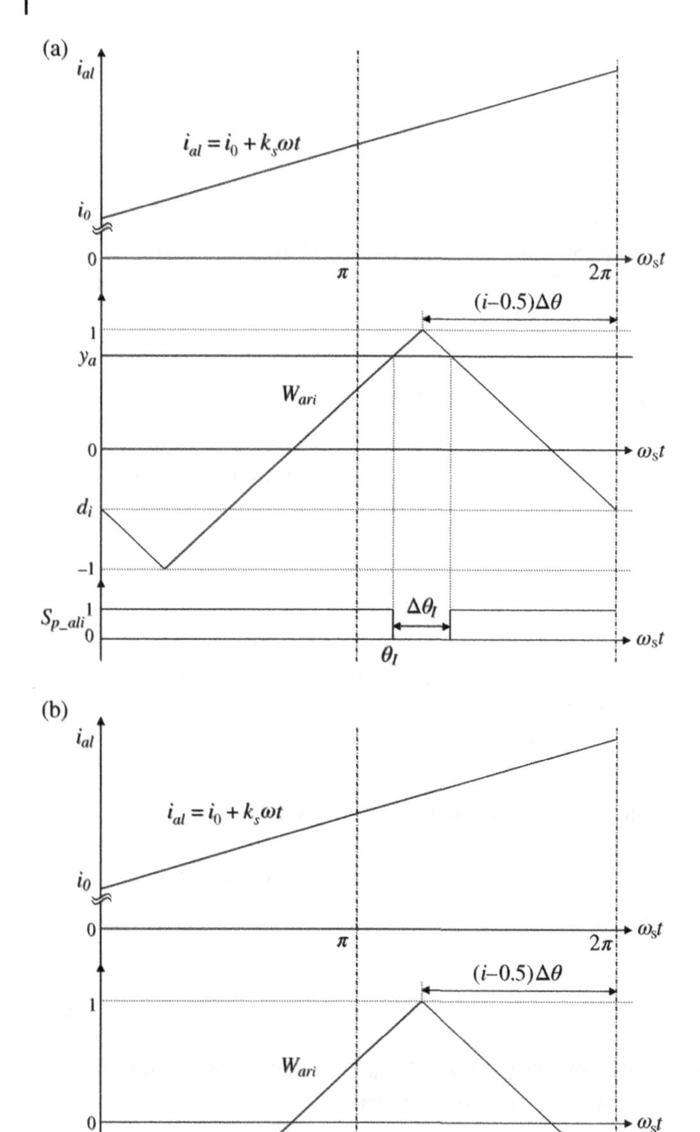

Figure 2.30 Lower arm switching signal in phase A. (a) Model I. (b) Model II.

2) *Mode-II*: $y_a < d_i$. The S_{p_ali} only has one on-time state in each carrier wave period, as shown in Figure 2.30b. The θ_{II} and $\Delta\theta_{II}$ in Figure 2.30b can be calculated as

$$\theta_{II} = \begin{cases} \frac{\pi}{2}\left(1 - y_a - \frac{2i-1}{\pi}\Delta\theta\right), & (n/2 \geq i > 0) \\ \frac{\pi}{2}\left(5 - y_a - \frac{2i-1}{\pi}\Delta\theta\right), & (n \geq i > n/2) \end{cases} \tag{2.55}$$

$$\Delta\theta_{II} = \pi\left(1 + y_a\right) \tag{2.56}$$

Suppose the carrier wave frequency is reasonably higher than that of the reference signal, as shown in Figure 2.30, the lower arm current i_{al} in each carrier wave period may be linearized as

$$i_{al} = i_0 + k_s\omega t \tag{2.57}$$

where i_0 is the initial value of the lower arm current in the carrier wave period, as shown in Figure 2.30. $k_s = di_{al}/d\omega t$ is the slope of the lower arm current, which may be positive ($k_s > 0$) or negative ($k_s < 0$).

During the "on" state of the SM under the lower arm switching signal S_{p_ali}, its capacitor voltage would be changed along with its charge transfer under the lower arm current i_{al}. The charge transfer Q_{ali} of the lower arm capacitor in each carrier wave period can be expressed as the integral of the lower arm current i_{al} over the on-time of S_{p_ali}, as

$$Q_{ali} = \int_0^{2\pi} S_{p\_ali} \cdot i_{al} d\left(\omega_s t\right) \tag{2.58}$$

Substituting equations (2.54)–(2.57) into equation (2.58), the lower arm capacitor charge transfer Q_{ali} corresponding to the lower arm switching signal S_{p_ali} can be rewritten as

$$Q_{ali} = \pi\left(1 + y_a\right)i_0 + \frac{\pi^2}{2}k_s\lambda_i, \quad \left(i = 1,2,\ldots n\right) \tag{2.59}$$

with

$$\lambda_i = \begin{cases} 4y_a + \left(1 - y_a\right)\frac{2i-1}{\pi}\Delta\theta, & \left(i = 1,2,\ldots n\right), \text{ Mode I} \\ \left(1 + y_a\right)\left(2 - \frac{2i-1}{\pi}\Delta\theta\right), & \left(i = 1,2,\ldots n/2\right), \text{ Mode II} \\ \left(1 + y_a\right)\left(6 - \frac{2i-1}{\pi}\Delta\theta\right), & \left(i = 2/n+1,\ldots n\right), \text{ Mode II} \end{cases} \tag{2.60}$$

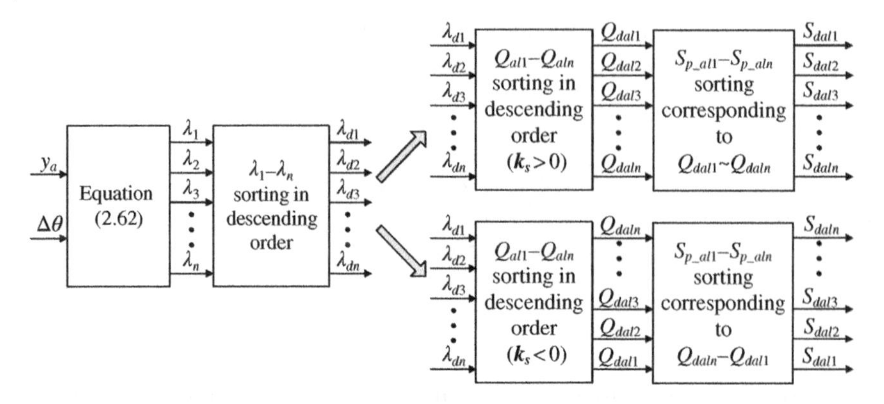

Figure 2.31 Sorting of the lower arm switching signals in phase A.

The n lower arm switching signals S_{p_al1}–S_{p_aln} in each carrier wave period can be sorted based on their corresponding contributions to capacitor charge transfer Q_{al1}–Q_{aln}, as shown in Figure 2.31. The first term on the right side of the equation (2.59) is the same for all the charge transfer, and the difference of the charge transfer only depends on the second term on the right side of the equation (2.59). As a result, the sorting of the capacitor charge transfer is related to λ and k_s. With the reference y_a and $\Delta\theta$, the λ_1–λ_n can be obtained by equation (2.60). And then, λ_1–λ_n are sorted in the descending order to obtain λ_{d1}–λ_{dn} ($\lambda_{d1} > \lambda_{d2} \ldots > \lambda_{dn}$).

- If $k_s > 0$, the charge transfers Q_{al1}–Q_{aln} produced by S_{p_al1}–S_{p_aln} in the lower arm can be sorted in descending order as Q_{dal1}–Q_{daln} ($Q_{dal1} > Q_{dal2} > \ldots > Q_{daln}$) corresponding to λ_{d1}–λ_{dn}. The switching signal sequence S_{al1}–S_{aln} can also be sorted based on their corresponding contributions to the capacitor charge transfer in descending order as S_{dal1}–S_{daln}.
- If $k_s < 0$, the sequence is reversed, and the charge transfers in descending order are Q_{daln}–Q_{dal1} ($Q_{daln} > \ldots > Q_{dal2} > Q_{dal1}$) corresponding to λ_{dn}–λ_{d1}. The switching signal sequence according to their corresponding contributions to the capacitor charge transfer in descending order is S_{daln}–S_{dal1}.

The upper arm of phase A can be analyzed with the same method as that for the lower arm of phase A, which is not repeated here.

2.4.5.3 Capacitor Voltage Analysis

Suppose that the lower arm capacitor voltages u_{cal1}–u_{caln} of phase A are sorted in descending voltage order, the difference of the maximum capacitor voltage and the minimum capacitor voltage is

$$\Delta u_{mm} = \mathrm{Max}\left[u_{cal1} - u_{caln}\right] - \mathrm{Min}\left[u_{cal1} - u_{caln}\right] \tag{2.61}$$

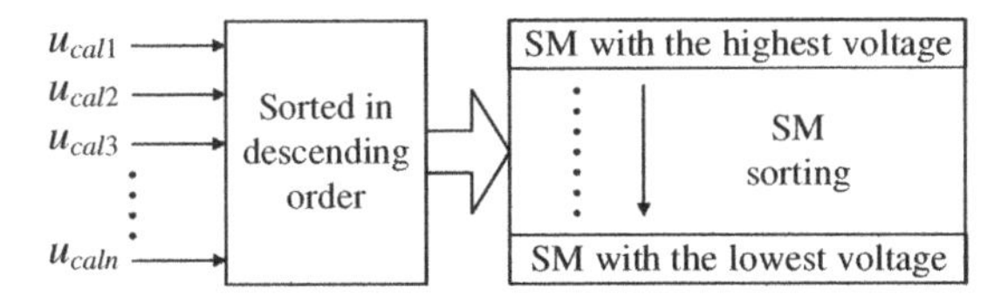

Figure 2.32 SMs sorting in lower arm of phase A.

Table 2.5 Capacitor voltage changing trend.

SMs sorting	k_s	Switching signal sequence	Charge transfer distribution	Δu_{mm}	Capacitor voltage trend
High voltage to low voltage	>0	S_{dal1}–S_{daln}	$Q_{dal1}>\ldots>Q_{daln}$	Increased	Unbalanced
		S_{daln}–S_{dal1}	$Q_{daln}<\ldots<Q_{dal1}$	Reduced	Balanced
	<0	S_{dal1}–S_{daln}	$Q_{dal1}<\ldots<Q_{daln}$	Reduced	Balanced
		S_{daln}–S_{dal1}	$Q_{daln}>\ldots>Q_{dal1}$	Increased	Unbalanced

The SMs in the lower arm of phase A can be sorted corresponding to the capacitor voltage in descending order, as shown in Figure 2.32.

Table 2.5 describes the changing trend of the lower arm capacitor voltage of phase A when the sorted SMs in Figure 2.32 are driven by the switching signal sequence S_{dal1}–S_{daln} or S_{daln}–S_{dal1} in Figure 2.31.

- When $k_s>0$, the sorted SMs in Figure 2.32 can be driven with the switching signal sequence S_{daln}–S_{dal1} in each carrier wave period to ensure Δu_{mm} decreasing and then the capacitor voltage tends to be balanced.
- When $k_s<0$, the sorted SMs in Figure 2.32 can be driven with the switching signal sequence S_{dal1}–S_{daln} in each carrier wave period to ensure Δu_{mm} decreasing for balancing the capacitor voltage.

Otherwise, the Δu_{mm} would be increased and the capacitor voltage tends to be unbalanced if the sorted SMs in Figure 2.32 are driven with the switching signal sequence S_{dal1}–S_{daln} under $k_s>0$, or driven with the switching signal sequence S_{daln}–S_{dal1} under $k_s<0$ in each carrier wave period. The voltage changing of the upper arm capacitor in phase A can be analyzed with the same method for that of the lower arm capacitor in phase A, which is not described here.

Table 2.5 shows that, in order to balance the capacitor voltage, the switching signals should be sorted in the appropriate direction S_{dal1}–S_{daln} or S_{daln}–S_{dal1} to make Δu_{mm} reduced. Otherwise, the Δu_{mm} would be increased and the SM capacitor voltage balancing does not work.

2.4.5.4 Voltage Balancing Control

Figure 2.33 shows the VBC for the MMC. In the control, the capacitor voltages u_{cal1}–u_{caln} in the lower arm of phase A are monitored and sampled at the frequency of f_s, which is the same to the carrier frequency. $T_s = 1/f_s$ is the sampling period. And then, the SMs in the lower arm are sorted corresponding to the capacitor voltage with descending order. In addition, the switching signal sequences S_{dal1}–S_{daln} and S_{daln}–S_{dal1} are calculated based on Figure 2.31. The appropriate selection of the switching signal sequence S_{dal1}–S_{daln} or S_{daln}–S_{dal1} for the sorted SMs in each carrier wave period can ensure capacitor voltage balancing. The VBC in Figure 2.33 is implemented in each carrier wave period, which is between two adjacent samplings.

In order to realize capacitor voltage balancing, the appropriate switching signal sequences S_{dal1}–S_{daln} or S_{daln}–S_{dal1} for the i-th ($i = 2,3\ldots$) carrier wave period can be determined by the (i–1)-th and i-th samplings. Supposing the α-th SM ($1 \le \alpha \le n$) and the β-th SM ($1 \le \beta \le n$) has, respectively, the maximum and minimum capacitor voltage $u_{cal}\alpha(t)$ and $u_{cal}\beta(t)$ at the (i–1)-th sampling and sorting, as shown in Figure 2.33, the $\Delta u_{mm(i-1)}$ is

$$\Delta u_{mm(i-1)} = u_{cal\alpha}(t) - u_{cal\beta}(t) \tag{2.62}$$

After the sorted SMs are driven by the switching signal sequence S_{dal1}–S_{daln} or S_{daln}–S_{dal1} in the (i–1)-th carrier wave period, the capacitor voltage of the α-th SM and the β-th SM will be changed to $u_{cal\_}\alpha(t+T_s)$ and $u_{cal\_}\beta(t+T_s)$ at the i-th sampling. In order to realize capacitor voltage balancing, the switching signal sequence

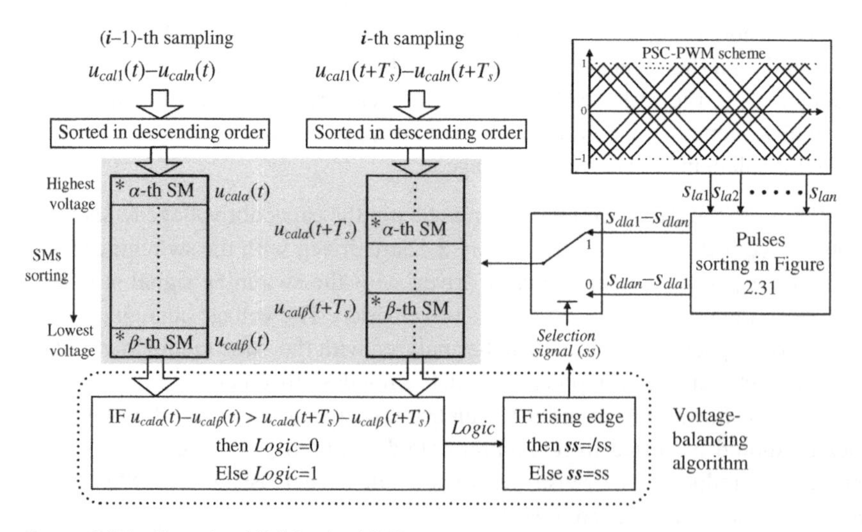

Figure 2.33 Capacitor VBC for the MMC.

for the i-th carrier wave period can be decided by the relationship between $\Delta u_{mm(i-1)}$ and $\Delta u_{mmi} = u_{cal}\alpha(t+T_s) - u_{cal}\beta(t+T_s)$ as follows.

- If $\Delta u_{mm(i-1)} > \Delta u_{mmi}$, which means the $\Delta u_{mm(i-1)}$ is reduced during the $(i-1)$-th carrier wave period and the capacitor voltage tends to be balanced with the given switching signal sequence for the $(i-1)$-th carrier wave period. If the *logic*, as shown in Figure 2.33, is 0 or 1 at the $(i-1)$-th sampling and sorting, the *logic* will be kept as 0 or step changed to 0 at the i-th sampling and sorting. Therefore, the selection signal *ss*, as shown in Figure 2.33, will be unchanged. The switching signal sequence for the i-th carrier wave period will be the same to that for the $(i-1)$-th carrier wave period.
- If $\Delta_{mm(i-1)} < \Delta_{mmi}$, which means that the $\Delta u_{mm(i-1)}$ is increased during the $(i-1)$-th carrier wave period and the capacitor voltage tends to be unbalanced with the given switching signal sequence for the $(i-1)$-th carrier wave period. At the i-th sampling and sorting, the *logic* will be step changed from 0 to 1. The selection signal *ss* will be changed on the rising edge of the *logic*, and the reversed switching signal sequence will be used for the i-th carrier wave period.

Figure 2.34 shows an example of the change of the selection signal *ss*. The *ss* is 1 in the $(i-1)$-th carrier wave period and is changed to 0 in the i-th carrier wave period, which results in that the switching signal sequence is changed from S_{dal1}–S_{dal4} to S_{dal4}–S_{dal1} in the i-th carrier wave period. The switching frequency is the same as the carrier wave frequency, as shown in Figure 2.34.

Figure 2.33 shows that the pulse sorting can be achieved with the linearization method for capacitor VBC and without the measurement of the arm current. In Figures 2.30 and 2.33, the linearization method achieves the reasonable accuracy at the cost of the switching frequency and the required switching frequency is suitable for some applications of MMCs. The different switching frequencies are discussed in Case Study 2.3 for the MMCs.

The upper arm capacitor voltage in phase A can be balanced with the same method for the lower arm in phase A and the capacitor voltage balancing in phases B and C can be realized with the same method for phase A, which is not repeated here.

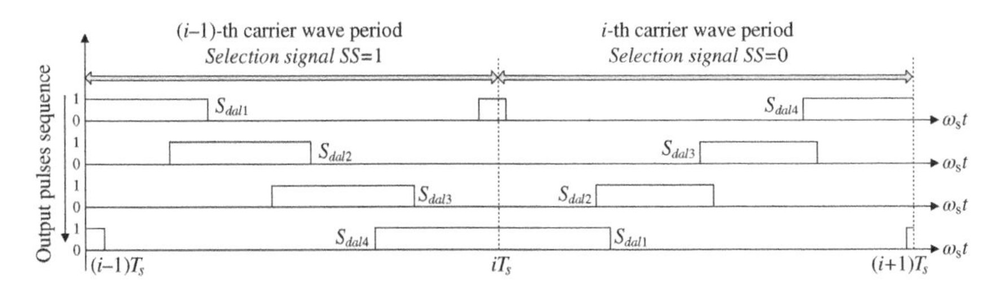

Figure 2.34 Example of selection signal change of MMCs with four SMs per arm.

Case Study 2.3 Analysis of the PSC-PWM Pulse Energy Sorting Based VBC

Objective: To verify the voltage balancing control, an MMC system, as shown in Figure 2.35, is simulated with the professional tool PSCAD/EMTDC. The circulating current elimination control in [1] is used here. The system parameters are shown in Table 2.6.

Simulation results and analysis:

Figure 2.36 shows the performance of the three-phase MMC under the voltage balancing control. In this situation, the active power P and reactive power Q of the MMC system are controlled as 1 MW and 0 MVar, respectively. The carrier wave frequency is 700 Hz. Figure 2.36a,b shows the voltage u_{ab}, u_{bc}, u_{ca} and current i_a, i_b, i_c. The arm current i_{au}, i_{al}, and circulating current i_{diff_a} of phase A is shown in Figure 2.36c. With the voltage balancing control, the capacitor voltage is kept balanced, as shown in Figure 2.36d.

Figure 2.37 shows the performance of the three-phase MMC under the voltage balancing control in another case. In this situation, the active power P and

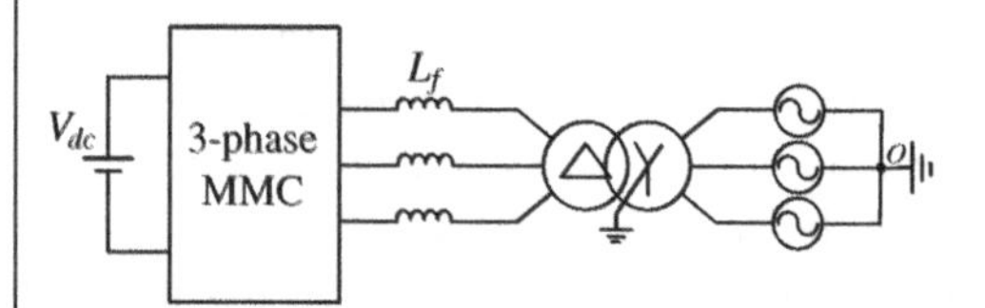

Figure 2.35 Block diagram of the simulation system.

Table 2.6 Simulation system parameters.

Parameter	Value
Rated active power P (MW)	1
Rated reactive power Q (Mvar)	0.16
Grid line-to-line voltage (kV)	11
Line frequency (Hz)	50
Transformer voltage rating	3 kV/11 kV
DC bus voltage V_{dc} (kV)	6
Number of SMs per arm n	10
SM capacitance C_{sm} (mF)	3.75
Arm inductance L_s (mH)	14
Inductance L_f (mH)	4

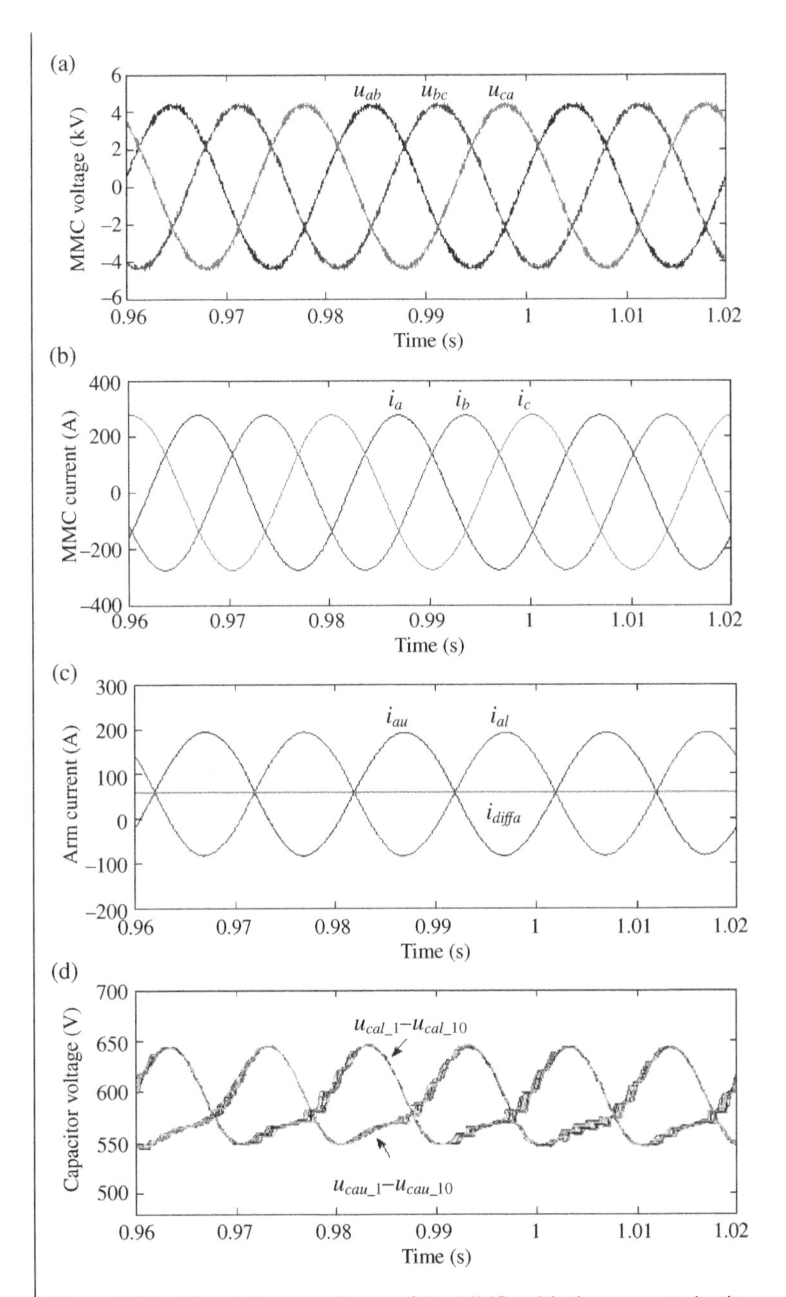

Figure 2.36 Simulation waveforms of the MMCs with the presented voltage balancing method. (a) Line-to-line voltage u_{ab}, u_{bc}, and u_{ca}. (b) Grid current i_a, i_b, and i_c. (c) Arm current i_{au}, i_{al}, and i_{diff_a} of phase A. (d) Upper and lower arm capacitor voltage of phase A.

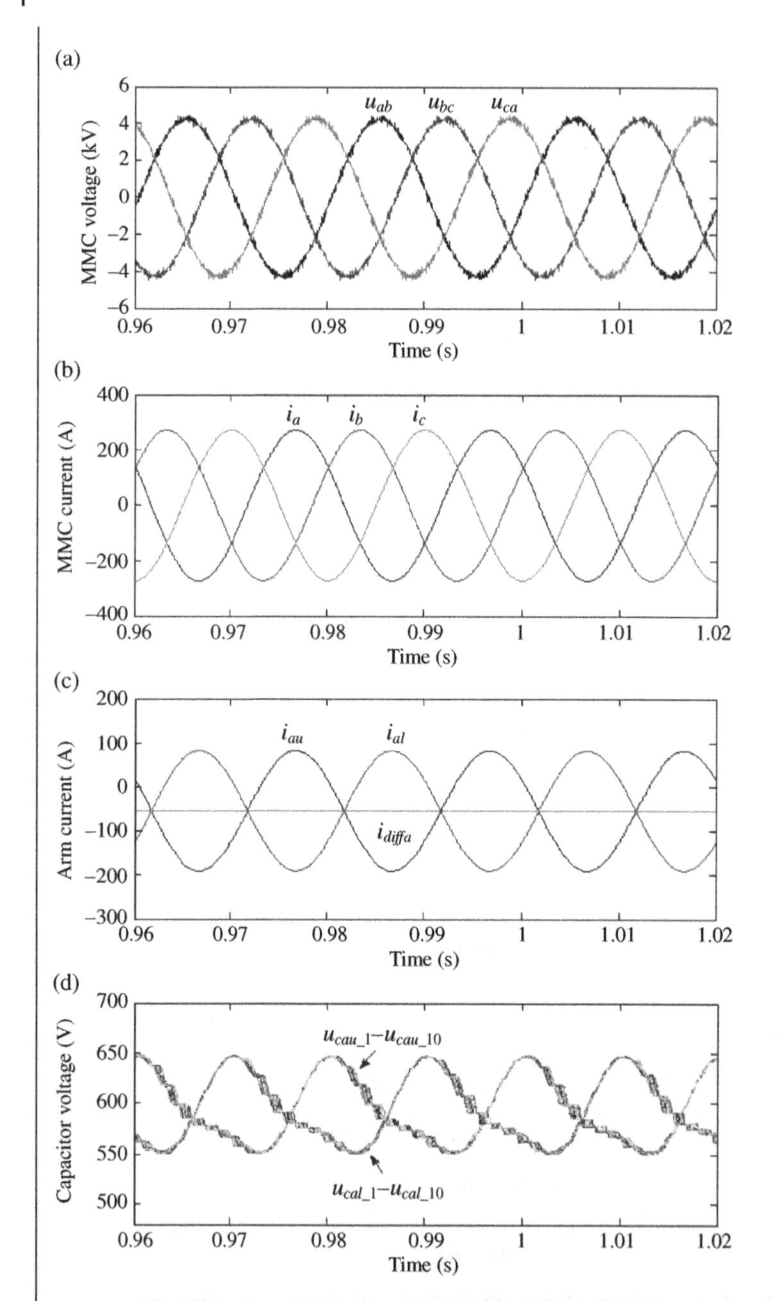

Figure 2.37 Simulation waveforms of the MMCs with the voltage balancing control. (a) Line-to-line voltage u_{ab}, u_{bc}, and u_{ca}. (b) Grid current i_a, i_b, and i_c. (c) Arm current i_{au}, i_{al}, and i_{diff_a} of phase A. (d) Upper and lower arm capacitor voltage of phase A.

reactive power Q of the three-phase MMC are −1 MW and 0 MVar, respectively. The carrier wave frequency is 700 Hz. Figure 2.37a, b show the voltage u_{ab}, u_{bc}, u_{ca} and current i_a, i_b, i_c. The arm current i_{au}, i_{al}, and circulating current i_{diff_a} of phase A is shown in Figure 2.37c. With the voltage balancing control, the capacitor voltage is kept balanced, as shown in Figure 2.37d.

Figures 2.38–2.41 show the lower arm capacitor voltage and lower arm current of phase A in the modular multilevel converter under the different carrier wave frequencies including 200, 300, 400, and 500 Hz, where the capacitor voltages are kept balanced and the harmonics of the arm current are improved in comparison with [17]. In this case, the active power P and reactive power Q of the MMC are controlled as 1 MW and 0.16 MVar, respectively. Along with the increase of the switching frequency, the capacitor voltages are more stable and the capacitor voltage ripple is reduced because the linearization becomes more accurate along with the increase of the switching frequency in the modular multilevel converter. The voltage balancing control achieves the reasonable linearization accuracy at the switching frequency no lower than 200 Hz. As a consequence, the voltage balancing

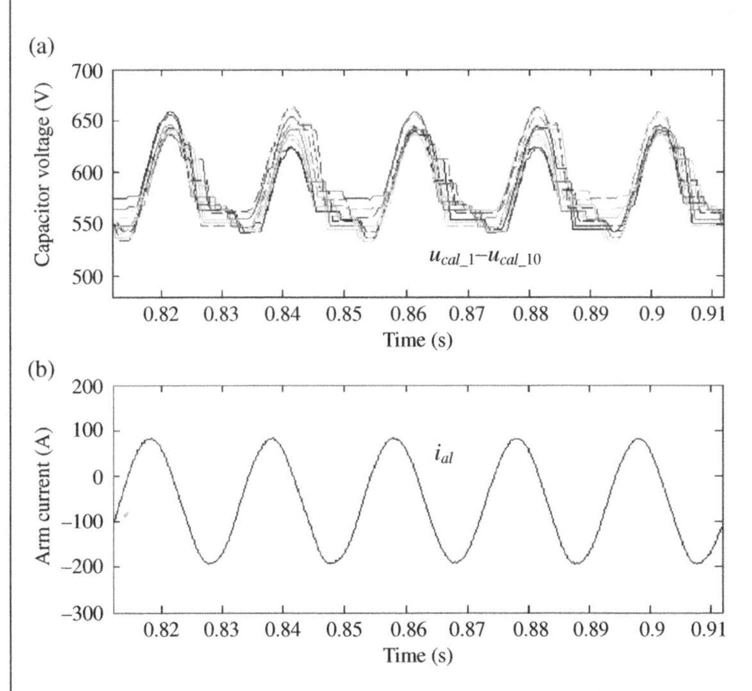

Figure 2.38 (a) Lower arm capacitor voltage u_{cal1}–u_{cal10}. (b) Lower arm current i_{al}. The MMC is operated under carrier wave frequency f_s as 200 Hz.

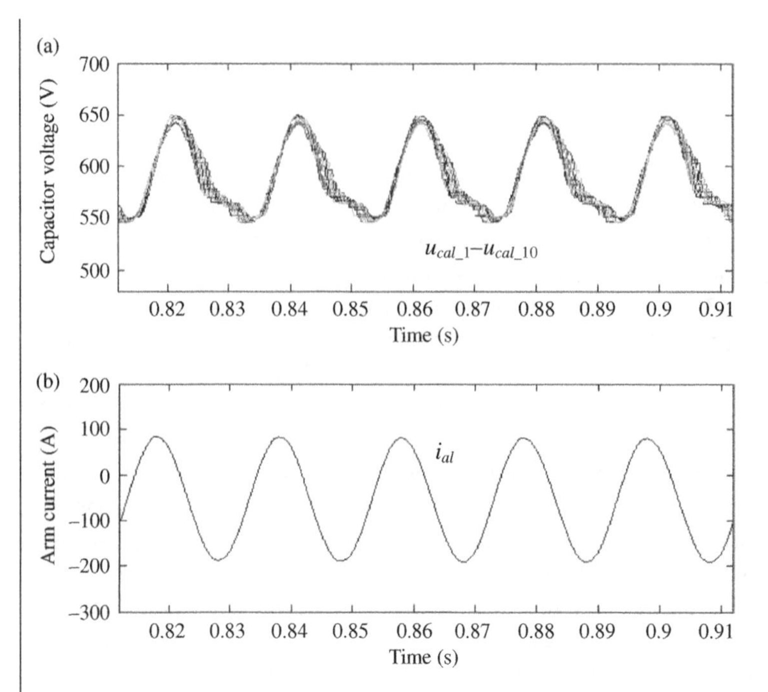

Figure 2.39 (a) Lower arm capacitor voltage $u_{cal1}-u_{cal10}$. (b) Lower arm current i_{al}. The MMC is operated under carrier wave frequency f_s as 300 Hz.

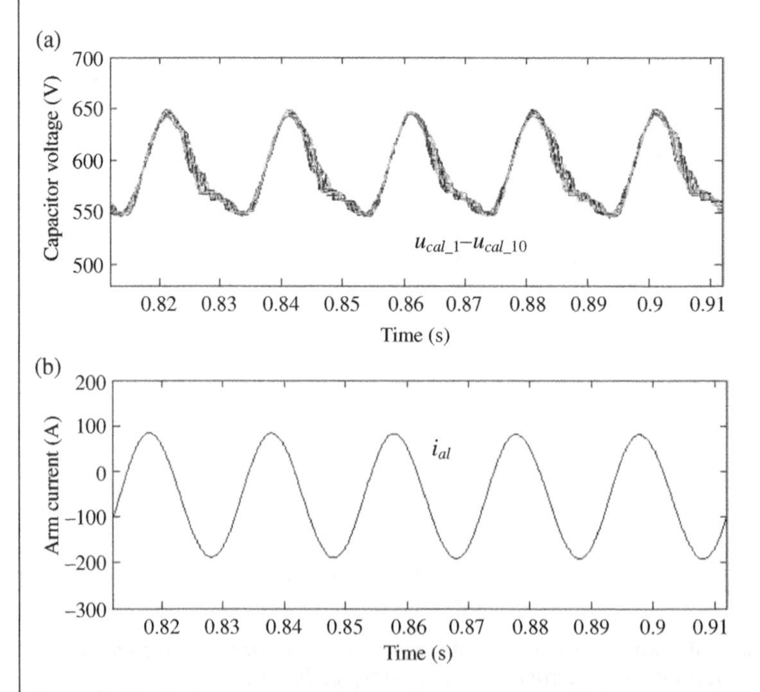

Figure 2.40 (a) Lower arm capacitor voltage $u_{cal1}-u_{cal10}$. (b) Lower arm current i_{al}. The MMC is operated under carrier wave frequency f_s as 400 Hz.

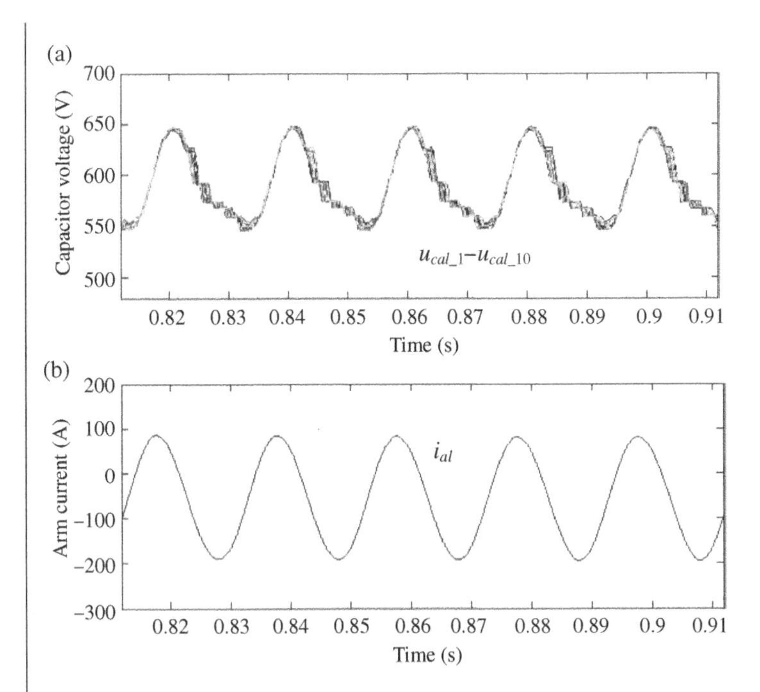

Figure 2.41 (a) Lower arm capacitor voltage u_{cal1}–u_{cal10}. (b) Lower arm current i_{al}. The MMC is operated under carrier wave frequency f_s as 500 Hz.

control can be used for some applications of the modular multilevel converter, such as motor drives and STATCOMs, where the modular multilevel converter is operated with a higher switching frequency [19–22].

2.5 Individual Capacitor Voltage Balancing Control

Compared with centralized capacitor VBC method, the individual capacitor VBC is implemented trough the closed-loop control for the individual SM capacitor voltage, which does not require a sorting technique to select the SMs, especially for the MMCs with a large number of SMs.

2.5.1 Average and Balancing Control Based VBC

Average and balancing control strategy consists of the average control and balancing control, which force the voltage of the capacitors in each arm to tend to average and follow the voltage reference [16].

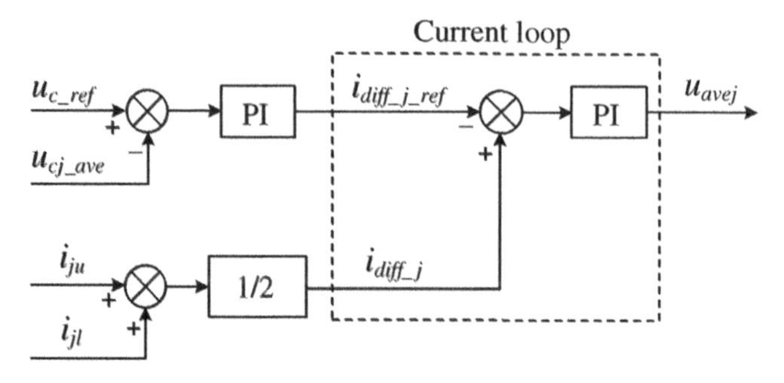

Figure 2.42 Average control.

2.5.1.1 Average Control

Figure 2.42 shows the averaging control, which forces the average value u_{cj_ave} of all capacitor voltages in phase j to follow the command value u_{c_ref}. In Figure 2.42, the average value u_{cj_ave} of all capacitor voltages in phase j is calculated as

$$u_{cj\_ave} = \frac{1}{2n}\left(\sum_{i=1}^{n} u_{cjui} + \sum_{i=1}^{n} u_{cjli}\right) \tag{2.63}$$

In Figure 2.42, when u_{c_ref} is higher than u_{cj_ave}, $i_{diff\_j\_ref}$ will increase through the PI controller, thus the function of the current loop will produce the reference voltage u_{avej} and force the actual current i_{diff_j} to increase to follow its command $i_{diff\_j\_ref}$. On the contrary, when u_{c_ref} is lower than u_{cj_ave}, $i_{diff\_j\_ref}$ will decrease through the PI controller, thus the function of the current loop will produce the reference voltage u_{avej} and force the actual current i_{diff_j} to decrease to follow its command $i_{diff\_j\_ref}$.

2.5.1.2 Balancing Control

Figure 2.43 shows the block diagram of balancing control for capacitor voltage balancing. Since the balancing control is based on the measurement of either i_{ju} or i_{jl}, the polarity of the produced voltage reference u_{cjui_bal} or u_{cjli_bal} should be changed according to that of i_{ju} or i_{jl}. The algorithm can be described below.

- When the capacitor voltage reference u_{c_ref} is higher than the actual capacitor voltage u_{cjui} (or u_{cjli}), a positive active power should be taken from the DC side into the SMs. When i_{ju} (or i_{jl}) is positive, the product of u_{cjui_bal} (or u_{cjli_bal}) and i_{ju} (or i_{jl}) forms the positive active power. When i_{ju} (or i_{jl}) is negative, the polarity of u_{cjui_bal} (or u_{cjli_bal}) should get inverse to take the positive active power.
- When the capacitor voltage reference u_{c_ref} is lower than the actual capacitor voltage u_{cjui} (or u_{cjli}), a positive active power should be taken from the SMs to the DC side. When i_{ju} (or i_{jl}) is negative, the product of u_{cjui_bal} (or u_{cjli_bal}) and i_{ju}

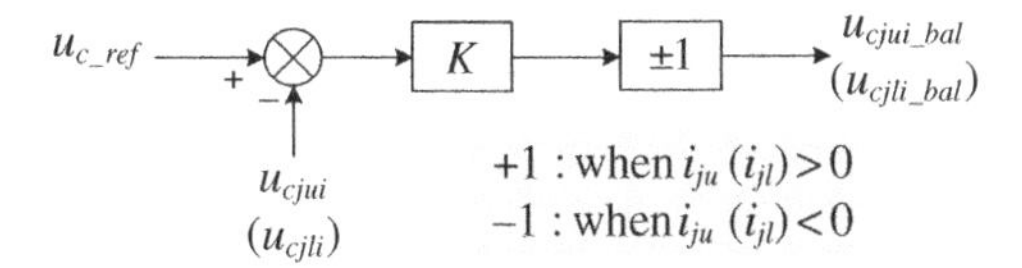

Figure 2.43 Balancing control.

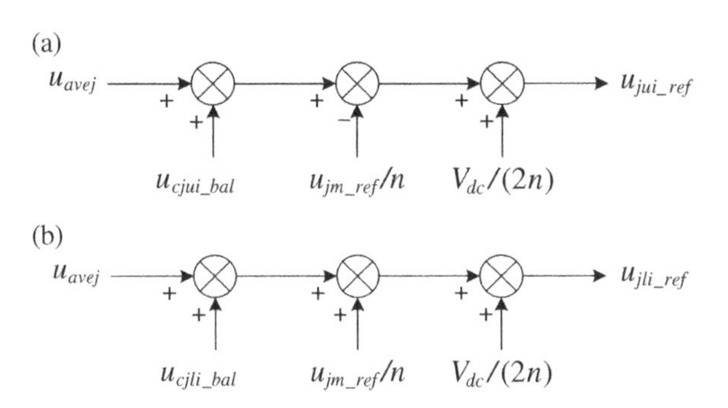

Figure 2.44 (a) The *i*-th SM's voltage reference in upper arm of phase *j*. (b) The *i*-th SM's voltage reference in lower arm of phase *j*.

(or i_{jl}) forms the positive active power. When i_{ju} (or i_{jl}) is positive, the polarity of u_{cjui_bal} (or u_{cjli_bal}) should get inverse to take the positive active power.

Figure 2.44a shows the voltage reference u_{jui_ref} for the i-th SM in the upper arm of phase j, and Figure 2.44b shows the voltage reference u_{jli_ref} for the i-th SM in the lower arm of phase j, as

$$\begin{cases} u_{jui\_ref} = u_{avej} + u_{cjui\_bal} - \dfrac{u_{jm\_ref}}{n} + \dfrac{V_{dc}}{2n} \\ u_{jli\_ref} = u_{avej} + u_{cjli\_bal} + \dfrac{u_{jm\_ref}}{n} + \dfrac{V_{dc}}{2n} \end{cases} \tag{2.64}$$

The voltage reference u_{jui_ref} and u_{jli_ref} is normalized by $V_{dc}/2$ and followed by the comparison with a triangular waveform under PSC-PWM mentioned in Section 1.4. And then, the switching function for each SM can be obtained, and accordingly each SM capacitor voltage can be individually controlled to be balanced.

2.5.2 Reference Modulation Index Based VBC

Reference modulation index based control can achieve the capacitor voltage control through controlling the modulation index to regulate the DC component in each SM capacitor current. This control can be realized without the knowledge of current, which reduces sensors, eliminates the adverse effects caused by the sensor noise, and improves the reliability.

2.5.2.1 Analysis of Capacitor Voltage

Suppose that the second-order harmonic circulating current is eliminated, the upper arm current in phase A is

$$i_{au} = \frac{i_{dc}}{3} + \frac{I_m}{2}\sin(\omega t + \theta) \tag{2.65}$$

the capacitor current i_{caui} in the i-th SM of the upper arm of phase A can be expressed as

$$i_{caui} = i_{au} \cdot \frac{1 + y_{au}}{2} \tag{2.66}$$

with

$$y_{au} = -m \cdot \sin(\omega t) \tag{2.67}$$

Substituting (2.65) and (2.67) into (2.66), the capacitor current i_{caui} can be rewritten as

$$i_{caui} = \underbrace{i_{cdc}}_{\text{DC component}} + \underbrace{\left[\frac{I_m}{4}\sin(\omega t + \theta) - \frac{mi_{dc}}{6}\sin(\omega t)\right]}_{\text{Fundamental component}} + \underbrace{\frac{mI_m}{8}\cos(2\omega t + \theta)}_{\text{Second-order component}} \tag{2.68}$$

with

$$i_{cdc} = \frac{i_{dc}}{6} - \frac{mI_m}{8}\cos(\theta) \tag{2.69}$$

The DC component i_{cdc} is zero in the steady-state operation of the MMC. In addition, the SM capacitor voltage can be regulated by the i_{cdc}, as

- capacitor voltage is increased by increase of i_{cdc} and
- capacitor voltage is reduced by reduction of i_{cdc}.

The i_{cdc} in each SM can be controlled by the corresponding modulation index m or the phase angle θ, which depends on the MMC operation mode. Figure 2.45 shows eight MMC operation modes, where $\vec{u_s}$ and $\vec{i_s}$ are the vectors of the MMC voltage u_{am}, u_{bm}, u_{cm} and current i_a, i_b, i_c, respectively, and $\vec{u_s}$ aligns along with the x-axis. The phase angle between $\vec{u_s}$ and $\vec{i_s}$ is θ.

- In mode 1, $\theta = 0$, the active power P is positive and the reactive power Q is 0.
- In mode 2, $0 < \theta < \pi/2$ and $P > 0$, $Q > 0$.

- In mode 3, $\theta = \pi/2$ and $P = 0, Q > 0$.
- In mode 4, $\pi/2 < \theta < \pi$ and $P < 0, Q > 0$.
- In mode 5, $\theta = \pi$ and $P < 0, Q = 0$.
- In mode 6, $\pi < \theta < 3\pi/2$ and $P < 0, Q < 0$.
- In mode 7, $\theta = 3\pi/2$ and $P = 0, Q < 0$.
- In mode 8, $3\pi/2 < \theta < 2\pi$ and $P > 0, Q < 0$.

2.5.2.2 Control of i_{cdc} by Modulation Index m

The control of i_{cdc} by m for the i-th SM in the MMC under different operation modes are shown in Table 2.7 and Figure 2.45, as follows.

- *Mode 1, 2, and 8*: $P > 0$. Here $i_{dc} > 0$ and $\cos(\theta) > 0$. According to equation (2.69), the i_{cdc} can be increased by the reduction of m and reduced by the increase of m.
- *Mode 3 and 7*: $P = 0$. Here, $i_{dc} = 0$ and $\cos(\theta) = 0$. According to equation (2.69), the i_{cdc} cannot be controlled by m.
- *Mode 4, 5, and 6*: $P < 0$. Here, $i_{dc} < 0$ and $\cos(\theta) < 0$. According to equation (2.69), the i_{cdc} can be increased by the increase of m and reduced by the reduction of m.

Table 2.7 Control of i_{cdc} by m.

Mode	P	i_{dc}	cos(θ)	m	i_{cdc}
1, 2, 8	>0	>0	>0		Inversely proportional
3, 7	0	0	0		Uncontrollable
4, 5, 6	<0	<0	<0		Proportional

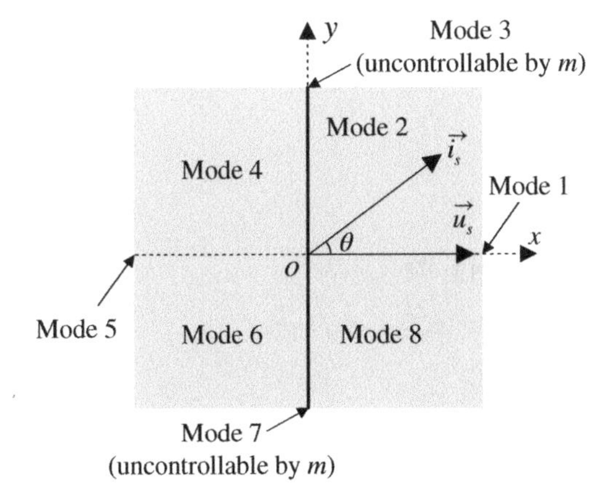

Figure 2.45 Control in the MMC by modulation index m.

Table 2.7 shows that the i_{cdc} in each SM capacitor current can be controlled by the corresponding modulation index m of the reference signal for each SM when the MMC works in various operation modes except Modes 3 and 7. When the MMC works on the right side of the y-axis, the i_{cdc} is proportional to m; when the MMC works on the left side of the y-axis, the i_{cdc} is inversely proportional to m.

2.5.2.3 Voltage Balancing Control by *m*

The central control for the MMC is shown in Figure 2.46, based on e_d, e_q and i_d, i_q of the dq-axis components of the grid voltage e_a, e_b, e_c and current i_a, i_b, i_c, respectively. Based on the control objective of the three-phase MMC system such as active power control, reactive power control, and DC-link voltage control, the current references i_{d_ref} and i_{q_ref} can be obtained [1, 2]. The vector control method is adopted in Figure 2.46, which regulates the i_d, i_q to follow the current references i_{d_ref} and i_{q_ref}, respectively, and generates the dq-axis voltage references u_{dm_ref} and u_{qm_ref}, respectively. Afterwards, the angle compensation component θ_m and the peak value U_m of the voltage reference can be obtained as

$$\begin{cases} \theta_m = \tan^{-1}\left(u_{qm\_ref} / u_{dm\_ref}\right) \\ U_m = \sqrt{u_{dm\_ref}^2 + u_{qm\_ref}^2} \end{cases} \tag{2.70}$$

The modulation index m_{ref} of the reference signal is $m_{ref} = 2U_m/V_{dc}$. In Figure 2.46, the phase angle θ_e of the grid voltage is obtained by the PLL. The phase angle of the reference signal is $\theta_{ref} = \theta_m + \theta_e$.

Figure 2.47 shows the reference modulation index based control for the i-th SM in the upper arm of phase A. For each SM, the PI controller is used to regulate its modulation index m to ensure the capacitor voltage balancing. Figure 2.47 shows the PI controller is used to produce the compensation modulation index m_{caui} for the i-th SM in the upper arm of phase A. The modulation index m_{aui} for the i-th SM in the upper arm of phase A is

$$m_{aui} = m_{ref} - m_{caui} \tag{2.71}$$

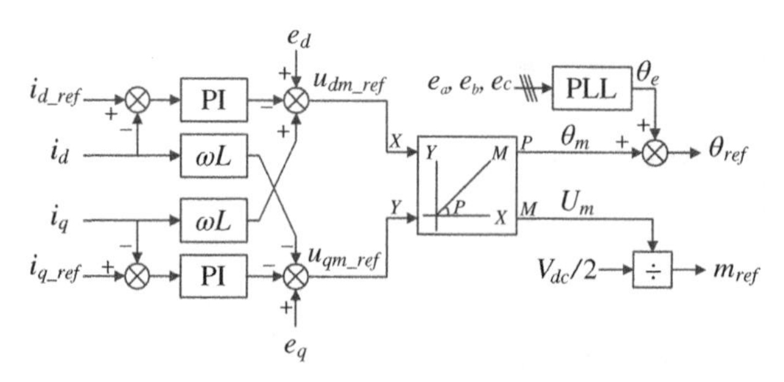

Figure 2.46 Current inner loop control for the MMC.

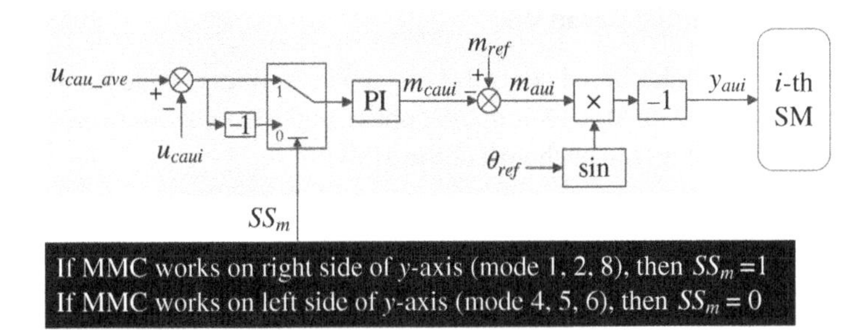

Figure 2.47 The reference modulation index based control for the *i*-th SM in the upper arm of phase A.

The reference for the *i*-th SM in the upper arm of phase A is

$$y_{aui} = -m_{aui} \cdot \sin\left(\theta_{ref}\right) \tag{2.72}$$

The implementation of the reference modulation index based control is related to the operation mode of the MMC, where the selection signal SS_m is 1 when the MMC works in Mode 1, 2, 8; the selection signal SS_m is 0 when the MMC works in Mode 4, 5, 6, as shown in Figure 2.47.

- $SS_m = 1$: If the capacitor voltage u_{caui} in the *i*-th SM is less than the average voltage u_{cau_ave} in the upper arm of phase A, the PI controller would increase the m_{caui} and reduce the m_{aui}. As a result, the DC component in the capacitor current would be increased according to Table 2.7 to increase u_{caui} to follow u_{cau_ave}. If $u_{caui} > u_{cau\_ave}$, the PI controller would reduce the m_{caui} and increase the m_{aui}. As a result, the DC component in the capacitor current would be reduced according to Table 2.7 to reduce u_{caui} to follow u_{cau_ave}.
- $SS_m = 0$: If $u_{caui} < u_{cau\_ave}$, the PI controller would reduce m_{caui} and increase m_{aui}. As a result, the DC component in the capacitor current would be increased according to Table 2.7 to increase u_{caui} to follow u_{cau_ave}. If $u_{caui} > u_{cau\_ave}$, the PI controller would increase m_{caui} and reduce m_{aui}. As a result, the DC component in the capacitor current would be reduced according to Table 2.7 to reduce u_{caui} to follow u_{cau_ave}.

The reference modulation index based control can be applied to the MMC in some applications. For the MMC works with the power transferring from DC side to AC side such as medium-voltage motor drive [23, 24] and grid integration of photovoltaic system [25], $SS_m = 1$ can be adopted for the MMC. For the MMC works with the power transferring from AC side to DC side such as the active rectifier of the medium-voltage motor drive [23], $SS_m = 0$ can be adopted for the MMC.

2.5.3 Reference Phase Angle Based VBC

The reference modulation index based control is not appliable under Mode 3 and 7, as analyzed in Section 2.5.2. In Mode 3 and 7, the phase angle θ can be controlled to balance the capacitor voltages in each arm of the MMC.

2.5.3.1 Control of i_{cdc} by Phase Angle θ

Figure 2.48 and Table 2.8 show the control of i_{cdc} by θ for the i-th SM in the MMC under different operation modes according to (2.69), as follows.

- *Mode 1*: $P > 0$ and $Q = 0$. Here, the $\cos(\theta)$ reaches its maximum 1. According to equation (2.69), the i_{cdc} cannot be regulated by θ.
- *Mode 2*: $P > 0$ and $Q > 0$. Here, $i_{dc} > 0$ and $\cos(\theta) > 0$. According to equation (2.69), the i_{cdc} can be increased by the increase of θ and reduced by the reduction of θ.

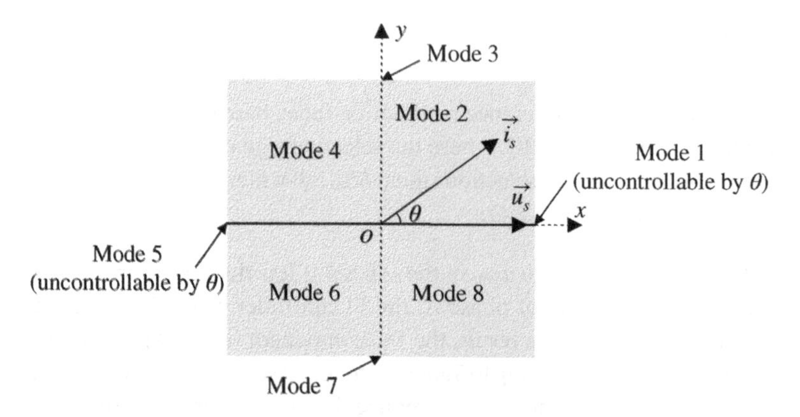

Figure 2.48 Control in the MMC by phase angle θ.

Table 2.8 Control of i_{cdc} by θ.

Mode	Q	i_{dc}	cos(θ)	θ	i_{cdc}
1	0	>0	1		Uncontrollable
5	0	<0	−1		
2	>0	>0	>0		Proportional
3		0	0		
4		<0	<0		
6	<0	<0	<0		Inversely proportional
7		0	0		
8		>0	>0		

- *Mode 3*: $P = 0$ and $Q > 0$. Here, $i_{dc} = 0$. According to equation (2.69), the i_{cdc} can be increased by the increase of θ and reduced by the reduction of θ.
- *Mode 4*: $P < 0$ and $Q > 0$. Here, $i_{dc} < 0$ and $\cos(\theta) < 0$. According to equation (2.69), the i_{cdc} can be increased by the increase of θ and reduced by the reduction of θ.
- *Mode 5*: $P < 0$ and $Q = 0$. Here, the $\cos(\theta)$ reaches its minimum -1. According equation (2.69), the i_{cdc} cannot be regulated by θ.
- *Mode 6*: $P < 0$ and $Q < 0$. Here, $i_{dc} < 0$ and $\cos(\theta) < 0$. According to equation (2.69), the i_{cdc} can be increased by the reduction of θ and reduced by the increase of θ.
- *Mode 7*: $P = 0$ and $Q < 0$. Here, $i_{dc} = 0$. According to equation (2.69), the i_{cdc} can be increased by the reduction of θ and reduced by the increase of θ.
- *Mode 8*: $P > 0$ and $Q < 0$. Here, $i_{dc} > 0$ and $\cos(\theta) > 0$. According to equation (2.69), the i_{cdc} can be increased by the reduction of θ and reduced by the increase of θ.

Table 2.8 shows that the DC component i_{cdc} in each SM capacitor current can be controlled by the corresponding phase angle θ of the reference signal for each SM when the MMC works in various modes except Modes 1 and 5. When the MMC works above the x-axis, the i_{cdc} is proportional to the θ; when the MMC works below the x-axis, the i_{cdc} is inversely proportional to the θ.

2.5.3.2 Voltage Balancing Control by θ

Figure 2.49 shows the reference phase angle θ for the i-th SM in the upper arm of phase A. For each SM, the PI controller is used to regulate its phase angle θ to ensure the capacitor voltage balancing. The PI controller is used to produce the compensation phase angle θ_{caui} for the i-th SM in the upper arm of phase A. The phase angle θ_{aui} for the i-th SM in the upper arm of phase A is

$$\theta_{aui} = \theta_{ref} + \theta_{caui} \tag{2.73}$$

The reference for the i-th SM in the upper arm of phase A is

$$y_{aui} = -m_{ref} \cdot \sin\left(\theta_{aui}\right) \tag{2.74}$$

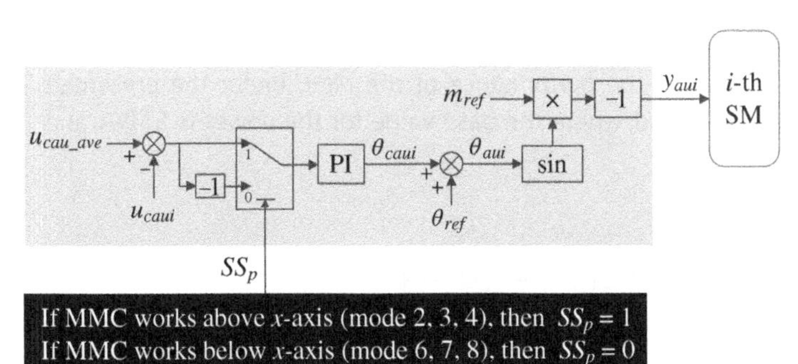

Figure 2.49 Reference phase angle based control for the i-th SM in upper arm of phase A.

The implementation of the reference phase angle based control is related to the operation mode of the MMC, where the selection signal SS_p is 1 when MMC works in Mode 2, 3, 4; the selection signal SS_p is 0 when MMC works in Mode 6, 7, 8, as shown in Figure. 2.49.

- $SS_p = 1$: If $u_{caui} < u_{cau\_ave}$, the PI controller would increase θ_{caui} and increase θ_{aui}. As a result, the DC component in the capacitor current would be increased according to Table 2.8 to increase u_{caui} to follow u_{cau_ave}. If $u_{caui} > u_{cau\_ave}$, the PI controller would reduce θ_{caui} and reduce θ_{aui}. As a result, the DC component in the capacitor current would be reduced according to Table 2.8 to reduce u_{caui} to follow the u_{cau_ave}.
- $SS_p = 0$: If $u_{caui} < u_{cau\_ave}$, the PI controller would reduce θ_{caui} and reduce θ_{aui}. As a result, the DC component in the capacitor current would be increased according to Table 2.8 to increase u_{caui} to follow u_{cau_ave}. If $u_{caui} > u_{cau\_ave}$, the PI controller would increase θ_{caui} and increase θ_{aui}. As a result, the DC component in the capacitor current would be reduced according to Table 2.8 to reduce u_{caui} to follow the u_{cau_ave}.

The reference phase angle based control can be applied to the MMC in some applications such as the MMC based STATCOM for reactive power regulation [23, 26], and $SS_p = 1$ is adopted for the MMC if sending reactive power to the AC grid; $SS_p = 0$ is adopted for the MMC if absorbing reactive power from the AC grid.

Case Study 2.4 Analysis of the Reference Modulation Index Based Control and Phase Angle Based Control

Objective: In order to verify the presented voltage balancing method, an MMC system shown in Figure 2.50 is simulated with the professional time-domain simulation tool PSCAD/EMTDC. The system parameters are shown in Table 2.9.

Simulation results and analysis:

Figures 2.51–2.58 show the performance of the MMC under the presented voltage balancing method, where the base value for the power is 5 MVA, and

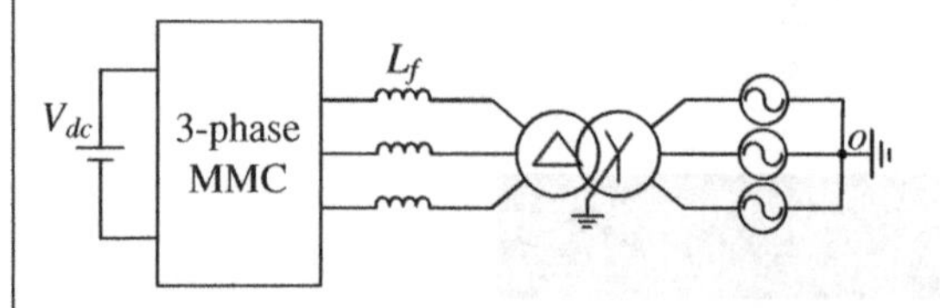

Figure 2.50 Block diagram of the simulation system.

Table 2.9 Simulation system parameters.

Parameter	Value
Grid line-to-line voltage (kV)	33
Line frequency (Hz)	50
Transformer voltage rating	3 kV/33 kV
DC bus voltage V_{dc} (kV)	6
Number of SMs per arm n	6
Rated capacitor voltage (kV)	1
SM capacitance C_{sm} (mF)	15
Arm inductance L_s (mH)	2
Inductance L_f (mH)	0.5
Switching frequency (kHz)	1

the base value for grid voltage is the peak value of the grid line-to-line voltage. The base value for the current is the peak value of the grid current when active power is 5 MW and reactive power Q is 0. The base value for the capacitor voltage is the rated capacitor voltage. Figure 2.51 shows the performance of the MMC working in Mode 1, and the reference modulation index based control with $SS_m = 1$ is adopted. Figure 2.51a shows that the grid line-to-line voltage e_{ab} leads grid current i_a by 30°. Here, the P is 1 p.u. and the Q is 0. Figure 2.51b shows the upper and lower arm current i_{au} and i_{al} in phase A. With the control, the upper and lower arm capacitor voltages $u_{cau1}-u_{cau6}$ and $u_{cal1}-u_{cal6}$ are kept balanced, as shown in Figure 2.51c.

Figure 2.52 shows the performance of the MMC working from Mode 8 to 2, and the reference modulation index based control with $SS_m = 1$ is adopted here. Figure 2.52a shows that the P is 1 p.u. and the Q is gradually changed from −0.4 to 0.4 p.u. With the presented control, the capacitor voltages $u_{cau1}-u_{cau6}$ and $u_{cal1}-u_{cal6}$ are kept balanced, as shown in Figure 2.52b.

Figure 2.53 shows the performance of the MMC in Mode 5, and the reference modulation index based control with $SS_m = 0$ is adopted. In this situation, the P is −1 p.u. and Q is 0. Figure 2.53a shows that the e_{ab} lags by i_a 150°. Figure 2.53b shows the i_{au} and i_{al} in phase A. With the presented control, the capacitor voltages $u_{cau1}-u_{cau6}$ and $u_{cal1}-u_{cal6}$ are kept balanced, as shown in Figure 2.53c.

Figure 2.54 shows the performance of the MMC working from Mode 6 to 4, and the reference modulation index based control with $SS_m = 0$ is adopted here. Figure 2.54a shows that the P is 1 p.u. and the Q is gradually changed

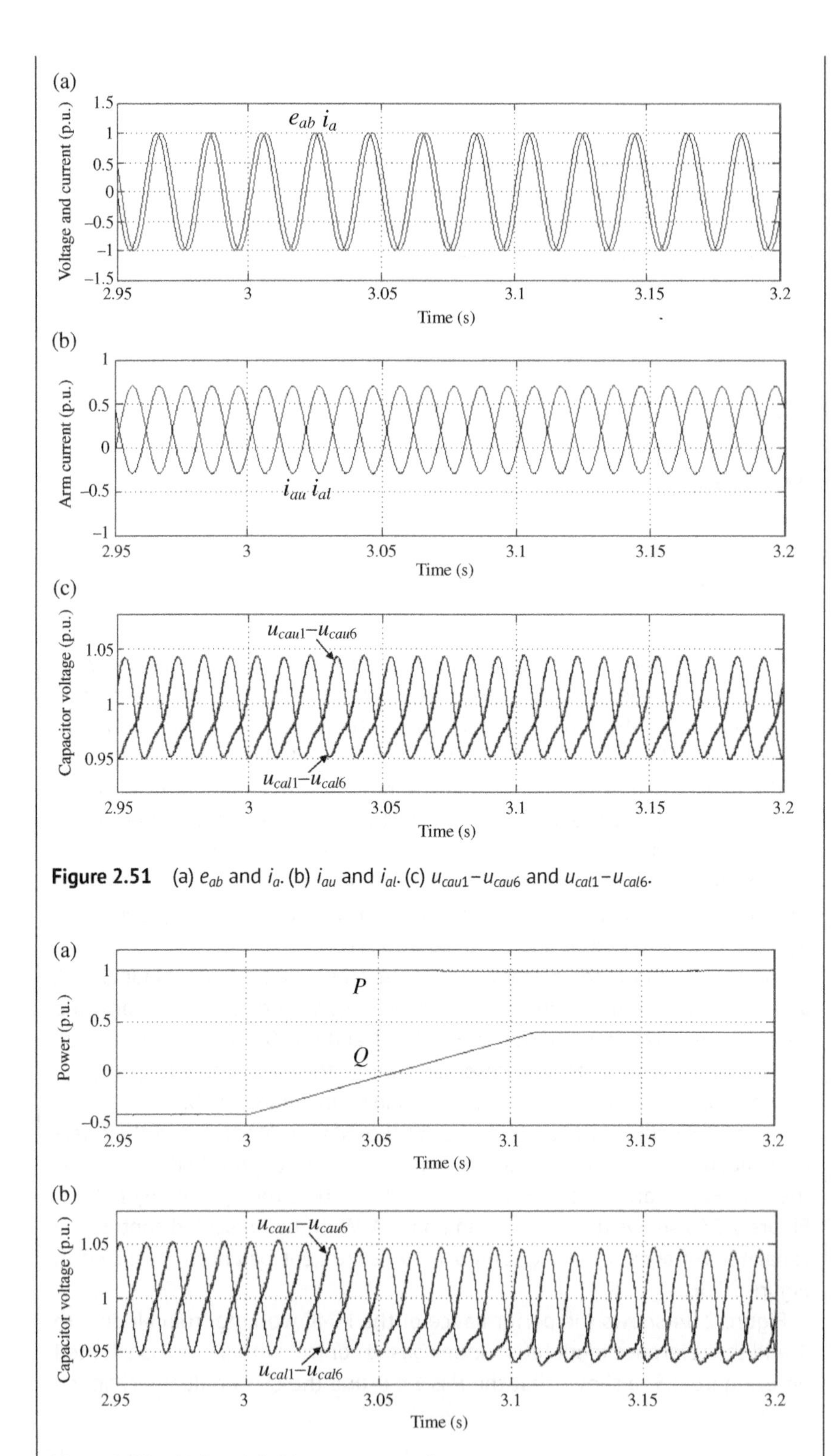

Figure 2.51 (a) e_{ab} and i_a. (b) i_{au} and i_{al}. (c) $u_{cau1}-u_{cau6}$ and $u_{cal1}-u_{cal6}$.

Figure 2.52 (a) P and Q. (b) $u_{cau1}-u_{cau6}$ and $u_{cal1}-u_{cal6}$.

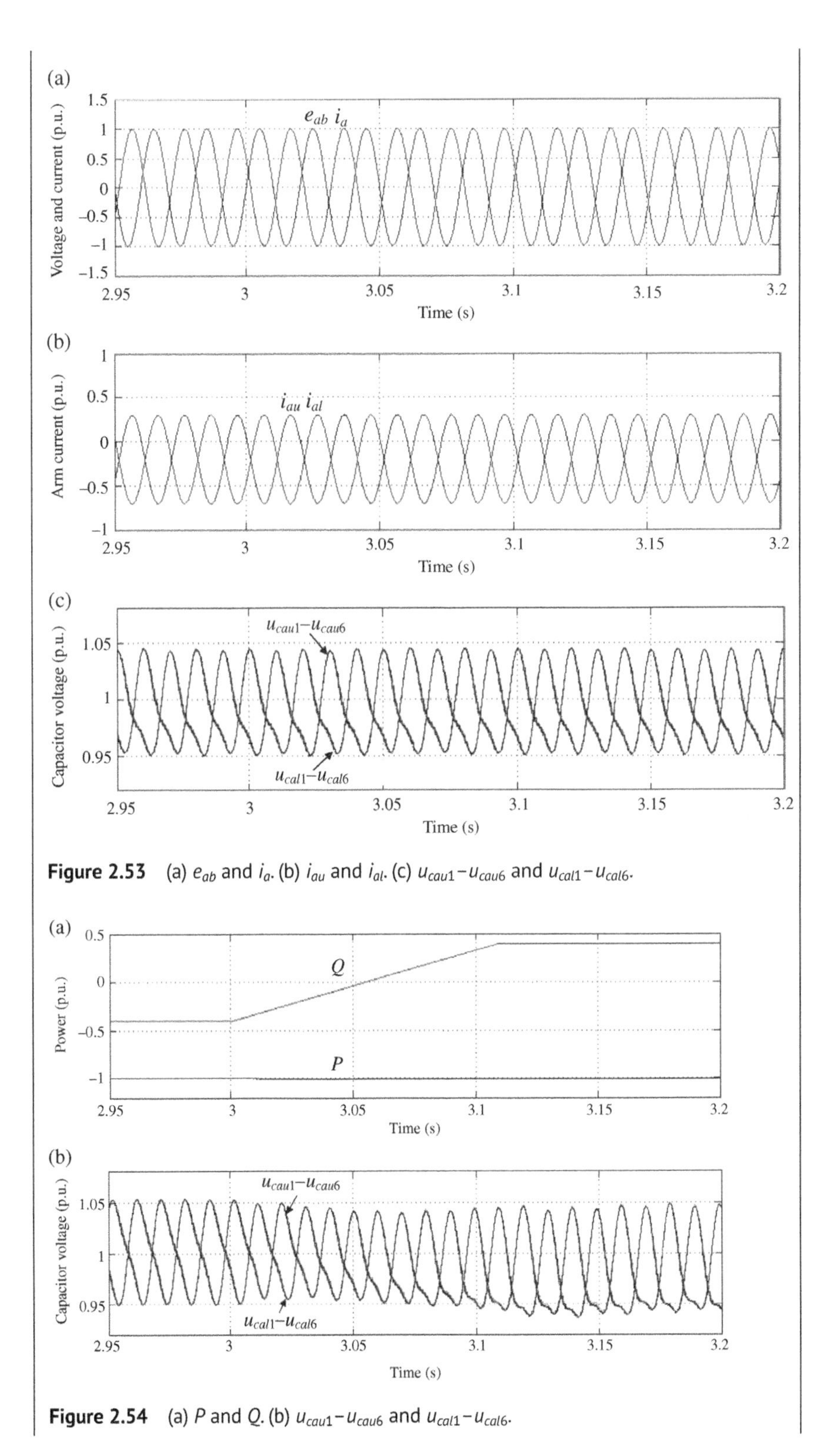

Figure 2.53 (a) e_{ab} and i_a. (b) i_{au} and i_{al}. (c) $u_{cau1}-u_{cau6}$ and $u_{cal1}-u_{cal6}$.

Figure 2.54 (a) P and Q. (b) $u_{cau1}-u_{cau6}$ and $u_{cal1}-u_{cal6}$.

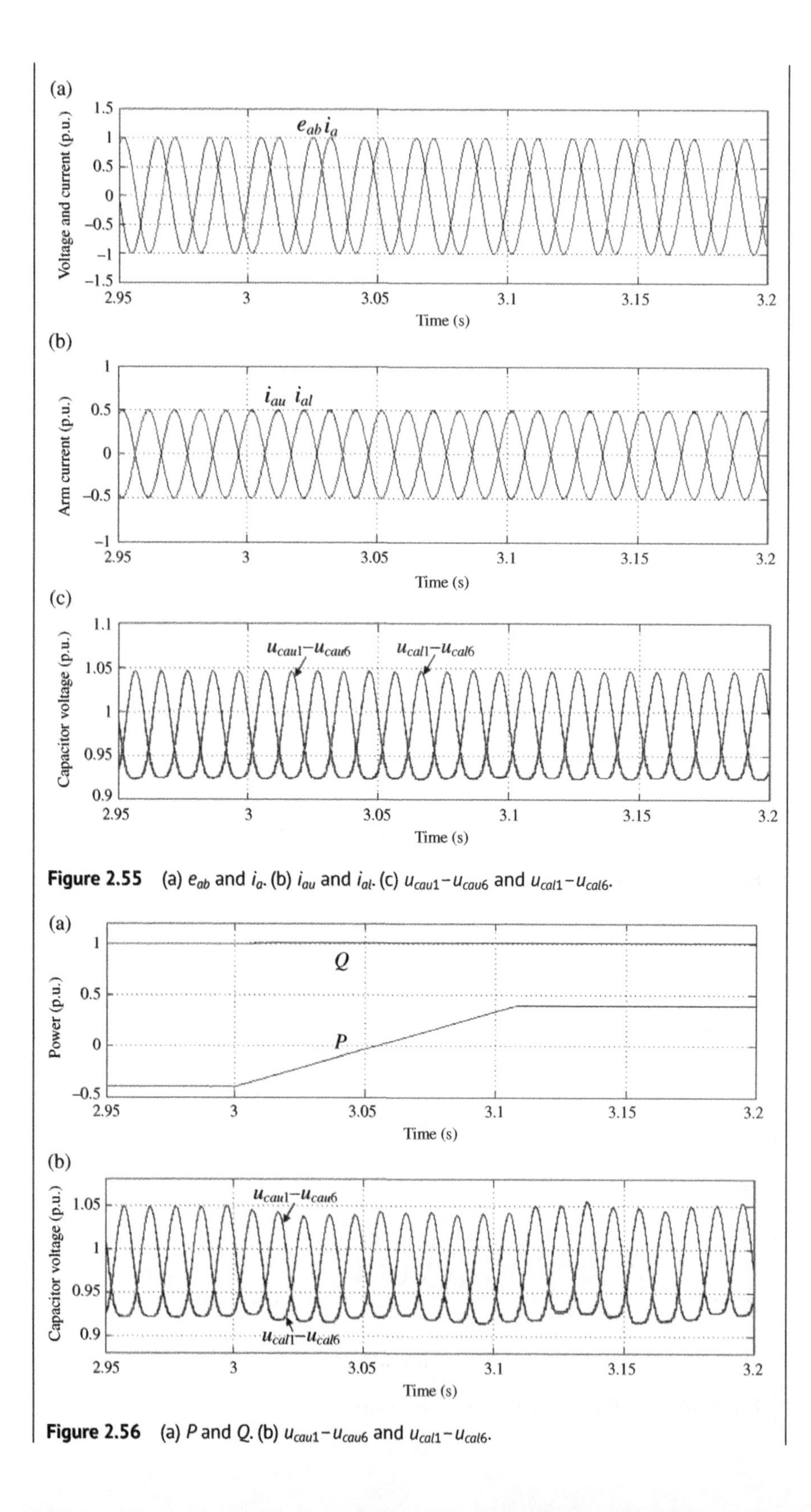

Figure 2.55 (a) e_{ab} and i_a. (b) i_{au} and i_{al}. (c) $u_{cau1}-u_{cau6}$ and $u_{cal1}-u_{cal6}$.

Figure 2.56 (a) P and Q. (b) $u_{cau1}-u_{cau6}$ and $u_{cal1}-u_{cal6}$.

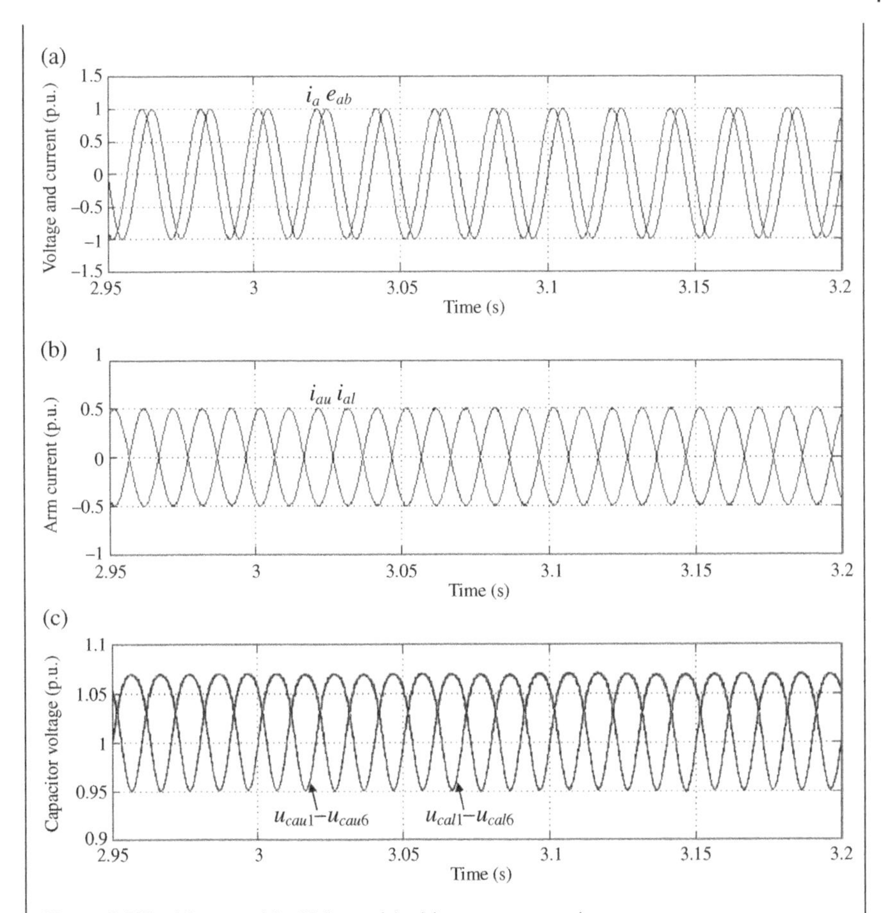

Figure 2.57 (a) e_{ab} and i_a. (b) i_{au} and i_{al}. (c) $u_{cau1}-u_{cau6}$ and $u_{cal1}-u_{cal6}$.

from −0.4 to 0.4 p.u. With the presented control, the capacitor voltages $u_{cau1}-u_{cau6}$ and $u_{cal1}-u_{cal6}$ are kept balanced, as shown in Figure 2.54b.

Figure 2.55 shows the performance of the MMC working in Mode 3 and the reference phase angle based control with $SS_p = 1$ is adopted here. Figure 2.55a shows that e_{ab} leads i_a by 120°. In this situation, the P is 0 and Q is 1 p.u. Figure 2.55b shows the upper and lower arm current i_{au} and i_{al} in phase A. With the presented control, the upper arm capacitor voltages $u_{cau1}-u_{cau6}$ and the lower arm capacitor voltages $u_{cal1}-u_{cal6}$ are kept balanced, as shown in Figure 2.55c.

Figure 2.56 shows the performance of the three-phase MMC working from Mode 4 to 2, and the reference phase angle based control with SS_p = 1 is adopted here. Figure 2.56a shows that the reactive power Q is 1 p.u. and the active power P is gradually changed from −0.4 to 0.4 p.u. With the voltage

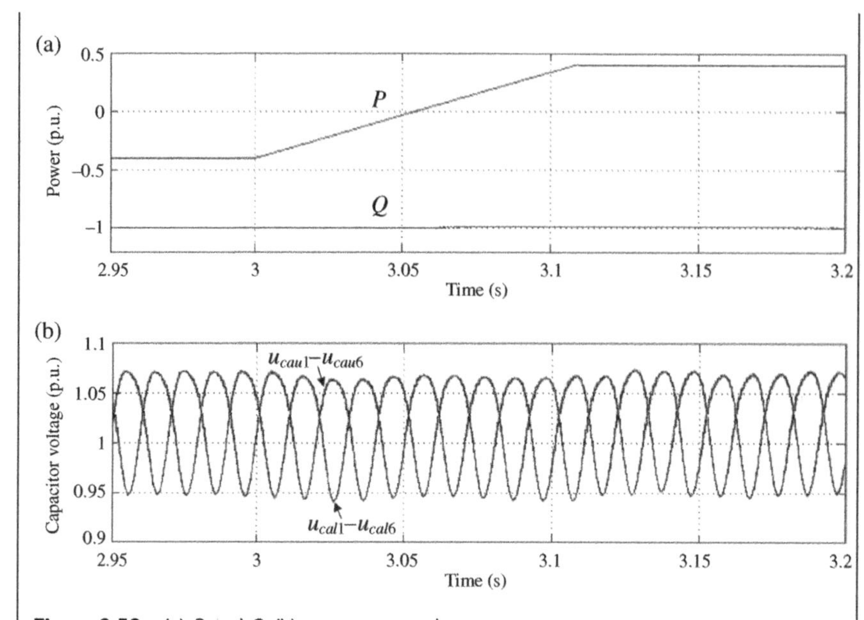

Figure 2.58 (a) P and Q. (b) $u_{cau1}-u_{cau6}$ and $u_{cal1}-u_{cal6}$.

balancing control, the capacitor voltages $u_{cau1}-u_{cau6}$ and $u_{cal1}-u_{cal6}$ in the MMC are kept balanced, as shown in Figure 2.56b.

Figure 2.57 shows the performance of the MMC working in Mode 7, and the reference phase angle based control with $SS_p = 0$ is adopted here. Figure 2.57a shows that e_{ab} lags i_a by 60°. In this situation, the P is 0 and Q is −1 p.u. Figure 2.57b shows the upper and lower arm current i_{au} and i_{al} in phase A. With the presented control, the capacitor voltages $u_{cau1}-u_{cau6}$ and $u_{cal1}-u_{cal6}$ are kept balanced, as shown in Figure 2.57c.

Figure 2.58 shows the performance of the MMC working from Mode 6 to 8, and the reference phase angle based control with $SS_p = 0$ is adopted here. Figure 2.58a shows that the Q is −1 p.u. and the P is gradually changed from −0.4 to 0.4 p.u. With the presented control, the $u_{cau1}-u_{cau6}$ and $u_{cal1}-u_{cal6}$ are kept balanced, as shown in Figure 2.58b.

2.6 Circulating Current Control

The circulating current in the MMC does not contribute to the AC-side current, but it distorts arm current and normally increases power losses in the arm. Therefore, the circulating current is normally expected to be suppressed. This section describes several CCC methods.

2.6.1 Proportional Integration Control

The three-phase circulating currents i_{diff_a}, i_{diff_b}, i_{diff_c} mainly contain the DC component $i_{dc}/3$ and the second-order harmonic component i_{2f_a}, i_{2f_b}, i_{2f_c}, which can be expressed as

$$\begin{cases} i_{diff\_a} = \frac{i_{dc}}{3} + i_{2f\_a} = \frac{i_{dc}}{3} + I_{2f}\left(2\omega t + \varphi\right) \\ i_{diff\_b} = \frac{i_{dc}}{3} + i_{2f\_b} = \frac{i_{dc}}{3} + I_{2f}\left(2\omega t + \varphi + \frac{2\pi}{3}\right) \\ i_{diff\_c} = \frac{i_{dc}}{3} + i_{2f\_c} = \frac{i_{dc}}{3} + I_{2f}\left(2\omega t + \varphi - \frac{2\pi}{3}\right) \end{cases} \tag{2.75}$$

Figure 2.59 shows the relationship of the circulating currents i_{diff_j} and voltages u_{diff_j} ($j = a, b, c$) imposed on the arm inductance L_s. According to Figure 2.59, the voltages u_{diff_j} can be expressed as

$$\begin{bmatrix} u_{diff\_a} \\ u_{diff\_b} \\ u_{diff\_c} \end{bmatrix} = 2L_s \frac{d}{dt} \begin{bmatrix} i_{diff\_a} \\ i_{diff\_b} \\ i_{diff\_c} \end{bmatrix} \tag{2.76}$$

The voltages u_{diff_j} ($j = a, b, c$) and circulating currents i_{diff_j} ($j = a, b, c$) can be transformed into d-axis and q-axis components in the negative-sequence rotational reference frame as

$$\begin{bmatrix} u_{diff\_d} \\ u_{diff\_q} \end{bmatrix} = 2L_s \frac{d}{dt} \begin{bmatrix} i_{diff\_d} \\ i_{diff\_q} \end{bmatrix} - 2 \begin{bmatrix} 0 & -2\omega L_s \\ 2\omega L_s & 0 \end{bmatrix} \begin{bmatrix} i_{diff\_d} \\ i_{diff\_q} \end{bmatrix} \tag{2.77}$$

where u_{diff_d} and u_{diff_q} are, respectively, the d-axis and q-axis components of the voltage u_{diff_a}, u_{diff_b}, u_{diff_c}; i_{diff_d} and i_{diff_q} are, respectively, the d-axis and q-axis components of the current i_{diff_a}, i_{diff_b}, i_{diff_c}.

Figure 2.60 shows the block diagram of the model of circulating currents [1]. The d-axis and q-axis components i_{diff_d} and i_{diff_q} are coupled. In order to realize the independent control of the d-axis and q-axis components, it is necessary to realize the decoupling of i_{diff_d} and i_{diff_q}.

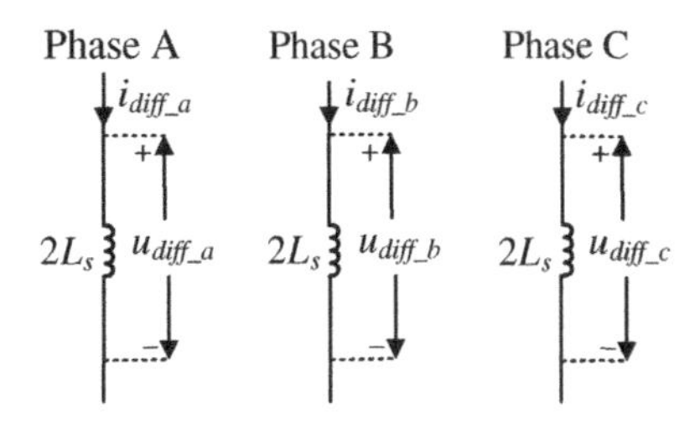

Figure 2.59 The relationship between the circulating currents i_{diff_j} and voltages u_{diff_j}.

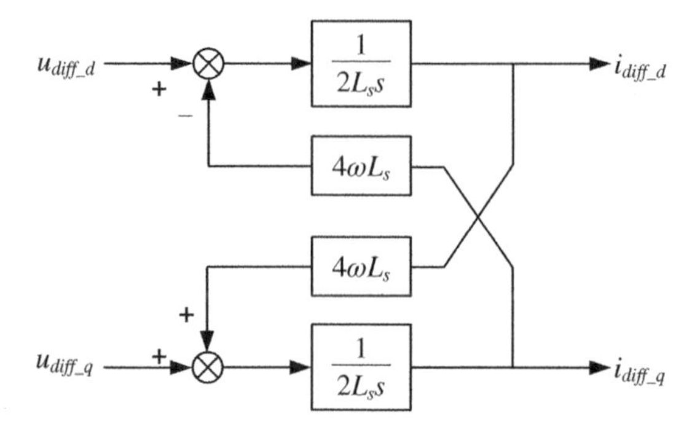

Figure 2.60 Block diagram of the model of circulating currents.

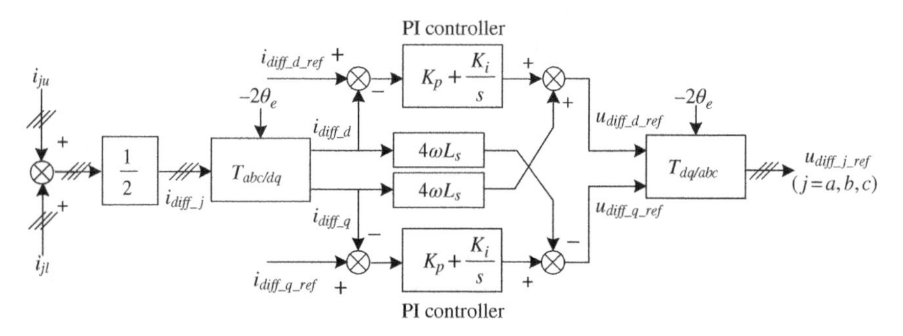

Figure 2.61 PI control of circulating currents.

Figure 2.61 shows the diagram of the PI control for circulating currents [1], which is explained in detail as follows.

- Based on equation (1.28), the circulating current i_{diff_j} ($j = a$, b, and c) can be obtained as half of the summation of the upper arm current i_{ju} and the lower arm current i_{jl}. Then, i_{diff_a}, i_{diff_b}, and i_{diff_c} are converted into i_{diff_d} and i_{diff_q} by the transformation matrix $T_{acb/dq}$ in equation (2.7), where i_{diff_d} and i_{diff_q} are the DC components in the negative sequence rotational reference frame.
- The PI controller is used to suppress i_{diff_d} and i_{diff_q} to zero. The i_{diff_d} and i_{diff_q} are coupling, which affects the dynamic performance of the PI control. Therefore, the feedforward decoupling strategy is used to achieve decoupling, and the control equations are as follows.

$$\begin{cases} u_{diff\_d\_ref} = K_p\left(i_{diff\_d\_ref} - i_{diff\_d}\right) + K_i\int\left(i_{diff\_d\_ref} - i_{diff\_d}\right)dt + 4\omega L_s i_{diff\_q} \\ u_{diff\_q\_ref} = K_p\left(i_{diff\_d\_ref} - i_{diff\_d}\right) + K_i\int\left(i_{diff\_d\_ref} - i_{diff\_d}\right)dt - 4\omega L_s i_{diff\_d} \end{cases} \tag{2.78}$$

where K_p and K_i are, respectively, the proportional coefficient and integral coefficient of the PI controller. $u_{diff\_d\_ref}$ and $u_{diff\_q\_ref}$ are, respectively, the d-axis and q-axis components of the reference voltage; $i_{diff\_d\_ref}$ and $i_{diff\_q\_ref}$ are, respectively, the d-axis and q-axis components of the circulating current reference, which should be set to 0.

- $u_{diff\_a\_ref}$, $u_{diff\_b\_ref}$, $u_{diff\_c\_ref}$ are calculated by the inverse transformation $T_{dq/abc}$. By adding $u_{diff\_j\_ref}$ $(j = a, b, c)$ to the upper arm reference voltage and the lower arm reference voltage in phase j, respectively, as shown in Figure 2.1, the suppression of harmonics of circulating currents can be achieved.

2.6.2 Multiple Proportional Resonant Control

In order to achieve no static error control of the even harmonic components of the circulating current, the proportional resonant (PR) controllers are introduced. In the real application, the quasi-PR controller with certain bandwidth is generally preferred for better digital realization and noise injection [27]. Therefore, quasi-PR control is more appropriate. The transfer function of quasi-PR controller is

$$G_{PR}(s) = K_p + \frac{2\omega_c K_r s}{s^2 + 2\omega_c s + \omega_h^2} \tag{2.79}$$

where ω_h is the resonant frequency. K_p and K_r are the proportional constant and the integral constant, respectively. ω_c is the cut-off frequency and mainly affects the bandwidth of the resonant controller at the resonant frequency.

Multiple even-order harmonics of circulating current need to be suppressed in the MMC [28]. Therefore, multiple quasi-PR controllers are used to suppress the circulating current harmonics [28], which can be described as

$$G_m(s) = K_p + \sum_{h=2,4,6,\cdots} \frac{2\omega_c K_r s}{s^2 + 2\omega_c s + (h\omega)^2} \tag{2.80}$$

Figure 2.62 shows the control block diagram of multiple quasi-PR control of circulating currents, where h is even. Based on equation (1.28), the circulating current components i_{diff_j} can be obtained by the upper arm current i_{ju} and the lower arm current i_{jl}. The multiple quasi-PR controllers in (2.80) are used to achieve small static error for tracking $i_{diff\_j\_ref}$ and produce reference voltage $u_{diff\_j\_ref}$, where $i_{diff\_j\_ref}$ is the reference of circulating currents and should be set to 0. By adding $u_{diff\_j\_ref}$ to the upper arm reference voltage and the lower arm reference voltage in phase j, respectively, as shown in Figure 2.1, the suppression of harmonics of circulating currents can be achieved.

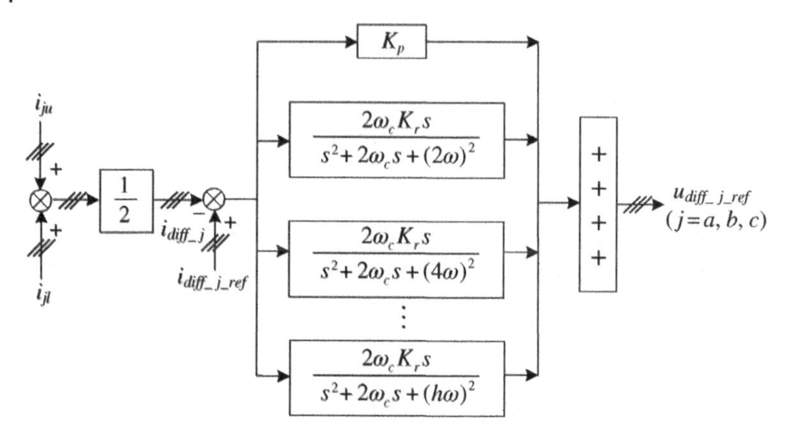

Figure 2.62 Multiple quasi-PR control of circulating currents.

2.6.3 Repetitive Control

In Section 2.6.2, multiple PR controllers are needed to suppress multiple even-order harmonics of circulating current, which increases the complexity of system design. In addition, higher order resonant controller will be useless because the low sampling and processing frequency in the MMC system causes the control bandwidth of the circulating current controller to be always low [29].

In order to realize effective suppression of multiple even-order harmonics of circulating current, the repetitive controller can be introduced. Figure 2.63a shows the control block diagram of the repetitive control in continuous domain, where e and u_o are the input and output of the repetitive controller, and T_f is the fundamental period as $T_f = \omega/2\pi$. According to Figure 2.63a, the transfer function of the repetitive controller in continuous domain can be calculated as

$$G_c(s) = \frac{u_o(s)}{e(s)} = \frac{e^{-T_f s}}{1-e^{-T_f s}} = -\frac{1}{2} + \frac{1}{T_f s} + \frac{2}{T_f}\sum_{n=1}^{\infty}\frac{s}{s^2+(n\omega)^2} \tag{2.81}$$

Equation (2.81) shows that the repetitive controller can be equivalent to the parallel connection of the proportional term, integral term, and the infinite resonant terms. Therefore, repetitive control can provide approximately infinite gain

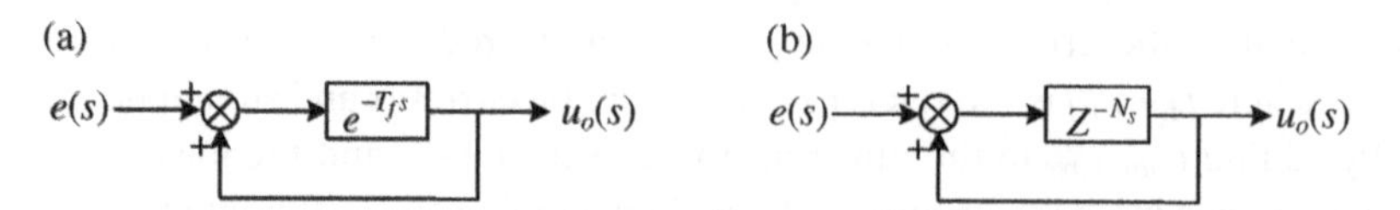

Figure 2.63 Repetitive control. (a) Continuous domain. (b) Discrete domain.

at the specified frequency and its harmonic frequencies, and it is widely used to achieve zero steady-state error at multiple harmonic frequencies.

Since the controller is usually implemented in a digital control, it is necessary to discretize the repetitive control in the discrete domain in order to actually use the repetitive control. According to Figure 2.63a, the control block diagram of the repetitive control in discrete domain can be obtained, which is shown in Figure 2.63b, where $N_s = T_f/T_s$.

Normally, the step of the repetitive control strategy is fundamental period, which results in a long dynamic tracking time for repetitive control and cannot achieve instant response. This section introduces a control method that combines repetitive control and PI control in parallel. On the one hand, it improves the system's ability to suppress multiple even-order harmonics, and on the other hand, it also has the characteristics of instant response of PI control.

Figure 2.64 shows the repetitive control and PI control of circulating currents. Based on (1.28), the circulating current component i_{diff_j} can be obtained by the upper arm current i_{ju} and the lower arm current i_{jl}. The repetitive controller is paralleled by the PI controller, which is used to achieve zero steady-state error for tracking $i_{diff\_j\_ref}$ and produce reference voltage $u_{diff\_j\_ref}$, where $i_{diff\_j\_ref}$ is the reference of circulating current and should be set to 0. By adding $u_{diff\_j\_ref}$ to the upper arm reference voltage and the lower arm reference voltage in phase j, respectively, as shown in Figure 2.1, the suppression of harmonics of circulating currents can be achieved.

For better stability, the repetitive controller in Figure 2.63 can be improved to the repetitive controller in Figure 2.65. In Figure 2.65, the K_{rc} is the gain of the repetitive controller, and $Q(z)$ is the stabilization filter. L_n is the number of samples for phase leading to compensate the lag of the plant and the digital controller [29]. According to Figure 2.65, the transfer function of the repetitive controller in the discrete domain can be calculated as

$$G_{rc}(z) = \frac{u_o(z)}{e(z)} = \frac{K_{rc} z^{-N_s + L_n}}{1 - Q(z) z^{-N_s}} \tag{2.82}$$

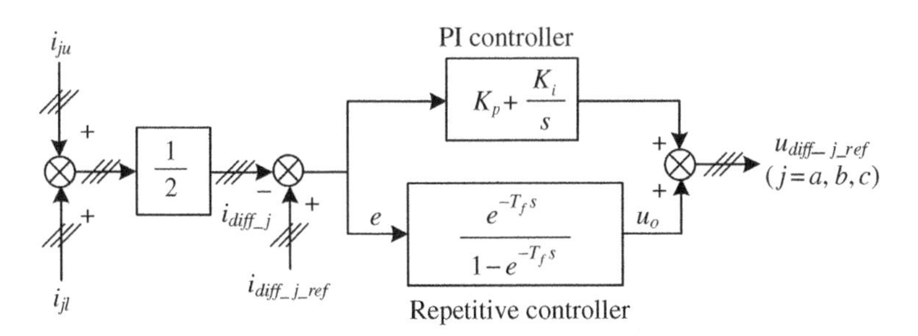

Figure 2.64 Repetitive control and PI control of circulating currents.

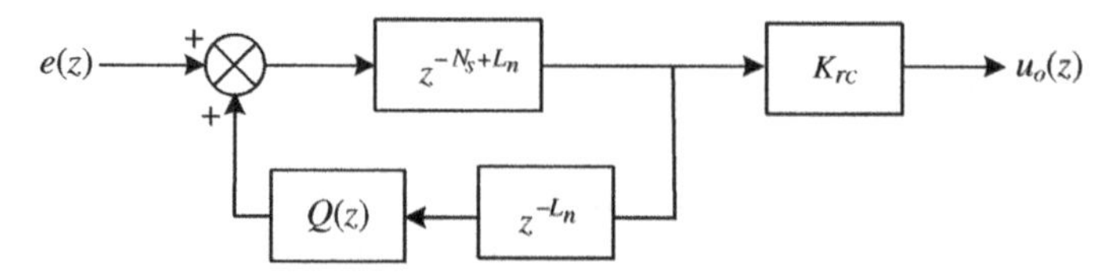

Figure 2.65 Improved repetitive control.

2.7 Summary

In this chapter, the recent advancements in the control of the MMC are introduced and analyzed. These control strategies are classified according to external and internal control objectives of the MMC.

In different application scenarios, the external control objectives are regulated by specific output control based on the current control. For the grid-connected applications, the DC-link voltage and power control is applied to regulate the power change between the MMC and grid. For the grid forming applications, the amplitude and frequency of AC-side voltage is regulated to form the AC grid.

As one of the internal control objectives, the capacitor voltage of each SM should keep balanced to ensure the safe operation of the MMC. Based on the sorting algorithm, several centralized capacitor VBC are illustrated. For each individual SM, several individual SM VBC are introduced. Circulating current is another typical internal control objective of the MMC, which should be suppressed to improve the efficiency of the MMC. In this chapter, several different CCC are introduced to achieve the circulating current suppression.

References

1 Q. Tu, Z. Xu, and L. Xu, "Reduced switching-frequency modulation and circulating current suppression for modular multilevel converters," *IEEE Trans. Power Delivery*, vol. 26, no. 3, pp. 2009–2017, Jul. 2011.

2 F. Deng and Z. Chen, "Voltage-balancing method for modular multilevel converters switched at grid frequency," *IEEE Trans. Ind. Electron.*, vol. 62, no. 5, pp. 2835–2847, May 2015.

3 M. Saeedifard and R. Iravani, "Dynamic performance of a modular multilevel back-to-back HVDC system," *IEEE Trans. Power Delivery*, vol. 25, no. 4, pp. 2903–2912, Oct. 2010.

4 H. Akagi, E. H. Watanabe, and M. Aredes, "The instantaneous power theory," in *Instantaneous Power Theory and Applications to Power Conditioning*, IEEE, 2017, pp. 37–109.

5 F. Mahr and J. Jaeger, "Advanced grid-forming control of HVDC systems for reliable grid restoration," in *2018 IEEE Power & Energy Society General Meeting (PESGM)*, 2018, pp. 1–5.
6 E. Sánchez-Sánchez, E. Prieto-Araujo, and O. Gomis-Bellmunt, "The role of the internal energy in MMCs operating in grid-forming mode," *IEEE J. Emerg. Sel. Top. Power Electron.*, vol. 8, no. 2, pp. 949–962, Jun. 2020.
7 Z. Li, C. Zang, P. Zeng, H. Yu, S. Li, and J. Bian, "Control of a grid-forming inverter based on sliding-mode and mixed H_2/H_∞ control," *IEEE Trans. Ind. Electron.*, vol. 64, no. 5, pp. 3862–3872, May 2017, doi: 10.1109/TIE.2016.2636798.
8 M. Raza, E. Prieto-Araujo, and O. Gomis-Bellmunt, "Small-signal stability analysis of offshore AC network having multiple VSC-HVDC systems," *IEEE Trans. Power Delivery*, vol. 33, no. 2, pp. 830–839, Apr. 2018.
9 F. Deng, Y. Lv, C. Liu, Q. Heng, Q. Yu, and J. Zhao, "Overview on submodule topologies, modeling, modulation, control schemes, fault diagnosis, and tolerant control strategies of modular multilevel converters," *Chin. J. Elect. Eng.*, vol. 6, no. 1, pp. 1–21, Mar. 2020.
10 Z. Li, F. Gao, F. Xu, X. Ma, *et al.*, "Power module capacitor voltage balancing method for a ±350kV/1000 MW modular multilevel converter," *IEEE Trans. Power Electron.*, vol. 31, no. 6, pp. 3977–3984, Jun. 2016.
11 L. G. Franquelo, J. Rodriguez, J. I. Leon, S. Kouro, R. Portillo, and M. A. M. Prats, "The age of multilevel converter arrives," *IEEE Ind. Electron. Mag.*, vol. 2, no. 2, pp. 28–39, Jun. 2008.
12 K. Ilves, A. Antonopoulos, L. Harnefors, S. Norrga, and H. P. Nee, "Circulating current control in modular multilevel converters with fundamental switching frequency," in *Proceedings of ECCE*, 2012, pp. 249–256.
13 K. Ilves, A. Antonopoulos, S. Norrga, and H. P. Nee, "A new modulation method for the modular multilevel converter allowing fundamental switching frequency," *IEEE Trans. Power Electron.*, vol. 27, no. 8, pp. 3482–3494, Aug. 2012.
14 J. Qin and M. Saeedifard, "Predictive control of a modular multilevel converter for a back-to-back HVDC system," *IEEE Trans. Power Delivery*, vol. 27, no. 3, pp. 1538–1547, Jul. 2012.
15 J. Qin and M. Saeedifard, "Reduced switching-frequency voltage-balancing strategies for modular multilevel HVDC- converters," *IEEE Trans. Power Delivery*, vol. 28, no. 4, pp. 2403–2410, Oct. 2013.
16 M. Hagiwara and H. Akagi, "Control and experiment of pulsewidth-modulated modular multilevel converters," *IEEE Trans. Power Electron.*, vol. 24, no. 7, pp. 1737–1746, Jul. 2009.
17 F. Deng and Z. Chen, "A control method for voltage balancing in modular multilevel converters," *IEEE Trans. Power Electron.*, vol. 29, no. 1, pp. 66–67, Jan. 2014.

18 K. Wang, Y. Li, Z. Zheng, and Lie Xu, "Voltage balancing and fluctuation-suppression methods of floating capacitors in a new modular multilevel converter," *IEEE Trans. Ind. Electron.*, vol. 60, no. 5, pp. 1943–1954, May 2013.

19 H. P. Mohammadi and M. T. Bina, "A transformerless medium-voltage STATCOM topology based on extended modular multilevel converters," *IEEE Trans. Power Electron.*, vol. 26, no. 5, pp. 1534–1545, May 2011.

20 M. Hagiwara, K. Nishimura, and H. Akagi, "A medium-voltage motor drive with a modular multilevel PWM inverter," *IEEE Trans. Power Electron.*, vol. 25, no. 7, pp. 1786–1799, Jul. 2010.

21 A. Antonopoulos, K. Ilves, L. Ängquist, and H. Nee, "On interaction between internal converter dynamics and current control of high-performance high-power ac motor drives with modular multilevel converters," in *Proceedings of IEEE ECCE*, Atlanta, 2010, pp. 4293–4298.

22 K. Ilves, L. Harnefors, S. Norrga, and H. Nee, "Predictive sorting algorithm for modular multilevel converters minimizing the spread in the submodule capacitor voltages," in *Proceedings of IEEE ECCE-Asia*, 2013, pp. 325–331.

23 A. Dekka, B. Wu, R. L. Fuentes, M. Perez, and N. R. Zargari, "Evolution of topologies, modeling, control schemes, and applications of modular multilevel converters," *IEEE J. Emerg. Sel. Top. Power Electron.*, vol. 5, no. 4, pp. 1631–1656, Dec. 2017.

24 BENSHAW. M21 3000 Series Medium Voltage Motor Drive. 2013. [Online]. Available: http://benshaw.com/uploadedFiles/Literature/Benshaw_M2L_MVFD_2.3-6.6kV.pdf.

25 M. R. Islam, A. M. Mahfuz-Ur-Rahman, M. M. Islam, Y. G. Guo, and J. G. Zhu, "Modular medium-voltage grid-connected converter with improved switching techniques for solar photovoltaic systems," *IEEE Trans. Ind. Electron.*, vol. 64, no. 11, pp. 8887–8896, Nov. 2017.

26 M. Hagiwara, R. Maeda, and H. Akagi, "Negative-sequence reactive power control by a PWM STATCOM based on a modular multilevel cascade converter(MMCC-SDBC)," *IEEE Trans. Ind. Appl.*, vol. 48, no. 2, pp. 720–729, Mar./Apr. 2012.

27 M. Hagiwara and H. Akagi, "Control and analysis of the modular multilevel cascaded converter based on double-star chopper-cells (MMCC-DSCC)," *IEEE Trans. Power Electron.*, vo1. 26, no. 6, pp. 1649–1658, Jun. 2011.

28 X. She, A. Huang, X. Ni, and R. Burgos, "AC circulating currents suppression in modular multilevel converter," in *IECON 2012 - 38th Annual Conference on IEEE Industrial Electronics Society*, 2012, pp. 191–196.

29 M. Zhang, L. Huang, W. Yao, and Z. Lu, "Circulating harmonic current elimination of a CPS-PWM-based modular multilevel converter with a plug-in repetitive controller," *IEEE Trans. Power Electron.*, vol. 29, no. 4, pp. 2083–2097, April 2014.

3

Fault Detection of MMCs under IGBT Faults

3.1 Introduction

Reliability and uninterruptable operation are major concerns for the modular multilevel converter (MMC) systems. The power semiconductor devices are one of the most fragile components and affect the reliability of the MMC. It is reported that the failure of power semiconductor devices accounts for about 21% of failures in the converter system [1]. The MMC system usually consists of a large number of insulated gate bipolar transistors (IGBTs), which makes IGBT a potential failure source threatening the reliable operation of the overall system.

Generally, the failure of the IGBT can be categorized as the short-circuit fault and the open-circuit fault. The short-circuit fault may cause destructive consequences. The time between the fault occurrence and the device failure is very short, about several microseconds. Therefore, the protection circuit is typically integrated into the IGBT gate drivers to shut down the switches when a short-circuit fault is detected. The open-circuit fault, on the other hand, can be tolerated for a long time. However, it will distort the voltage and current waveforms of the MMC, may lead to overcurrent or overvoltage, and may even cause secondary damages. Therefore, it is essential to develop methods to detect the IGBT faults in the MMC.

This chapter deals with the recent advancements in IGBT faults detection and protection methods for MMCs. The IGBT short-circuit fault and open-circuit fault are briefly introduced in Section 3.2. Section 3.3 introduces the detection and protection methods under IGBT short-circuit faults. Section 3.4 introduces the features of the submodule's IGBT open-circuit faults. Based on the features depicted in Section 3.4, a Kalman filter (KF) based fault detection method is

Modular Multilevel Converters: Control, Fault Detection, and Protection, First Edition.
Fujin Deng, Chengkai Liu, and Zhe Chen.

Published 2023 by John Wiley & Sons, Inc.

presented for the MMC in Section 3.5. In addition, an integrator-based fault detection method is presented in Section 3.6 to detect the IGBT faults. Section 3.7 presents a sliding-time window (STW) based IGBT open-circuit fault detection method. Section 3.8 presents an isolation forest (IF) based IGBT open-circuit fault detection method. Finally, Section 3.9 gives the summary of this chapter.

3.2 IGBT Faults

IGBT is the core component of the MMC system and is widely used in high-power applications. According to a survey, the power semiconductor device IGBT is considered to be one of the most fragile components in power electronic converters [2]. Factors such as environmental condition, electrical loading, and mechanical vibration will all exert stresses on the IGBTs and affect their reliability, which are closely related to the faults of the IGBT modules [3].

Figure 3.1 shows the structure of a widely used wire-bond IGBT module package. The weak points in the wire-bond IGBT modules are the bond wires, the chip solder, and the substrate solder joints [4]. Long-term operation and power fluctuations may cause degradation of the IGBT modules, which may lead to the wear-out failure of the IGBT modules. Causal factors such as transient overvoltage, overcurrent, and electrical overstress could cause the catastrophic failures of the IGBT modules. It should be noted that both wear-out failures and catastrophic failures may cause the same failure characteristics.

The IGBT failures can be generally divided into two classes according to the fault characteristics, including short-circuit fault and open-circuit fault, as shown in Figure 3.2.

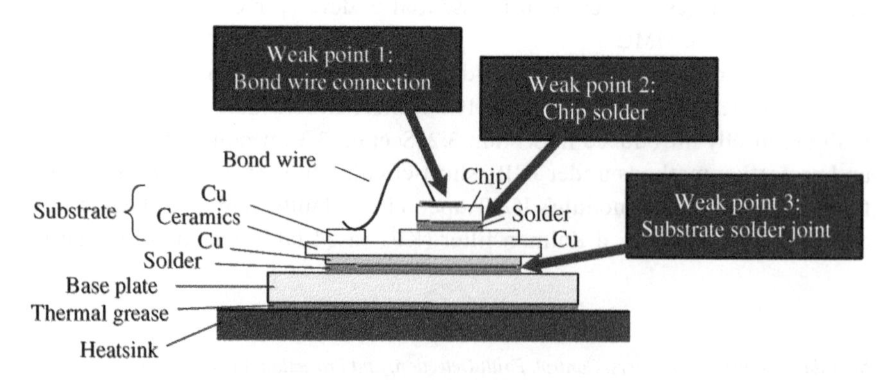

Figure 3.1 Structure of a wire-bond IGBT power module package.

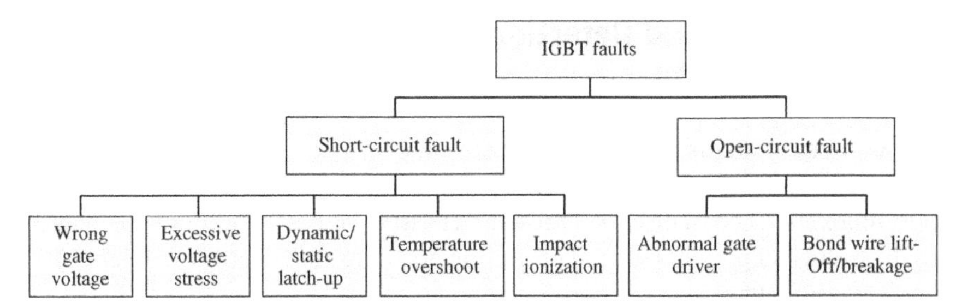

Figure 3.2 IGBT faults and main mechanisms.

3.2.1 IGBT Short-Circuit Fault

The IGBT short-circuit fault is mainly caused by incorrect driving signals or device overstress. Specifically, it may happen due to either of the following reasons [5].

- *Wrong gate voltage*, which may be caused by drive circuit failure, auxiliary power supply failure, or dv/dt interference.
- *Excessive voltage stress*, which may break down the isolation gate zone of the module case.
- *Dynamic/static latch-up*, which will increase the collector current of the IGBT, resulting in excessive loss and damage to the device.
- *Temperature overshoot*, where extremely uneven junction temperature will produce thermal stress and mechanical stress to make the IGBT fail.
- *Impact ionization*, where impact ionization and current crowding will cause excess heat generation and ultimately lead to device breakdown.

The short-circuit fault of IGBT is usually destructive, and the time between fault occurrence and IGBT failure is relatively short.

3.2.2 IGBT Open-Circuit Fault

The IGBT open-circuit failure is mainly caused by the failure of the switch gate driver and the failures of the bond wire [6], as follows.

- *Abnormal gate driver*, which may be caused by device failure in gate driver or short-circuit/open-circuit of the driver board.
- *Lift-off and breakage of the bond wire*, which may be caused by the aging of the IGBT and the bond wire breakage.

The IGBT open-circuit fault will cause an unbalanced state in the circuit, which produces distorted waveforms and voltage/current stresses and obviously deteriorates the system performance. Although this type of failure usually does not cause the immediate system shutdown, uneven stress and thermal cycling may result in a secondary fault within the converter [7] and ultimately cause the entire system to fail.

3.3 Protection and Detection Under IGBT Short-Circuit Faults

The short-circuit faults of IGBT are catastrophic for IGBT because they will lead to an abnormal overcurrent and cause serious damage to other components in a short time. This section discusses the characteristics of the SM under IGBT short-circuit faults and presents several protection and detection strategies under IGBT short-circuit faults.

3.3.1 SM Under IGBT Short-Circuit Fault

For the MMC with half-bridge (HB) SM, the IGBT short-circuit faults mainly have two cases, i.e. T_1 short-circuit fault and T_2 short-circuit fault. The characteristics of the SM under these two cases are listed below.

1) *T_1 short-circuit fault*: Figure 3.3 shows the short-circuit fault in T_1.
   - When the switching function S is 1, where drive signal u_{g1} for T_1 is high-level voltage and the drive signal u_{g2} for T_2 is low-level voltage, the T_1 is short-circuit and T_2 is turned off. Here, the SM is inserted into the arm regardless of the direction of the arm current i_{arm}.
   - When the switching function S is 0, where u_{g1} is low-level drive voltage, u_{g2} is high-level drive voltage, the T_1 is short-circuit and T_2 is turned on. Due to the short-circuit fault of T_1, the SM capacitor is shorted and the surge current through T_1 and T_2 will be caused, which will destroy the devices.

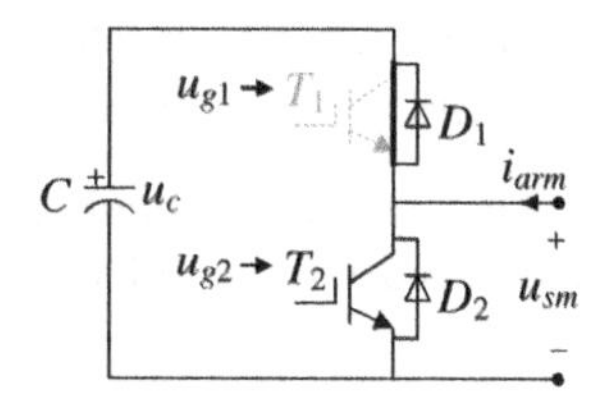

Figure 3.3 Short-circuit fault in T_1.

2) *T_2 short-circuit fault*: Figure 3.4 shows the short-circuit fault in T_2.
   - When the switching function S is 0, where u_{g1} is low-level drive voltage and u_{g2} is high-level drive voltage, the T_1 is turned off and T_2 is short-circuit. Here, the SM is bypassed from the arm regardless of the direction of the arm current i_{arm}.
   - When the switching function S is 1, where u_{g1} is high-level drive voltage and u_{g2} is low-level drive voltage, the T_1 is turned on and T_2 is short-circuit. Due to the short-circuit fault of T_2, the SM capacitor is shorted and the surge current through T_1 and T_2 will destroy the devices.

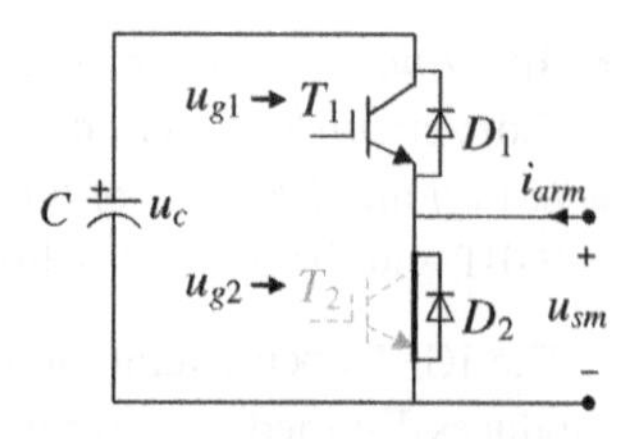

Figure 3.4 Short-circuit fault in T_2.

3.3.2 Protection and Detection Under IGBT Short-Circuit Fault

The detection time for the IGBT short-circuit fault must be within 10 μs to save the power devices from destruction and to avoid the shoot-through fault between the complementary devices [5]. Therefore, most of the short-circuit detection is implemented by using a hardware circuit with additional sensors and the protection is typically integrated in the gate drivers of the IGBT, which can guarantee the fast response and immediate shutdown of the IGBTs when the short-circuit fault occurs [6]. Some typical detection and protection methods are introduced below and the comparison among these methods are listed in the Table 3.1.

1) *De-saturation detection* [8]: This method detects the collector voltage rising from the low saturation value to the bus voltage level when the gate drive signal is high. If the short-circuit fault has occurred, the gate signals will be blocked to protect the device. This method uses very simple sensing circuit but is not suitable for high-speed IGBT switching, because it requires a blanking time of around 1–5 μs.

Table 3.1 Comparison of short-circuit detection methods.

Methods	Parameters required	Turn-off	Implementation effort	Reliability	Drawbacks
De-saturation detection [8]	Collector voltage	Abrupt	Low	Medium	Device turn-off not assured
di/*dt* feedback control method [9]	Device current	Soft	High	Medium	Stray inductance difficult to control
Gate voltage comparison [10]	Gate voltage	Soft	Low	Low	Requires complex circuitry
Current mirror method [11]	Device current	Abrupt	Low	Medium	Expensive
Gate voltage monitoring method [12]	Gate voltage	Abrupt	Low	Low	Requires complex circuitry
Protection by snubber and clamp circuit [13]	Device voltage	N/A	High	Low	Expensive
Protection by slow turn-off of IGBT [13]	Gate voltage	Soft	High	Low	Requires complex circuitry

2) *di/dt feedback control method* [9]: This method measures the induced voltage across the stray inductance between the Kelvin emitter and the power emitter based on the current change rate *di*/*dt*. The measured voltage is compared with the preset threshold value to determine whether a short-circuit fault has occurred. If the short-circuit fault happens, this method will limit the fault current change rate to a safe range and block the gate signals. After then, a slow turn-off mechanism is applied to reduce the voltage overshoot. This method does not require any blanking time and provides dynamic control of the gate voltage. But additional complex circuitry is needed.
3) *Gate voltage comparison* [10]: The gate voltage under short-circuit fault is different from that under normal conditions. This fault detection method is conducted by identifying the changes of gate voltage. Normally, when the device turns on, the gate voltage of IGBT with Miller effect is smaller than the gate input voltage from the drive circuits. However, if the short-circuit fault occurs, the gate voltage will be increased up to or even above the gate input voltage. Therefore, the short-circuit fault can be detected by comparing the gate voltage and the gate input voltage. This method requires complicated protection circuitry.
4) *Current mirror method* [11]: This method integrates a second IGBT into the main IGBT. The second IGBT carries the scaled-down current of the main IGBT, which is named as mirror current. The collector current of the main IGBT is obtained based on the mirror current of the second IGBT. This method is expensive because it requires a special device with a current mirror circuit.
5) *Gate voltage monitoring method* [12]: There is a variable capacitance in the equivalent IGBT gate circuit, which affects the gate current and gate voltage during fault conditions. This method identifies the short-circuit fault based on the variations of gate voltage characteristics during turn-on transient. This method requires complicated circuitry.
6) *Protection by snubber and clamp circuit* [13]: The collector-emitter voltage will increase rapidly when the IGBT is turned off due to the existence of the stray inductance in the circuit. This method uses snubber capacitor to absorb the energy trapped in the stray inductance to avoid the voltage overshoot of the device. However, these circuits are not efficient for the protection against turn-off transients caused by rapid suppression of the gate drive, because it requires high-capacity high-voltage snubber capacitors that are bulky and costly.
7) *Protection by slow turn-off of IGBT using additional passive components* [13]: This method addresses the IGBT fault current caused by the generated overshoot voltage during the rapid turn-off of the device. When a fault current is detected, this method can slow down the falling rate of the gate voltage by additional passive components like resistor or capacitor. In the scheme with resistor, a big value of gate resistor is switched in series with the IGBT gate to consume the discharging energy. In the scheme with capacitor, a big value of

capacitor is switched in parallel with the IGBT gate input capacitor to increase the discharge time constant. Therefore, the fault current falling rate is reduced, preventing the device from breakdown.

3.4 MMC Features Under IGBT Open-Circuit Faults

This section analyses the MMC performance under the IGBT open-circuit faults, including T_1 open-circuit fault and T_2 open-circuit fault, respectively.

3.4.1 Faulty SM Features Under T_1 Open-Circuit Fault

Figure 3.5 shows T_1 open-circuit fault in the SM. Table 3.2 shows the operation modes of the SM with T_1 open-circuit fault, where the fault only affects the Mode 3 and does not affect Modes 1, 2, and 4, as follows.

- *Mode 1*: $i_{arm} > 0$ and $S = 1$: the SM is inserted into the arm, the arm current i_{arm} flows through D_1 and C, the C is charged, the capacitor voltage u_c is increased, and the SM output voltage is $u_{sm} = u_c$.
- *Mode 2*: $i_{arm} > 0$ and $S = 0$: the SM is bypassed from the arm, the arm current i_{arm} flows through T_2, the C is bypassed, the capacitor voltage u_c is unchanged, and the SM output voltage is $u_{sm} = 0$.
- *Mode 3*: $i_{arm} < 0$ and $S = 1$: owing to open-circuit T_1, the arm current i_{arm} cannot flow through T_1, the i_{arm} flows through D_2, the capacitor C is bypassed, the capacitor voltage u_c is unchanged, and the SM output voltage is $u_{sm} = 0$.
- *Mode 4*: $i_{arm} < 0$ and $S = 0$: the SM is bypassed from the arm, the arm current i_{arm} flows through D_2, the C is bypassed, the capacitor voltage u_c is unchanged, and the SM output voltage is $u_{sm} = 0$.

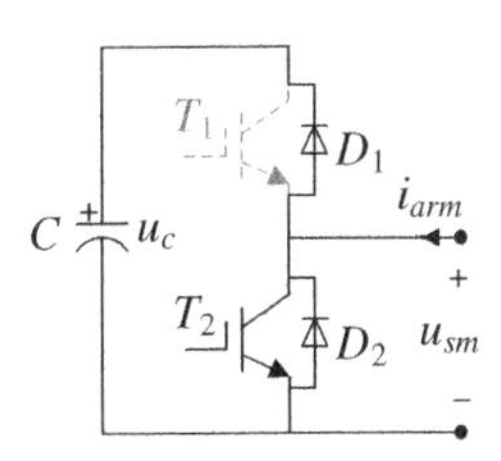

Figure 3.5 T_1 open-circuit fault.

Table 3.2 SM under T_1 open-circuit fault.

Mode	i_{arm}	Switching function S	SM state	i_{arm} path	u_{sm}	u_c
1	>0	1	Insert	D_1 and C	u_c	Increased
2		0	Bypass	T_2	0	Unchanged
3	<0	1	Bypass	D_2	0	Unchanged
4		0	Bypass	D_2	0	Unchanged

Table 3.2 shows that the capacitor in the faulty SM maybe charged and its capacitor voltage is increased when $i_{arm}>0$. However, the capacitor in the faulty SM is bypassed from the arm and its capacitor voltage is unchanged when $i_{arm}<0$.

3.4.2 Faulty SM Features Under T_2 Open-Circuit Fault

3.4.2.1 Operation Mode of Faulty SM

Figure 3.6 shows T_2 open-circuit fault in the SM. Table 3.3 shows the operation modes of the SM with T_2 open-circuit fault, where the fault only affects the Mode 2 and does not affect Modes 1, 3, and 4, as follows.

- *Mode 1*: $i_{arm}>0$ and $S=1$: the SM is inserted into the arm, the arm current i_{arm} flows through D_1 and C, the C is charged, the capacitor voltage u_c is increased, and the SM output voltage is $u_{sm}=u_c$.
- *Mode 2*: $i_{arm}>0$ and $S=0$: owing to open-circuit T_2, the i_{arm} cannot flow through T_2, the i_{arm} flows through D_1, the SM is inserted into the arm, capacitor C is charged, the capacitor voltage u_c is increased, and the SM output voltage is $u_{sm}=u_c$.
- *Mode 3*: $i_{arm}<0$ and $S=1$: the SM is inserted into the arm, the i_{arm} flows through T_1, the C is discharged, the capacitor voltage u_c is reduced, and SM output voltage is $u_{sm}=u_c$.
- *Mode 4*: $i_{arm}<0$ and $S=0$: the SM is bypassed from the arm, the i_{arm} flows through D_2, the C is bypassed, the capacitor voltage u_c is unchanged, and the SM output voltage is $u_{sm}=0$.

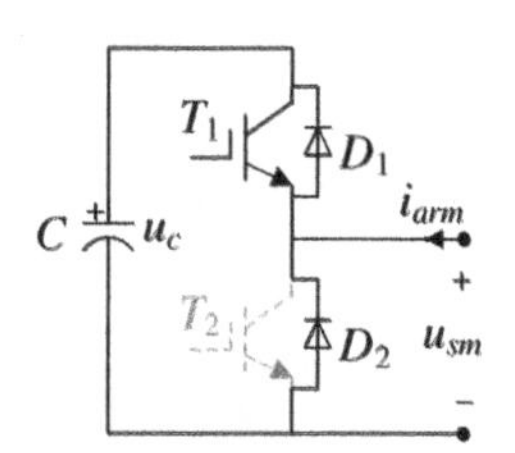

Figure 3.6 T_2 open-circuit fault.

Table 3.3 shows that the capacitor in the faulty SM is always inserted into the arm and its capacitor voltage is increased when $i_{arm}>0$. The capacitor in the faulty SM maybe inserted into the arm and its capacitor voltage is decreased or bypassed from the arm and its capacitor

Table 3.3 SM under T_2 open-circuit fault.

Mode	i_{arm}	Switching function S	SM state	i_{arm} path	u_{sm}	u_c
1	>0	1	Insert	D_1 and C	u_c	Increased
2		0	Insert	D_1 and C	u_c	Increased
3	<0	1	Insert	C and T_1	u_c	Decreased
4		0	Bypass	D_2	0	Unchanged

voltage is unchanged when $i_{arm} < 0$. As a result, the faulty SM capacitor voltage may be balanced or unbalanced as follows.

- *Balanced capacitor voltage*: if the increased capacitor voltage of faulty SM in the period when $i_{arm} > 0$ can be decreased to zero in the period when $i_{arm} < 0$.
- *Unbalanced capacitor voltage*: if the increased capacitor voltage of faulty SM in the period when $i_{arm} > 0$ cannot be decreased to zero in the period when $i_{arm} < 0$.

3.4.2.2 Faulty SM Capacitor Voltage of MMCs in Inverter Mode

Figure 3.7 shows the upper arm current i_{au} in one period T_f where the MMC works in inverter mode and the i_{au} can be expressed as

$$i_{au} = \frac{i_{dc}}{3} + \frac{I_m}{2}\sin(\omega t + \theta), \quad (i_{dc} > 0) \tag{3.1}$$

During the period when $i_{au} > 0$, the capacitor in faulty SM would be charged all the time by i_{au} according to Table 3.3, and the charge transferred to the capacitor in the faulty SM is

$$\begin{aligned} Q_{i1} &= \int_{t_{i1}}^{t_{i2}} i_{au} dt \\ &= \frac{I_m}{\omega}\sqrt{1-(2i_{dc}/3I_m)^2} + \frac{2i_{dc}}{3\omega}\left(\arcsin(2i_{dc}/3I_m) + \frac{\pi}{2}\right) \end{aligned} \tag{3.2}$$

The capacitor voltage increment in the faulty SM during the period when $i_{au} > 0$ can be expressed as

$$\Delta u_{i1} = \frac{Q_{i1}}{C} \tag{3.3}$$

During the period when $i_{au} < 0$, the capacitor in faulty SM would be reduced to keep the capacitor voltage balancing. Suppose that the capacitor in the faulty SM

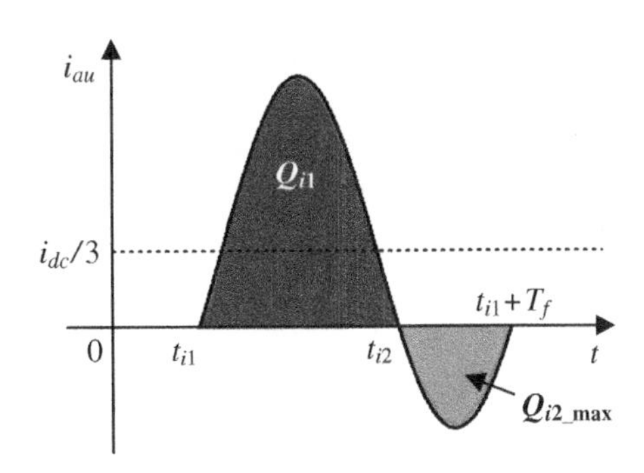

Figure 3.7 Upper arm current i_{au} of MMCs in inverter mode.

Table 3.4 Capacitor voltage in inverter mode under T_2 fault.

Fault type	MMC mode	Trend of capacitor voltage
T_2	Inverter	Increase

is discharged all the time during the period when $i_{au}<0$, the maximum charge transferred away from the capacitor in the faulty SM would be

$$\begin{aligned} Q_{i2\_max} &= \int_{t_{i2}}^{t_{i1}+T_f} |i_{au}|dt \\ &= \frac{I_m}{\omega}\sqrt{1-(2i_{dc}/3I_m)^2} + \frac{2i_{dc}}{3\omega}\left(\arcsin(2i_{dc}/3I_m)-\frac{\pi}{2}\right) \end{aligned} \tag{3.4}$$

The corresponding maximum capacitor voltage decrement in the faulty SM during the period when $i_{au}<0$ is

$$\Delta u_{i2\_max} = \frac{Q_{i2\_max}}{C} \tag{3.5}$$

Owing to $i_{dc}>0$ when the MMC works in inverter mode, there is

$$\Delta u_{i1} > \Delta u_{i2\_max} \tag{3.6}$$

The maximum capacitor voltage decrement in the faulty SM during the period when $i_{au}<0$ is less than the capacitor voltage increment in the faulty SM during the period when $i_{au}>0$, which results in that the capacitor voltage in the faulty SM cannot be kept balanced. The capacitor voltage in the faulty SM would be gradually increased over the time, as shown in Table 3.4 [14].

3.4.2.3 Faulty SM Capacitor Voltage of MMCs in Rectifier Mode

Figure 3.8 shows the upper arm current in phase A of the MMC in rectifier mode, which can be expressed as

$$i_{au} = \frac{i_{dc}}{3} + \frac{I_m}{2}\sin(\omega t + \theta), \quad (i_{dc}<0) \tag{3.7}$$

During the period when $i_{au}>0$, the capacitor in the faulty SM would be charged all the time by i_{au} according to Table 3.3. The charge transferred to the capacitor in the faulty SM is

$$\begin{aligned} Q_{r1} &= \int_{t_{r1}}^{t_{r2}} i_{au}dt \\ &= \frac{I_m}{\omega}\sqrt{1-(2i_{dc}/3I_m)^2} + \frac{2i_{dc}}{3\omega}\left(\arcsin(2i_{dc}/3I_m)+\frac{\pi}{2}\right) \end{aligned} \tag{3.8}$$

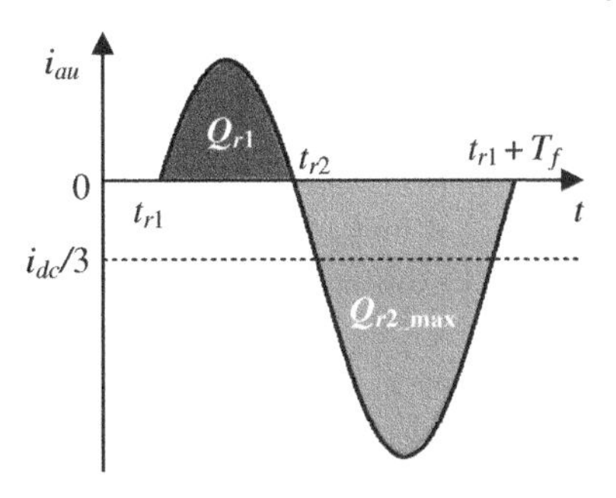

Figure 3.8 Upper arm current i_{au} of MMCs in rectifier mode.

The capacitor voltage increment in the faulty SM during the period when $i_{au} > 0$ can be expressed as

$$\Delta u_{r1} = \frac{Q_{r1}}{C} \tag{3.9}$$

During the period when $i_{au} < 0$, the capacitor in the faulty SM would be reduced to keep the capacitor voltage balancing. Assuming that the capacitor in the faulty SM is discharged all the time during the period when $i_{au} < 0$, the maximum charge transferred away from the capacitor in the faulty SM would be

$$\begin{aligned} Q_{r2\_\max} &= \int_{t_{r2}}^{t_{r1}+T_f} |i_{au}|dt \\ &= \frac{I_m}{\omega}\sqrt{1-(2i_{dc}/3I_m)^2} + \frac{2i_{dc}}{3\omega}\left(\arcsin(2i_{dc}/3I_m) - \frac{\pi}{2}\right) \end{aligned} \tag{3.10}$$

The corresponding maximum capacitor voltage decrement is

$$\Delta u_{r2\_\max} = \frac{Q_{r2\_\max}}{C} \tag{3.11}$$

Owing to $i_{dc} < 0$ when the MMC works in the rectifier mode, there is

$$\Delta u_{r1} < \Delta u_{r2\_\max} \tag{3.12}$$

The charge transferred to the capacitor in the faulty SM is less than the maximum charge transferred away from the capacitor in the faulty SM. As a result, the increased capacitor voltage during the period when $i_{au} > 0$ can be reduced to zero during the period when $i_{au} < 0$, and the capacitor voltage in the faulty SM can be balanced, as shown in Table 3.5 [15].

Table 3.5 Capacitor voltage in rectifier mode under T_2 fault.

Fault type	MMC mode	Trend of capacitor voltage
T_2	Rectifier	Balance

All capacitor voltages in the faulty arm can be kept balanced in each period according to Table 3.5. The operation of the faulty arm in one period is [15]

1) $[t_{r1}, t_{r2}]$: $i_{au} > 0$. In the upper arm, the faulty SM is charged all the time from t_{r1} to t_{r2}, shown in Figure 3.8, and the faulty SM is with the highest capacitor voltage in the arm. Hence, the faulty SM would be assigned with the switching function as 0 because $i_{au} > 0$. However, the faulty SM cannot be bypassed because of T_2 open-circuit fault, as shown in Table 3.3. Here, the faulty SM is inserted into the arm, as shown in Figure 3.9. For the lower arm, the SMs work normally. Suppose that the reference for phase A is y_a, during the period $[t_{r1}, t_{r2}]$, the upper arm voltage u_{au} and the lower arm voltages u_{al} are

$$\begin{cases} u_{au} = \dfrac{1-y_a}{2} \cdot nu_{cau} + u_{cau} \\ u_{al} = \dfrac{1+y_a}{2} \cdot nu_{cal} \end{cases} \quad t \in [t_{r1}, t_{r2}] \tag{3.13}$$

where u_{cau} and u_{cal} are the average capacitor voltages in the upper arm and lower arm of phase A, respectively.

2) $[t_{r2}, t_{r1} + T_f]$: $i_{au} < 0$. All SMs in the upper and lower arms work normally, as shown in Table 3.3, and therefore the u_{au} and u_{la} are

$$\begin{cases} u_{au} = \dfrac{1-y_a}{2} \cdot nu_{cau} \\ u_{al} = \dfrac{1+y_a}{2} \cdot nu_{cal} \end{cases} \quad t \in [t_{r2}, t_{r1} + T_f] \tag{3.14}$$

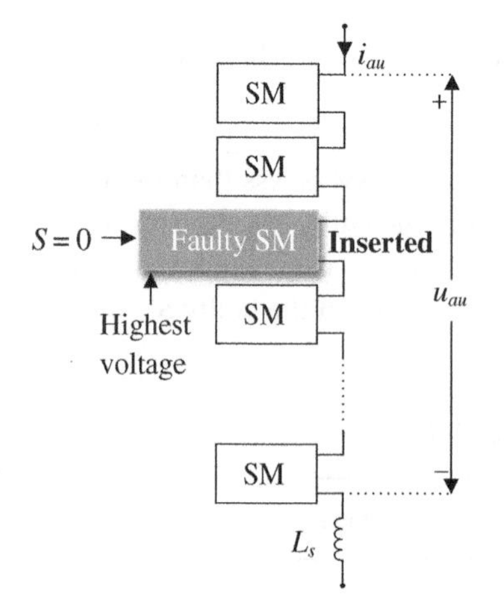

Figure 3.9 Upper arm voltage u_{au} when $i_{au} > 0$.

Based on equations (3.7), (3.13), and (3.14), the upper and lower arm capacitor voltage can be obtained as equation (3.15) in the steady-state operation of the MMC.

$$\begin{cases} u_{cau} = \dfrac{V_{dc}}{n} \cdot \dfrac{1}{1 + \dfrac{2}{\pi n} \arccos\left(-2i_{dc} / 3I_m\right)} \\ u_{cal} = \dfrac{V_{dc}}{n} \end{cases} \tag{3.15}$$

Although the capacitor voltages in the upper arm are kept balanced, they are reduced under faults according to equation (3.15). The capacitor voltages in the lower arm are nearly not affected and kept balanced. In the upper arm, the capacitor voltages are reduced less along with the increase of the number of the SMs in the arm. The capacitor voltages are reduced more along with the reduction of the number of the SMs in the arm.

Figure 3.10 shows the capacitor voltages in the upper arm of phase A, where the MMC works in rectifier mode with rated power. Under the T_2 open-circuit fault occurring to SM1 in the MMC, the peak value of the capacitor voltage in the faulty SM is a little higher than those of the healthy SMs and all capacitor voltages in the arm are kept balanced, which is consistent with the theoretical analysis. Along with the increase of the number of SMs in the arm, the capacitor voltages under faults are increased. Figure 3.11 shows the upper arm capacitor voltage and the lower arm capacitor voltages under various number of SMs in the arm. The capacitor voltages in the lower arm are nearly not affected by the fault, while the capacitor voltages in the upper arm are reduced under the fault. Along with the reduction of the number of SMs in the arm, the reduction of the capacitor voltages in the upper arm will be increased.

3.5 Kalman Filter Based Fault Detection Under IGBT Open-Circuit Faults

The IGBT open-circuit fault may distort the voltage and current in the MMC, cause overvoltage on the power devices, and lead to secondary damages to other power devices and failure of system. In order to ensure the safe operation of the MMC, it is essential to detect and locate the fault within a short time after the fault occurrence in the MMC. However, the MMC usually is composed of a large number of SMs, which may result in a very complex diagnosis algorithm and heavy calculation burden. Besides, the real-life characteristics (e.g. drift and stability) and quality (e.g. accuracy and repeatability) of large number of sensors used in the MMC bring lots of unwanted sensor noises and may affect critically

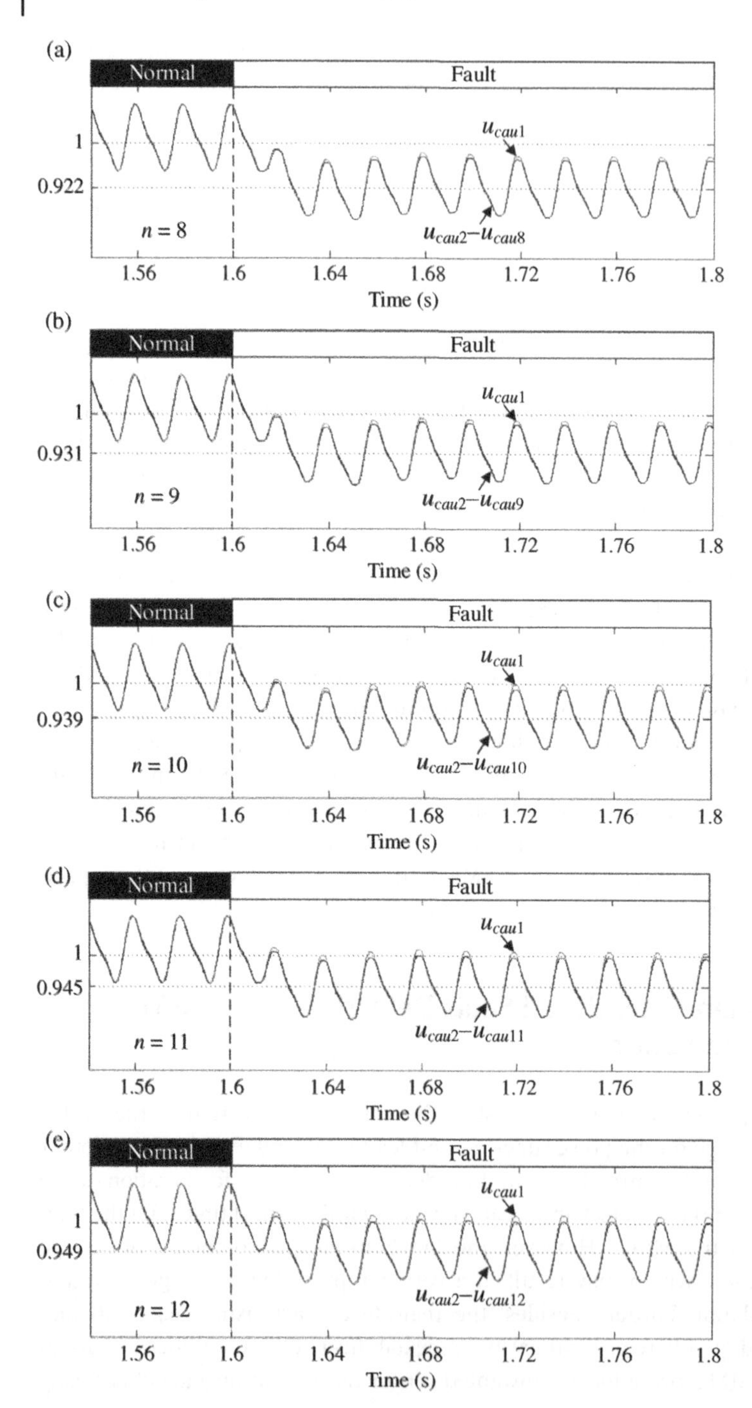

Figure 3.10 Upper arm capacitor voltages under T_2 open-circuit fault in SM1. (a) $n = 8$. (b) $n = 9$. (c) $n = 10$. (d) $n = 11$. (e) $n = 12$.

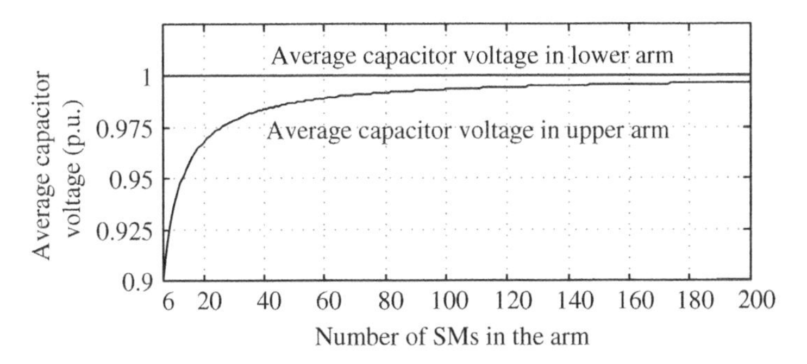

Figure 3.11 Upper and lower arm capacitor voltage of the MMC under various *n*.

the fault detection if the sensor noise is not taken into account [16], eventually leading to a problematic and erroneous result [17]. This section presents an IGBT open-circuit fault detection and localization method for the MMC based on KF.

3.5.1 Kalman Filter Algorithm

The KF is a commonly used method to optimally estimate the state of a dynamic system from a series of imperfect noisy measurements, especially in presence of uncertainties caused by sensor noise [18]. For the MMC system, KF can be used for optimally estimating the circulating current and fault detection because of the employment of a number of sensors associated with different level of uncertainties for the MMC. The KF can exploit the dynamics of the power electronics system iteratively and therefore remove the effects of the measurement noise and get the optimal estimates of the states.

Generally, a discrete-time linear dynamic system can be described as

$$\begin{cases} \mathbf{x}(k) = \mathbf{F}(k)\mathbf{x}(k-1) + \mathbf{G}(k)\mathbf{u}(k) + \mathbf{w}(k) \\ \mathbf{y}(k) = \mathbf{H}(k)\mathbf{x}(k) + \mathbf{v}(k) \end{cases} \tag{3.16}$$

where $\mathbf{x}$, $\mathbf{y}$, $\mathbf{u}$, $\mathbf{F}$, $\mathbf{H}$, and $\mathbf{G}$ are the state vector, measurement vector, control input vector, state transition matrix, measurement gain matrix, and control gain matrix, respectively. $\mathbf{w}$ is the process noise and $\mathbf{v}$ is the measurement noise, respectively.

According to equation (3.16) and [19], the typical KF state equations can be expressed as

$$\begin{cases} \mathbf{x}_\mathrm{p}(k) = \mathbf{F}(k)\mathbf{x}_\mathrm{c}(k-1) + \mathbf{G}(k)\mathbf{u}(k) \\ \mathbf{P}_\mathrm{p}(k) = \mathbf{F}(k)\mathbf{P}_\mathrm{c}(k-1)\mathbf{F}^\mathrm{T}(k) + \mathbf{Q}(k) \\ \mathbf{K}(k) = \dfrac{\mathbf{P}_\mathrm{p}(k)\mathbf{H}^\mathrm{T}(k)}{\mathbf{H}(k)\mathbf{P}_\mathrm{p}(k)\mathbf{H}^\mathrm{T}(k) + \mathbf{R}(k)} \\ \mathbf{x}_\mathrm{c}(k) = \mathbf{x}_\mathrm{p}(k) + \mathbf{K}(k)\left[\mathbf{y}(k) - \mathbf{H}(k)\mathbf{x}_\mathrm{p}(k)\right] \\ \mathbf{P}_\mathrm{c}(k) = \left[\mathbf{I} - \mathbf{K}(k)\mathbf{H}(k)\right]\mathbf{P}_\mathrm{p}(k) \end{cases} \tag{3.17}$$

where $\mathbf{x}_\mathrm{p}$ and $\mathbf{x}_\mathrm{c}$ are the predicted state estimate vector and the optimal state estimate vector, respectively. $\mathbf{P}_\mathrm{p}$ and $\mathbf{P}_\mathrm{c}$ are the auto-covariances of the predicted state estimate error vector and the optimal state estimate error vector, respectively. $\mathbf{K}_\mathbf{kf}$ is the KF gain. $\mathbf{Q}_\mathbf{kf}$ and $\mathbf{R}_\mathbf{kf}$ are the covariance matrices of the process noise $\mathbf{w}$ and measurement noise $\mathbf{v}$, respectively.

3.5.2 Circulating Current Estimation

The circulating current i_{diff_j} exists in phase j of the MMC, which is half of the sum of the upper arm current i_{ju} and the lower arm current i_{jl} in phase j of the MMC, as

$$i_{diff\_j} = \frac{i_{ju} + i_{jl}}{2} \tag{3.18}$$

The i_{diff_j} can also be expressed as

$$\frac{di_{diff\_j}(t)}{dt} = -\frac{R_s}{L_s} \cdot i_{diff\_j}(t) + \frac{1}{L_s}\left[\frac{V_{dc}}{2} - \frac{u_{ju}(t) + u_{jl}(t)}{2}\right] \tag{3.19}$$

with

$$\begin{cases} u_{ju} = \dfrac{1 + y_{ju}}{2} \cdot \displaystyle\sum_{i=1}^{n} u_{cjui} \\ u_{jl} = \dfrac{1 + y_{jl}}{2} \cdot \displaystyle\sum_{i=1}^{n} u_{cjli} \end{cases} \tag{3.20}$$

where u_{cjui} and u_{cjli} ($i = 1, 2..., n$) are the upper arm capacitor voltage of the i-th SM and the lower arm capacitor voltage of the i-th SM in phase j, respectively. R_s is the equivalent arm resistor.

Substituting equation (3.20) into equation (3.19) and discretizing the continuous-time mode, the system state variable i_{diff_j} can be calculated as

$$i_{diff\_j}(k) = F \cdot i_{diff\_j}(k-1) + G \cdot u(k-1) \tag{3.21}$$

with

$$\begin{cases} F = 1 - \dfrac{R_s}{L_s f_s} \\ G = \dfrac{1}{L_s f_s} \\ u = \dfrac{V_{dc}}{2} - \dfrac{1 + y_{ju}}{4} \displaystyle\sum_{i=1}^{n} u_{cju\_i} - \dfrac{1 + y_{jl}}{4} \displaystyle\sum_{i=1}^{n} u_{cjl\_i} \end{cases} \tag{3.22}$$

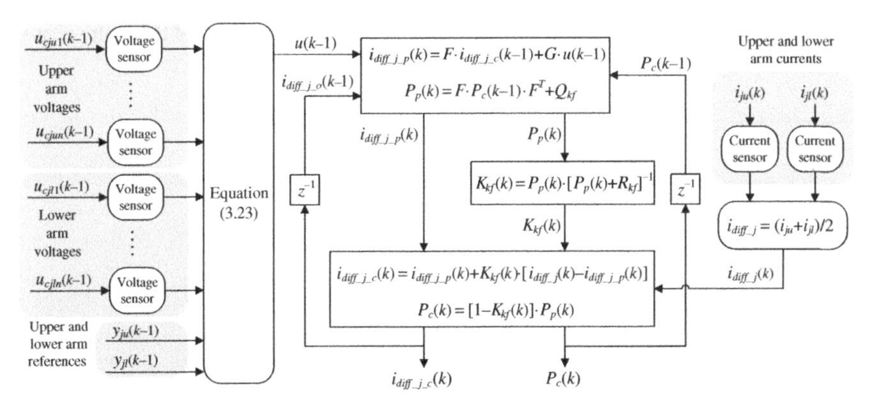

Figure 3.12 Circulating current estimation based on KF for phase *j* of the MMC.

where f_s is sampling frequency. The circulating current $i_{diff\_j}(k-1)$ can be obtained based on equation (3.18).

Based on the KF in equation (3.17), the circulating current value $i_{diff\_j\_c}$ can be estimated by the KF as

$$\begin{cases} i_{diff\_j\_p}(k) = F \cdot i_{diff\_j\_c}(k-1) + G \cdot u(k-1) \\ P_p(k) = F \cdot P_c(k-1) \cdot F^T + Q_{kf} \\ K(k) = \dfrac{P_p(k)}{P_p(k) + R_{kf}} \\ i_{diff\_j\_c}(k) = i_{diff\_j\_p}(k) + K_{kf}(k) \cdot \left[i_{diff\_j}(k) - i_{diff\_j\_p}(k) \right] \\ P_c(k) = \left[1 - K_{kf}(k)\right] \cdot P_p(k) \end{cases} \tag{3.23}$$

where $i_{diff\_j\_p}$ is the predicted state estimate of circulating current in phase *j*. P_p is the auto-covariance of predicted state estimate error. P_c is the auto-covariance of optimal state estimate error.

Figure 3.12 shows the KF algorithm for each phase of the MMC [20], where the arm capacitor voltage, the upper arm current, and the lower arm current are sampled at the sampling frequency. With the real-time monitoring and sampling of capacitor voltage and arm current values obtained by voltage and current sensors in each phase, the circulating current value $i_{diff\_j\_c}$ can be calculated by the implementation of the KF algorithm shown in Figure 3.12.

3.5.3 Faulty Phase Detection

In the normal operation of the MMC, the estimated circulating current value $i_{diff\_j\_c}$ is close to the measured circulating current value i_{diff_j}. However, once the

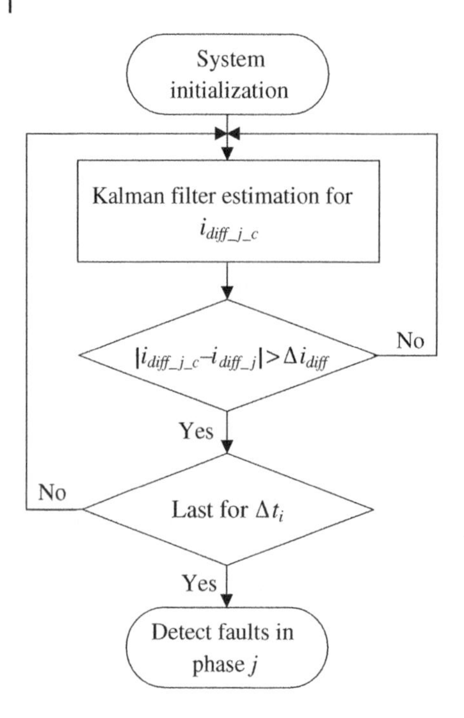

Figure 3.13 Flowchart of the faulty phase detection method.

fault occurs, the $i_{diff\_j\_c}$ estimated by KF will start deviating from the measured values i_{diff_j}, and the error between $i_{diff\_j\_c}$ and i_{diff_j} will be increased under persistent faults. Based on this fact, Figure 3.13 shows the faulty phase detection method [20], which employs a direct comparison of the estimated value $i_{diff\_j\_c}$ to the measured value i_{diff_j}. If $|i_{diff\_j\_c} - i_{diff\_j}| > \Delta i_{diff}$, where Δi_{diff} is the threshold value of the circulating current error, and it lasts at least for a period of Δt_i, then a fault is detected in phase j of the MMC. Otherwise, the system is considered as normal and the comparison will continue. The fault detection algorithm is implemented in each carrier wave period, which can reduce the calculation burden and lower the demand for processor.

3.5.4 Capacitor Voltage

The capacitor voltages of the MMC working in inverter mode are analyzed, where two types of IGBT open-circuit faults are considered, including T_1 open-circuit fault and T_2 open-circuit fault.

The power p_{sm} of one SM can be described with the SM output voltage u_{sm} and the arm current i_{arm} as

$$p_{sm} = u_{sm} \cdot i_{arm} \tag{3.24}$$

Based on the IGBT open-circuit faults characteristics of the MMC depicted in Tables 3.2 and 3.3, the SM power p_{sm_n}, p_{sm_T1}, and p_{sm_T2}, respectively, under normal operation, T_1 fault, and T_2 fault can be expressed, respectively, as

$$p_{sm\_n} = S \cdot u_c \cdot i_{arm} \tag{3.25}$$

$$p_{sm\_T1} = \begin{cases} S \cdot u_c \cdot i_{arm} & i_{arm} \geq 0 \\ 0 & i_{arm} < 0 \end{cases} \tag{3.26}$$

$$p_{sm\_T2} = \begin{cases} u_c \cdot i_{arm} & i_{arm} \geq 0 \\ S \cdot u_c \cdot i_{arm} & i_{arm} < 0 \end{cases} \tag{3.27}$$

From equations (3.25)–(3.27), the capacitor voltage characteristics of the SM under different open-circuit faults with the MMC operating in inverter mode can be obtained as shown in Table 3.6.

- Under T_1 fault, when $i_{arm} > 0$, $p_{sm\_n} = p_{sm\_T1}$; when $i_{arm} < 0$, $p_{sm\_n} \leq p_{sm\_T1}$, which results in that the capacitor voltage in healthy SMs are lower than that in SM under the T_1 fault.
- Under T_2 fault, when $i_{arm} < 0$, $p_{sm\_n} = p_{sm\_T2}$; when $i_{arm} > 0$, $p_{sm\_n} \leq p_{sm\_T2}$, which results in the capacitor voltage in healthy SMs are lower than that in the SM under the T_2 fault.

As a result, the capacitor voltage in the faulty SMs will be higher than that in the healthy SMs.

3.5.5 Faulty SM Detection

Figure 3.14 shows the faulty SM localization method [20]. After the fault is detected in phase j, the fault localization algorithm started to execute simultaneously in both upper and lower arms of phase j, where each capacitor voltage u_{cjki} ($k = u, l$) in the arm will be compared with the minimum capacitor

Table 3.6 Capacitor voltage characteristics under different faults.

Fault type	i_{arm}	Power relationship	Capacitor voltage trend
T_1	>0	$p_{sm\_n} = p_{sm\_T1}$	Faulty SM voltage > healthy SM voltage
	<0	$p_{sm\_n} \leq p_{sm\_T1}$	
T_2	>0	$p_{sm\_n} \leq p_{sm\_T2}$	
	<0	$p_{sm\_n} = p_{sm\_T2}$	

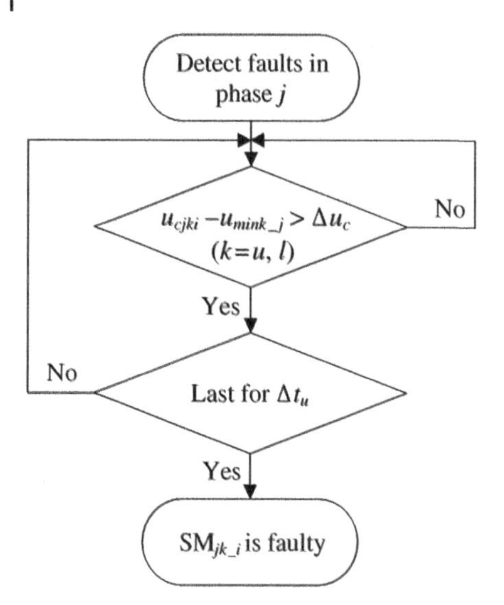

Figure 3.14 Flowchart of the faulty SM localization method.

voltage value for a given time instant $u_{mink\_j}(t) = \text{Min}[u_{cjk1}(t), u_{cjk2}(t), \ldots u_{cjkn}(t)]$, to determine which SM may contain fault in this arm. If the difference between the capacitor voltage u_{cjki} and the minimum capacitor voltage value $u_{mink\_j}(t)$ is over the threshold value Δu_c as $u_{cjki} - u_{mink\_j} > \Delta u_c$, and it lasts for a period of Δt_u, then it indicates that the SM_{jk_i} (which denotes the i-th SM in arm k of phase j) contains the fault.

Case Study 3.1 Analysis of Kalman Filter Based Fault Detection Under IGBT Open-Circuit Faults

Objective: In this case study, a three-phase MMC system is built with the time-domain simulation tool power systems computer aided design/electromagnetic transients including DC (PSCAD/EMTDC) to verify the KF based fault detection under IGBT open-circuit faults, as shown in Figure 3.15. The AC side of the three-phase MMC is connected with *RL* load.

Parameters: The main system parameters are listed in Table 3.7.

Simulation results and analysis:
Figure 3.16 shows the performance of the MMC under normal operation. Figure 3.16a shows the reference y_{au} and y_{al} for upper and lower arms of phase A. Moreover, Figure 3.16b shows the arm current i_{au}, i_{al}, and circulating

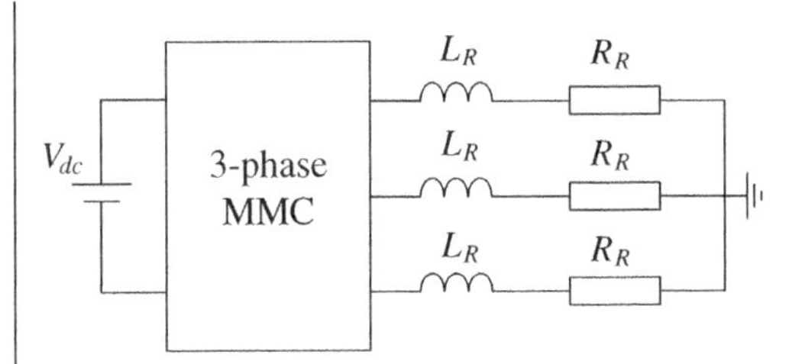

Figure 3.15 Block diagram of the simulation system.

Table 3.7 Parameters of the simulated three-phase MMC.

Parameter	Value
DC-link voltage V_{dc} (kV)	3.6
Load frequency (Hz)	50
Number of SMs per arm n	4
SM capacitance C (mF)	3.5
Arm inductance L_s (mH)	3
Load inductance L_R (mH)	10
Load resistance R_R (Ω)	10
Carrier frequency f_s (Hz)	1650
Q_{kf} for KF	4e-5
R_{kf} for KF	5e-4
Δi_{diff} (A)	11
Δt_i (ms)	1.5
Δu_c (V)	60
Δt_u (ms)	20

current i_{diff_a} of phase A. Besides, Figure 3.16c shows the capacitor voltage in phase A. According to these measured current and voltage, the optimally estimated circulating current $i_{diff\_a\_c}$ of phase A can be calculated in real time by KF, as shown in Figure 3.16d, where the $i_{diff\_a\_c}$ is well matched with the measured circulating current i_{diff_a} in phase A.

Figure 3.17 shows the performance of the MMC with an open-circuit fault in SM_{al_1}, T_1 at 1 second. Figure 3.17a shows the currents i_{au}, i_{al}, and i_{diff_a}. Figure 3.17b shows $i_{diff\_a\_c}$ and i_{diff_a}, where the $i_{diff\_a\_c}$ cannot match i_{diff_a} under the fault. According to the fault detection method, the fault can be detected in

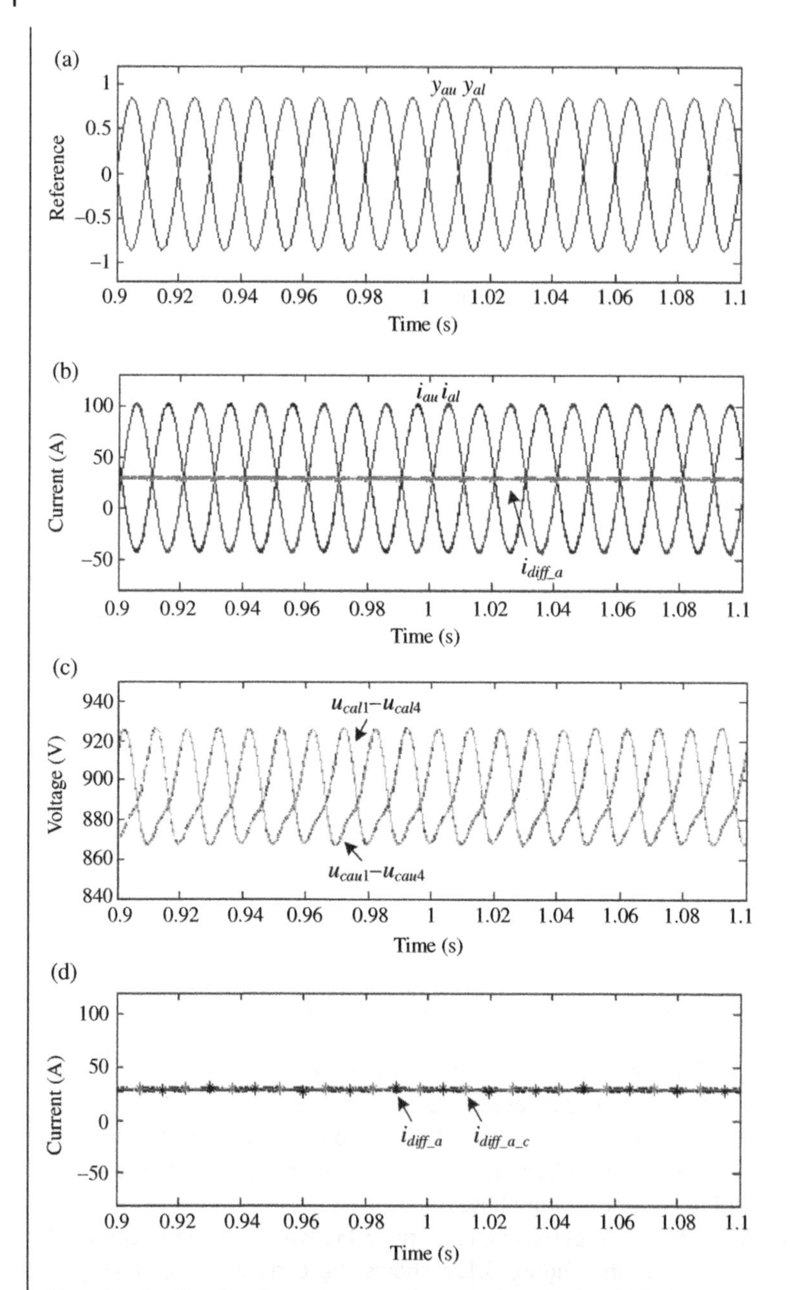

Figure 3.16 Simulated waveforms under normal operation. (a) References y_{au} and y_{al} for phase A. (b) Arm current i_{au}, i_{al}, and circulating current i_{diff_a}. (c) Capacitor voltage of phase A. (d) Measured circulating current i_{diff_a} and estimated circulating current $i_{diff\_a\_c}$.

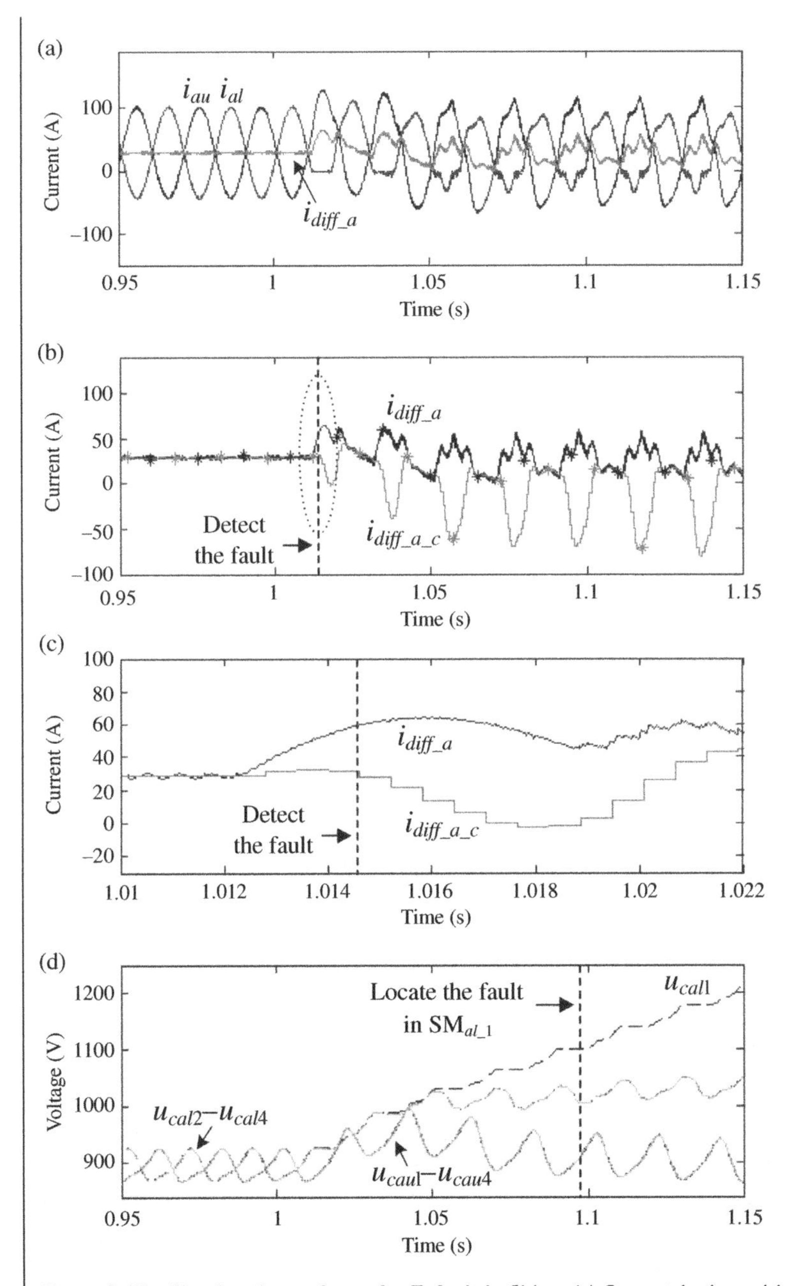

Figure 3.17 Simulated waveforms for T_1 fault in SM_{al_1}. (a) Current i_{au}, i_{al}, and i_{diff_a}. (b). Current i_{diff_a} and $i_{diff\_a\_c}$. (c) i_{diff_a} and $i_{diff\_a\_c}$ in small time scale. (d) Capacitor voltage of phase A.

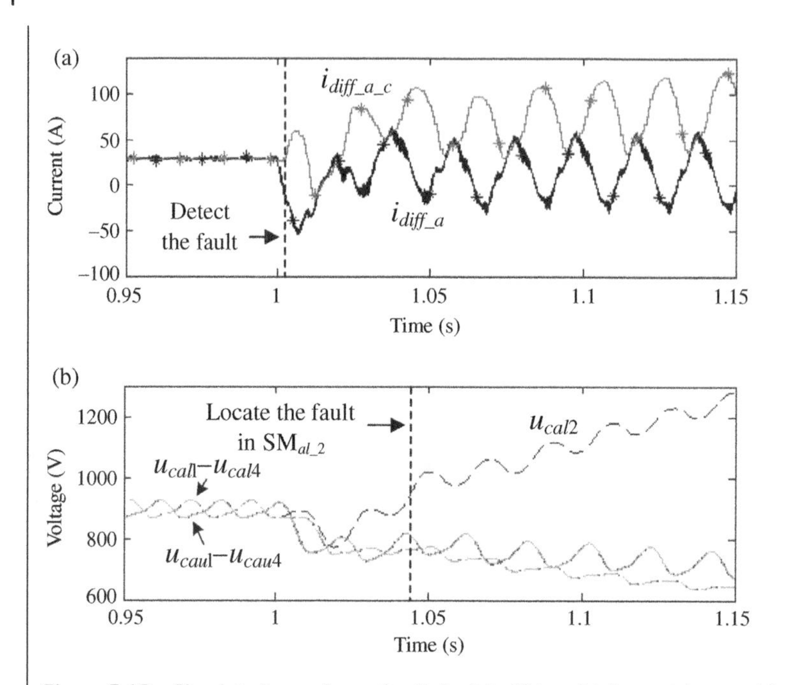

Figure 3.18 Simulated waveforms for T_2 fault in SM_{al_2}. (a) Current i_{diff_a} and $i_{diff\_a\_c}$. (b) Capacitor.

phase A at 1.0145 seconds, as shown in Figure 3.17c. The voltage u_{cal_1} in the SM_{al_1} is higher than that of the other SMs in the lower arm of phase A. With the fault localization method, the fault is located in SM_{al_1} at 1.096 seconds, as shown in Figure 3.17d.

Figure 3.18 shows the performance of the MMC with an open-circuit fault in SM_{al_2}, T_2 at 1 second. Once the fault occurs, the estimated circulating current $i_{diff\_a\_c}$ cannot match the measured circulating current i_{diff_a}, as shown in Figure 3.18a. The fault can be detected in phase A at 1.002 seconds. During the fault, the voltage u_{cal2} in the SM_{al_2} is higher than that of the other SMs in the lower arm of phase A, as shown in Figure 3.18b. An open-circuit fault is located in the SM_{al_2} at 1.044 seconds based on the fault detection method.

In this case, the faults in SM_{al_1}, T_1 and SM_{al_2}, T_2 occur at 1 and 1.05 seconds, respectively. Figure 3.19a shows $i_{diff\_a\_c}$ and i_{diff_a}, where the fault is detected in phase A at 1.014 seconds. Figure 3.19b shows the capacitor voltage in phase A, where the faulty SM_{al_1} and SM_{al_2} are located at 1.094 and 1.123 seconds, respectively.

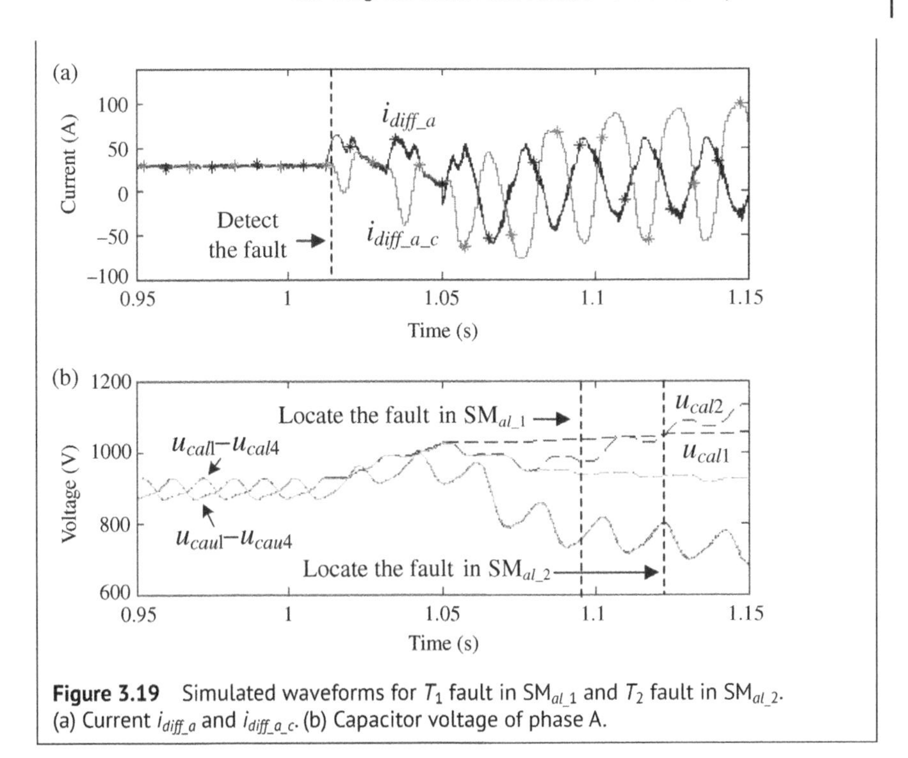

Figure 3.19 Simulated waveforms for T_1 fault in SM_{al_1} and T_2 fault in SM_{al_2}. (a) Current i_{diff_a} and $i_{diff\_a\_c}$. (b) Capacitor voltage of phase A.

3.6 Integrator Based Fault Detection Under IGBT Open-Circuit Faults

The KF based method can locate the faulty SM if its capacitor voltage in the faulty SM is unbalanced, where the KF based method can locate the faulty SM when its capacitor voltage is over the threshold value. However, the capacitor voltage in the faulty SM is kept balanced if the MMC works in rectifier mode as the analysis in Table 3.5, which increases the difficulty for fault localization. In this section, an integrator-based fault detection method [15] is presented no matter whether the MMC works in inverter mode or rectifier mode.

According to the analysis in Section 3.4, the capacitor in the faulty SM is charged all the time during the period when the arm current is positive, which results in that the capacitor voltage increment in the faulty SM is the highest in the arm during the period when arm current is positive. Based on the common fault characteristics of the MMC in rectifier mode and inverter mode, the integrator-based fault localization method for the MMC [15] is shown in Figure 3.20. After the

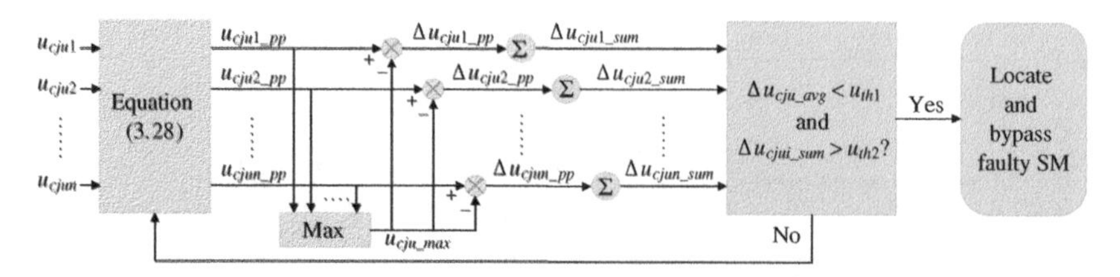

Figure 3.20 Fault localization strategy for MMCs under IGBT open-circuit fault.

open-circuit fault is detected in phase *j* of the MMC, the fault localization strategy will be implemented in the upper arm and lower arm of phase *j* to locate the faulty SM, respectively. Figure 3.20 shows the fault localization strategy for the upper arm of phase *j*. The fault localization strategy for the lower arm of phase *j* is the same as for the upper arm of phase *j*, which is not repeated here.

In Figure 3.20, all capacitor voltages u_{cju1}–u_{cjun} in the upper arm of phase *j* are real-time monitored. The capacitor voltages are increased during the period when $i_{au} > 0$. The capacitor voltage increment of the *i*-th SM in each period is

$$u_{cjui\_pp} = u_{cjui}(t_2) - u_{cjui}(t_1) \tag{3.28}$$

where $u_{cjui}(t_2)$ is the instantaneous capacitor voltage at the negative zero-crossing point of the *i*-th SM in each period. $u_{cjui}(t_1)$ is the instantaneous capacitor voltage at the positive zero-crossing point of the *i*-th SM in each period, as shown in Figures 3.7 and 3.8. The obtained capacitor voltage increment u_{cju1_pp}–u_{cjun_pp} can be used for MMC fault localization under IGBT open-circuit fault, as follows.

1) *Fault occurrence in arm*: If fault occurs in the *i*-th SM in the upper arm, the capacitor voltage increment u_{cjui_pp} in the faulty SM would be the maximum one among u_{cju1_pp}–u_{cjun_pp}. In Figure 3.20, the Δu_{cju1_pp}–Δu_{cjun_pp} can be obtained as

$$\begin{cases} \Delta u_{cju1\_pp} = u_{cju1\_pp} - \text{Max}\left[u_{cju1\_pp}, u_{cju2\_pp} \cdots u_{cjun\_pp}\right] \\ \Delta u_{cju2\_pp} = u_{cju2\_pp} - \text{Max}\left[u_{cju1\_pp}, u_{cju2\_pp} \cdots u_{cjun\_pp}\right] \\ \vdots \\ \vdots \\ \Delta u_{cjun\_pp} = u_{cjun\_pp} - \text{Max}\left[u_{cju1\_pp}, u_{cju2\_pp} \cdots u_{cjun\_pp}\right] \end{cases} \tag{3.29}$$

For the faulty *i*-th SM, its corresponding Δu_{cjui_pp} would be 0. For the normal SMs, their Δu_{cju1_pp}–$\Delta u_{cju\,(i-1)\_pp}$ and $\Delta u_{cju\,(i+1)\_pp}$–Δu_{cjun_pp} would be negative values.

In Figure 3.20, the Δu_{cju1_pp}–Δu_{cjun_pp} in each fundamental period is summed up, respectively, as

$$\begin{cases} \Delta u_{cju1\_sum} = \sum \Delta u_{cju1\_pp} \\ \Delta u_{cju2\_sum} = \sum \Delta u_{cju2\_pp} \\ \quad\vdots \\ \quad\vdots \\ \Delta u_{cjun\_sum} = \sum \Delta u_{cjun\_pp} \end{cases} \tag{3.30}$$

The Δu_{cju_avg} is the average value of Δu_{cju1_sum}–Δu_{cjun_sum}, which can be obtained as

$$\Delta u_{cju\_avg} = \frac{\Delta u_{cju1\_sum} + \Delta u_{cju2\_sum} + \cdots + \Delta u_{cjun\_sum}}{n} \tag{3.31}$$

After accumulation over a number of fundamental periods, the Δu_{cjui_sum} of the faulty SM is quite small. The Δu_{cju1_sum}–$\Delta u_{cju\,(i-1)\_sum}$ and $\Delta u_{cju(i+1)\_sum}$–Δu_{cjun_sum} of the healthy SMs would be decreased. Consequently, the average value Δu_{cju_avg} would be decreased too. According to the above characteristics of Δu_{cju1_sum}–Δu_{cjun_sum} and Δu_{cju_avg}, the faulty *i*-th SM can be located. If the Δu_{cju_avg} is less than the threshold value u_{th1} as $\Delta u_{cju\_avg} < u_{th1}$ and the Δu_{cjui_sum} of the *i*-th SM is greater than the threshold value u_{th2} as $\Delta u_{cjui\_sum} > u_{th2}$, the faulty SM can be located. The u_{th1} is smaller than u_{th2}. Afterwards, the faulty SM is bypassed from the faulty arm.

2) *No fault occurrence in arm*: If no fault occurs in the arm, the SM capacitor voltage increments u_{cju1_pp}–u_{cjun_pp} in the arm are close to each other, the Δu_{cju1_pp}–Δu_{cjun_pp} would be small, the Δu_{cju1_sum}–Δu_{cjun_sum} would be decreased slowly and close to each other as the average value Δu_{cju_avg}. As a result, Δu_{cju_avg} will decrease slowly and takes a long time to reach u_{th1}. Even though Δu_{cju_avg} reaches u_{th1}, Δu_{cju1_sum}–Δu_{cjuN_sum} are close to u_{th1} and less than u_{th2}, and therefore no SM would be false alarmed.

Case Study 3.2 Analysis of Integrator Based Fault Detection Under Lower IGBT Open-Circuit Faults

Objective: In this case study, the performance of the integrator-based fault detection method under T_2 open-circuit faults is studied for the MMC with the professional tool PSCAD/EMTDC.

Parameters: The simulation system parameters are shown in Table 3.8.

Simulation results and analysis:
Figure 3.21 shows the MMC performance under the fault localization method, where the MMC works in rectifier mode and transfers 120 MW (1.0 p.u.) active power and 0 MVar reactive power from AC side to DC side. At 1.6 seconds, the T_2 open-circuit fault happens to SM1 in the upper arm of phase A.

Table 3.8 Simulation system parameters.

Parameters	Value
DC-link voltage V_{dc} (kV)	120
Number of SMs per arm n	60
SM capacitance C (mF)	6
Arm inductance L_s (mH)	10
Carrier frequency (kHz)	1
u_{th1} (V)	−200
u_{th2} (V)	−40

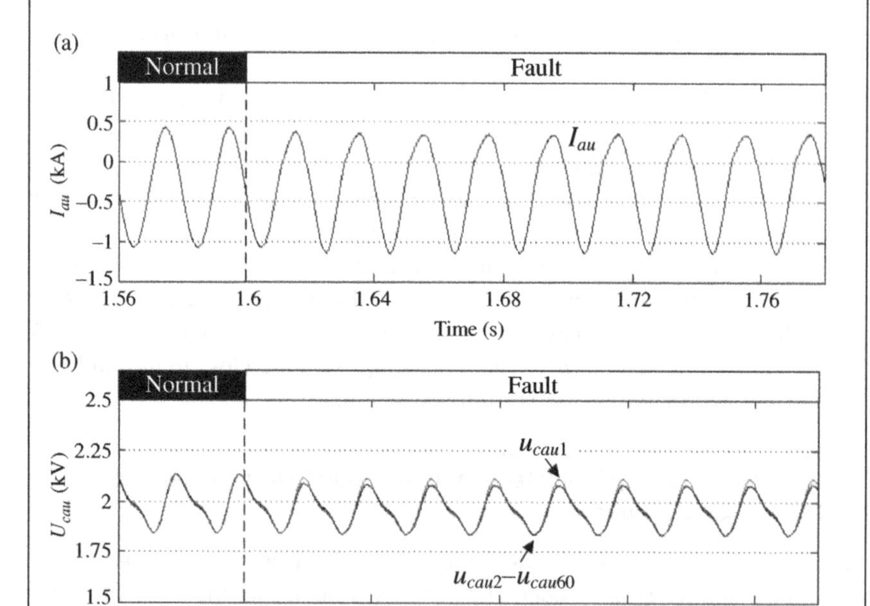

Figure 3.21 Simulation waveforms. (a) Upper arm current i_{au} in phase A. (b) Capacitor voltages u_{cau1}−u_{cau60} in phase A. (c) Δu_{cau1_sum}−Δu_{cau60_sum} in phase A. (d) Lower arm current i_{al} in phase A. (e) Capacitor voltages u_{cal1}−u_{cal60} in phase A. (f) Δu_{cal1_sum}−Δu_{cal60_sum} in phase A.

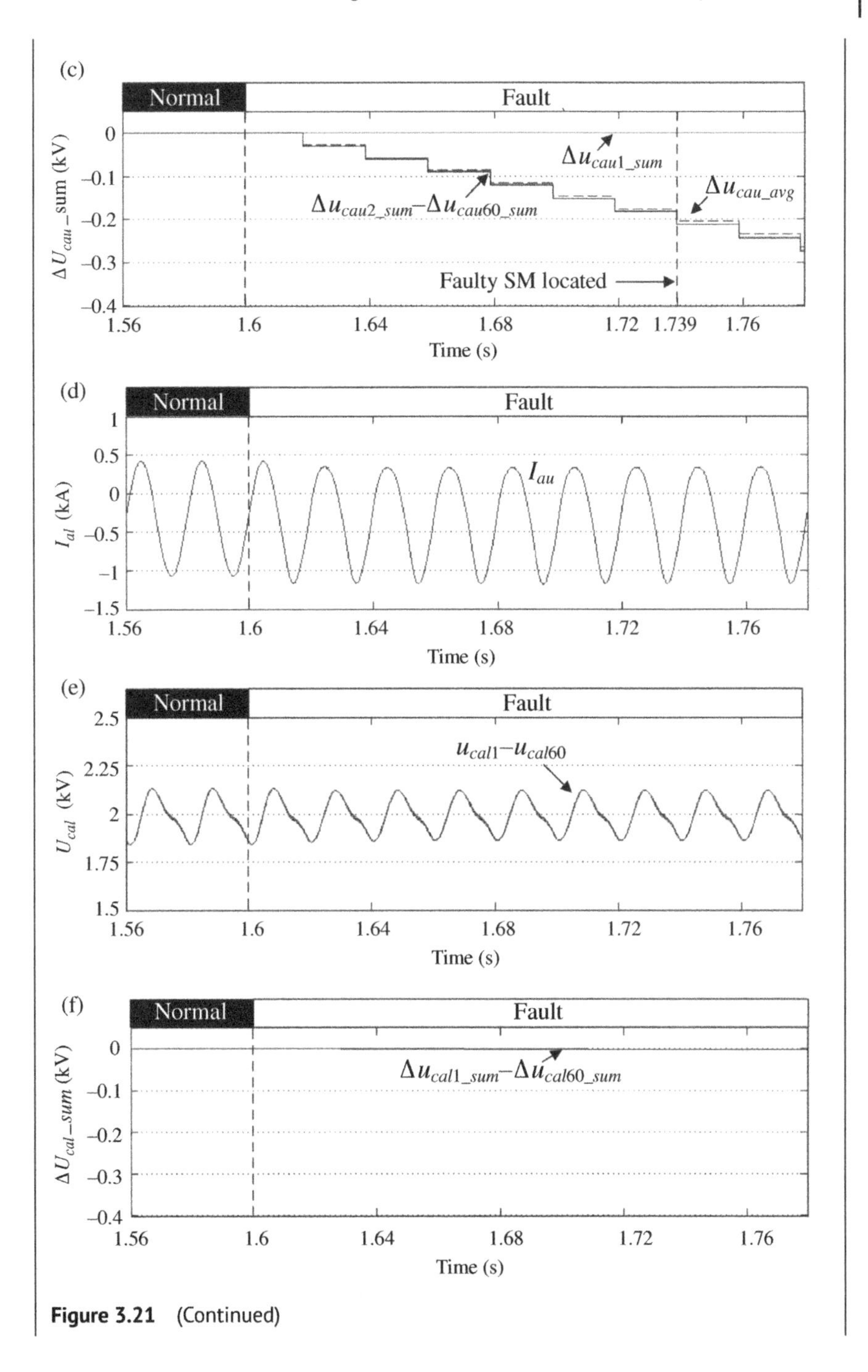

Figure 3.21 (Continued)

Figure 3.21a shows the upper arm current i_{au} in phase A, which is distorted under faults. Although the T_2 open-circuit fault occurs to the upper arm of phase A, the capacitor voltages in the upper arm are reduced slightly and still kept balanced, as shown in Figure 3.21b. Figure 3.21c shows Δu_{cau1_sum}–Δu_{cau60_sum} for the MMC in the upper arm of phase A, where the Δu_{cau2_sum}–Δu_{cau60_sum} and Δu_{cau_avg} are gradually decreased over a number of periods and only the Δu_{cau1_sum} is quite small and close to 0 because of the T_2 open-circuit fault occurrence to the SM1. Consequently, the faulty SM1 is located at 1.739 seconds. Figure 3.21d shows the lower arm current i_{al} in phase A. Figure 3.21e shows the capacitor voltages u_{cal1}–u_{cal60} in the lower arm are kept balanced and nearly not affected by the fault. Figure 3.21f shows Δu_{cal1_sum}–Δu_{cal60_sum} for the MMC in the lower arm of phase A, which are close to each other because the SMs in the lower arm are healthy.

3.7 STW Based Fault Detection Under IGBT Open-Circuit Faults

This section presents another switch open-circuit fault localization strategy for the MMC, where an STW based features extraction algorithm (FEA) is applied to extract the features of the MMC based on the feature relationship between neighboring STWs. Based on the extracted features, the fault in the MMC can be easily located with the two-dimensional convolutional neural networks (2D-CNNs). Figure 3.22 shows the STW-FEA based fault localization method for the MMC [21]. The STW-FEA based fault localization strategy can construct concise low-data-volume features samples for the MMC in both time domain and frequency domain, and accordingly it can locate the fault with short time and high accuracy for the MMC. In addition, it does not require the creation of complex mathematical models and manual setting of empirical thresholds.

3.7.1 MMC Data

Take the MMC with 6 SMs per arm as an example, where 72 kinds of switch open-circuit fault cases are considered and marked as F_1–F_{72}, as shown in Table 3.9.

Figure 3.22 STW-FEA based fault localization method for MMCs.

Table 3.9 Switch open-circuit fault cases.

Fault case	Fault phase	Fault arm	No. of fault SM	Fault type
F_{1-6}	A	Upper	1–6	I
F_{7-12}		Lower		
F_{13-18}		Upper		II
F_{19-24}		Lower		
F_{25-30}	B	Upper	1–6	I
F_{31-36}		Lower		
F_{37-42}		Upper		II
F_{43-48}		Lower		
F_{49-54}	C	Upper	1–6	I
F_{55-60}		Lower		
F_{61-66}		Upper		II
F_{67-72}		Lower		

The data in MMCs are sampled at the frequency f_s and the sampling period is $T_s = 1/f_s$. Here, 53 kinds of data related to the switch open-circuit faults in the MMC are acquired from 53 channels, as listed in Table 3.10. The data from the j-th channel is marked as $X_j(n)$ ($n = 0, 1...$) and all data are

$$X(n)=\left[X_1(n),X_2(n),\ldots,X_j(n),\ldots,X_{53}(n)\right]^T \tag{3.32}$$

To reduce the redundancy and features correlation of these data, the $X(n)$ are whitened and regularized as

$$X_z(n)=\left[X(n)\cdot X(n)^T\right]^{-1/2}\cdot X(n)=\left[X_{z1}(n),X_{z2}(n),\ldots,X_{z53}(n)\right]^T \tag{3.33}$$

3.7.2 Sliding-Time Windows

The STW $S(n)$, an $N \times 53$ data matrix, is presented to sample the data $X_z(n)$, as shown in Figure 3.23a. The sliding interval and frequency of the $S(n)$ is ΔN and $f_s/\Delta N$, respectively, where $\Delta N < N$. The i-th $S_i(n)$ can be expressed as

$$S_i(n)=X_Z(n)\cdot\omega(n) \tag{3.34}$$

with

$$\omega(n)=rect(n)=\begin{cases}1, & 0\le n<N\\ 0, & \text{other cases}\end{cases} \tag{3.35}$$

Table 3.10 53 kinds of data from 53 channels.

Channel	Channel name	Signal	Data $X(n)$
1	Time	$T(n)$	$X_1(n)$
2–4	MMC voltage	$u_a(n)$, $u_b(n)$, $u_c(n)$	$X_2(n)$–$X_4(n)$
5–7	MMC currents	$i_a(n)$, $i_b(n)$, $i_c(n)$	$X_5(n)$–$X_7(n)$
8–10	Phase A–C differential currents	$i_{diff\_a}(n)$, $i_{diff\_b}(n)$, $i_{diff\_c}(n)$	$X_8(n)$–$X_{10}(n)$
11–13	Phase A–C upper arm currents	$i_{au}(n)$, $i_{bu}(n)$, $i_{cu}(n)$	$X_{11}(n)$–$X_{13}(n)$
14–16	Phase A–C lower arm currents	$i_{al}(n)$, $i_{bl}(n)$, $i_{cl}(n)$	$X_{14}(n)$–$X_{16}(n)$
17	DC-side current	$i_{dc}(n)$	$X_{17}(n)$
18–29	Phase A SM 1–12 capacitor voltages	$u_{cau1}(n)$–$u_{cau6}(n)$, $u_{cal1}(n)$–$u_{cal6}(n)$	$X_{18}(n)$–$X_{29}(n)$
30–41	Phase B SM 1–12 capacitor voltages	$u_{cbu1}(n)$–$u_{cbu6}(n)$, $u_{cbl1}(n)$–$u_{cbl6}(n)$	$X_{30}(n)$–$X_{41}(n)$
42–53	Phase C SM 1–12 capacitor voltages	$u_{ccu1}(n)$–$u_{ccu6}(n)$, $u_{ccl1}(n)$–$u_{ccl6}(n)$	$X_{42}(n)$–$X_{53}(n)$

The $S_i(n)$ records N columns of data from $i{\cdot}\Delta N{\cdot}T_s$ to $(i{\cdot}\Delta N+N){\cdot}T_s$. Two neighboring STWs are overlapped with $(N-\Delta N)$ columns of data.

The $S_i(n)$ can be expressed as [22]

$$S_i(n) = P_J^{-1} \sum_{i,J\in Z} R_i^J \Phi_{J,i}(n) \tag{3.36}$$

with

$$\begin{cases} R_i^J = \sqrt{2}\sum_{p\in Z} \bar{f}_{p-2i} S_i(n), \Phi_{J-1,p} \\ \Phi_{J,i}(n) = \sqrt{2}\sum_{p\in Z} f_{p-2i} \Phi_{J-1,p}(n) \end{cases} \tag{3.37}$$

where R_i^J is the projection coefficient. P_J is the projection operator. $\Phi_{J,i}(n)$ is the basis function. $\bar{f}_{p-2i}$ and f_{p-2i} are the filter with length as L.

3.7.3 Feature of STW

The spectrum of $S_i(n)$ can be divided into 2^J number of different frequency bands (FBs) in the frequency domain, including FB_0– FB_{2^J-1}, as shown in Figure 3.23b.

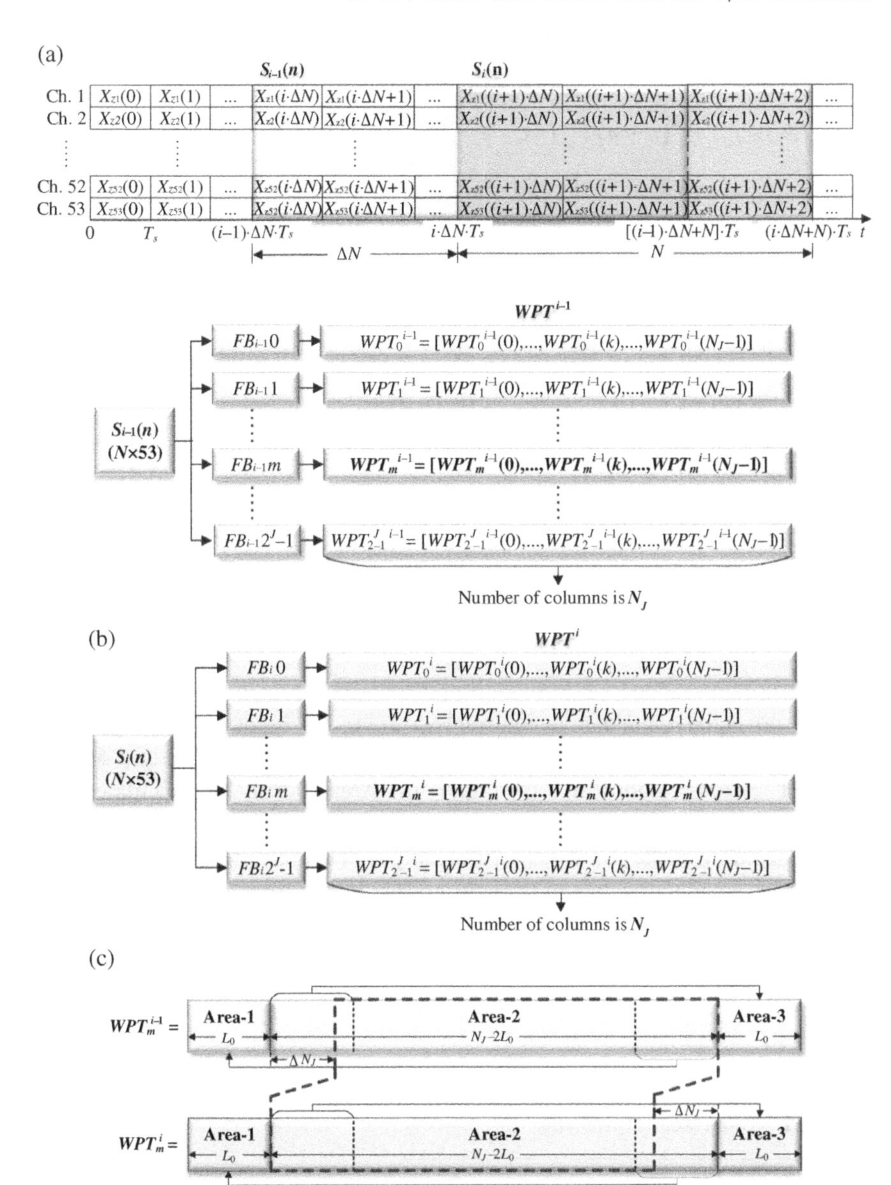

Figure 3.23 STW operation. (a) Sliding-time windows. (b) Features of STWs. (c) Feature relationship between neighboring STWs.

According to equation (3.36) and [22], the m-th FB ($0 \le m \le 2^J - 1$) can be expressed as

$$FB_m = P_J^{-1} R_i^J (n-m) \Phi_{J,i}(n) \tag{3.38}$$

According to equation (3.38) and Mallat wavelet porous algorithm, the feature $WPT_m^{\ i}$ corresponding to FB_m can be obtained as equation (3.39), which is a $53 \times N_J$ data matrix and $N_J = N/2^J$.

$$WPT_m^{\ i} = \left[WPT_m^{\ i}(0), WPT_m^{\ i}(1), \ldots, WPT_m^{\ i}(N_J - 1) \right] \tag{3.39}$$

with

$$WPT_m^i(k) = \begin{cases} WPT_m^i(k + N_J - 2L_0),\ 0 \le k < L_0 \\ \sum\limits_{h=-L/2}^{L/2} R_i^J(k-h) f(h), L_0 \le k < N_J - L_0 \\ WPT_m^i(k - N_J + 2L_0),\ N_J - L_0 \le k < N_J \end{cases} \tag{3.40}$$

The $WPT_m^i(k)$ is a 53×1 matrix, which can be divided into three areas, as shown in Figure 3.23c and Table 3.11.

- *Area-1*: column range $[0, L_0)$. According to data cycle principle [23], the data in Area-1 are equal to the data in column $[N_J - 2L_0$ to $N_J - L_0)$ of Area-2.
- *Area-2*: column range $[L_0, N_J - L_0)$. According to equation (3.39) and Mallat wavelet porous algorithm, the data in Area-2 can be calculated as equation (3.41).
- *Area-3*: column range $[N_J - L_0, N_J)$. According to data cycle principle, the data in Area-3 are equal to the data in the column $[L_0, 2L_0)$ of Area-2.

The feature WPT^i of $S_i(n)$ is a $53 \times N$ matrix, which is the combination of $WPT_0^i - WPT_{2^J-1}^i$ corresponding to $FB_0 - FB_{2^J-1}$ as

$$WPT^i = \left[WPT_0^i, \cdots, WPT_m^i, \cdots, WPT_{2^J-1}^i \right] \tag{3.41}$$

Table 3.11 Three areas of $WPT_m^{\ i}(k)$.

Area	No. of columns	Column range
1	L_0	$[0, L_0)$
2	N_J-$2L_0$	$[L_0, N_J$-$L_0)$
3	L_0	$[N_J$-$L_0, N_J)$

3.7.4 Features Relationships Between Neighboring STWs

The difference between $R_{i-1}^{J}(k)$ for $S_{i-1}(n)$ and $R_i^J(k)$ for $S_i(n)$ results in different $WPT_m^{i-1}(k)$ and $WPT_m^{i}(k)$ for $S_{i-1}(n)$ and $S_i(n)$, respectively, in Figure 3.23b, according to equations (3.36)–(3.40). Based on equation (3.37) and time-translational invariant, the relationship between $R_{i-1}^{J}(k)$ and $R_i^J(k)$ can be obtained as

$$R_i^J(k) = R_{i-1}^J(k - \Delta N_J), \qquad (0 \le k < N_J - \Delta N_J) \tag{3.42}$$

where $\Delta N_J = \Delta N/2^J$.

Based on equations (3.40) and (3.42), the relationship between $WPT_m^{i-1}(k)$ and $WPT_m^{i}(k)$ can be obtained as

$$WPT_m{}^{i}(k) = WPT_m{}^{i-1}(k - \Delta N_J), \qquad (L_0 \le k < N_J - \Delta N_J - L_0) \tag{3.43}$$

Hence, between neighboring $S_{i-1}(n)$ and $S_i(n)$, the data in column $[L_0+\Delta N_J, N_J-L_0)$ of WPT_m^{i-1} are the same to the data in column $[L_0, N_J-L_0-\Delta N_J)$ of WPT_m^{i}, as shown in Figure 3.23c.

3.7.5 Features Extraction Algorithm

Based on the feature relationships between neighboring STWs, an STW-FEA is presented to simplify the feature extraction [21]. Figure 3.24 shows the STW-FEA for WPT_m^{i}, which is divided into four zones, as shown in Table 3.12.

- *Zone I*: column range $[0, L_0)$. According to data cycle principle, the data in Zone I equals to the data in the column $[N_J-2L_0, N_J-L_0)$ of WPT_m^{i}.
- *Zone II*: column range $[L_0, N_J-L_0-\Delta N_J)$. According to equation (3.43), the data of Zone II in WPT_m^{i} can be directly obtained from WPT_m^{i-1}.
- *Zone III*: column range $[L_0, N_J-L_0-\Delta N_J)$. After calculation of $R_i^J(k)$ $(N_J-L_0-\Delta N_J \le k < N_J-L_0)$ with equations (3.36) and (3.37), the data in column $[N_J-L_0-\Delta N_J, N_J-L_0)$ can be calculated with equation (3.44), as

$$WPT_m^{i}(k) = \begin{cases} WPT_m^{i}(k + N_J - 2L_0), & k \in \text{Zone I} \\ WPT_m^{i-1}(k), & k \in \text{Zone II} \\ \sum_{h=-L/2}^{L/2} R_i^J(k-h) f(h), & k \in \text{Zone III} \\ WPT_m^{i}(k - N_J + 2L_0), & k \in \text{Zone IV} \end{cases} \tag{3.44}$$

- *Zone IV*: column range $[N_J-L_0, N_J)$. According to data cycle principle, the data in Zone IV equals to the data in the column $[L_0, 2L_0)$ of WPT_m^{i}.

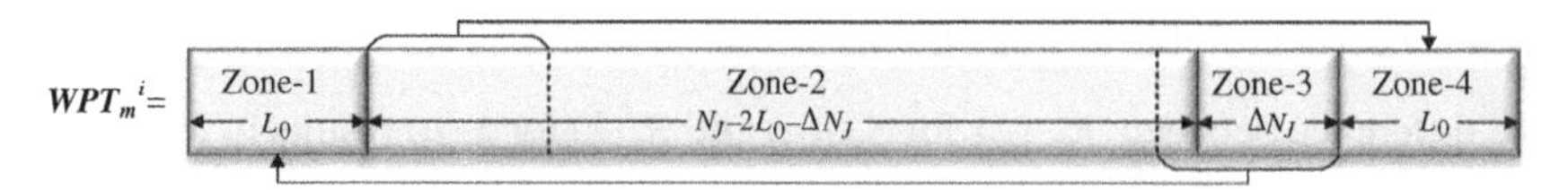

Figure 3.24 STW-FEA for $WPT_m{}^i(k)$.

Table 3.12 Four zones of $WPT_m{}^i$.

Zone	No. of columns	Column range
I	L_0	$[0, L_0)$
II	N_J-$2L_0$-ΔN_J	$[L_0, N_J$-L_0-$\Delta N_J)$
III	ΔN_J	$[N_J$-L_0-$\Delta N_J, N_J$-$L_0)$
IV	L_0	$[N_J$-$L_0, N_J)$

3.7.6 Energy Entropy Matrix

The energy entropy feature matrix E^i_{Total} of $S_i(n)$ can be calculated as equation (3.45), which is a 53 × 1 matrix and is only a low-data-volume features sample.

$$E^i{}_{Total} = \left[E^i{}_1, E^i{}_2, \ldots, E^i{}_{53}\right]^T \tag{3.45}$$

with

$$E^i_j = \sum_{m=0}^{2^J-1} E_{m,j}, \ \left(j = 1, 2, \ldots, 53\right) \tag{3.46}$$

$$E_{m,j} = \sum_{k=0}^{N_J-1} \left|WPT^i_m\left(k\right)_j\right|^2 \tag{3.47}$$

where $WPT^i_m\left(k\right)_j$ is the value in the j-th row of $WPT^i_m\left(k\right)$.

3.7.7 2D-CNN

The low-data-volume features sample E^i_{Total} for $S_i(n)$ is sent to the 2D-CNNs [24] for fault localization, as shown in Figure 3.25, where the 2D-CNNs are composed of input layer, l number of convolutional layers, pool layer, full connect layer, and output layer. After the input of the E^i_{Total} into the input layer, the *Relu* activation function is used to perform l times of the convolution operation in the convolutional layer as

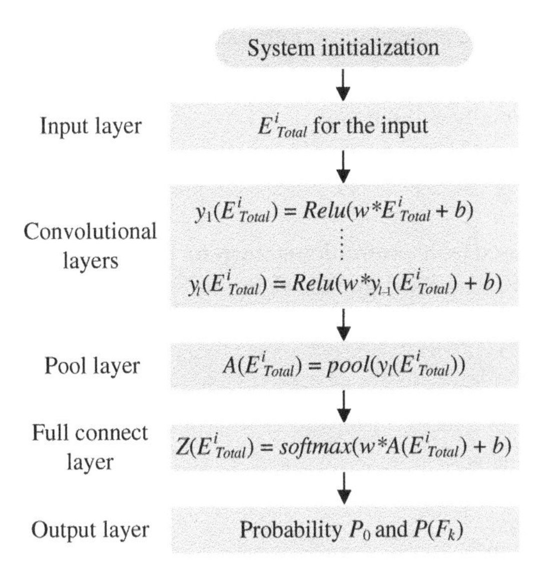

Figure 3.25 2D-CNNs for fault localization.

$$\begin{cases} y_1\left(E^{i}_{Total}\right) = Relu\left(w^{*}E^{i}_{Total} + b\right) \\ \vdots \\ y_l\left(E^{i}_{Total}\right) = Relu\left(w^{*}y_{l-1}\left(E^{i}_{Total}\right) + b\right) \end{cases} \tag{3.48}$$

where * represents the convolution operation. The $y_1(E^{i}_{Total})$ and $y_l(E^{i}_{Total})$ are the 1-th and l-th convolution operation output, respectively. The w and b are the weights and bias, respectively.

The output of the pool layer can be realized by the *pool* activation function as

$$A\left(E^{i}_{Total}\right) = pool\left(w^{*}y_l\left(E^{i}_{Total}\right) + b\right) \tag{3.49}$$

The output of the full connect layer can be realized by the *softmax* activation function as

$$Z\left(E^{i}_{Total}\right) = softmax\left(w^{*}A\left(E^{i}_{Total}\right) + b\right) \tag{3.50}$$

In the output layer, the probability P_0 corresponding to normal case without fault and the probability $P(F_k)$ (k = 1, 2..., 72) corresponding to the fault case F_k can be directly obtained from the $Z(E^{i}_{Total})$, which meets

$$P_0 + \sum_{k=1}^{72} P\left(F_k\right) = 100\% \tag{3.51}$$

The fault can be located based on the probability, as follows.

- If $P(F_k) > 50\%$ and lasts for several sliding interval ΔN, the fault F_k is alarmed.
- Otherwise, the MMC works normally without faults.

3.7.8 Fault Detection Method

Figure 3.26 shows the STW-FEA based fault localization strategy for the MMC [21]. The $S_i(n)$ is captured from the MMC data $X_z(n)$, and the algorithm is implemented with the period $\Delta N/f_s$.

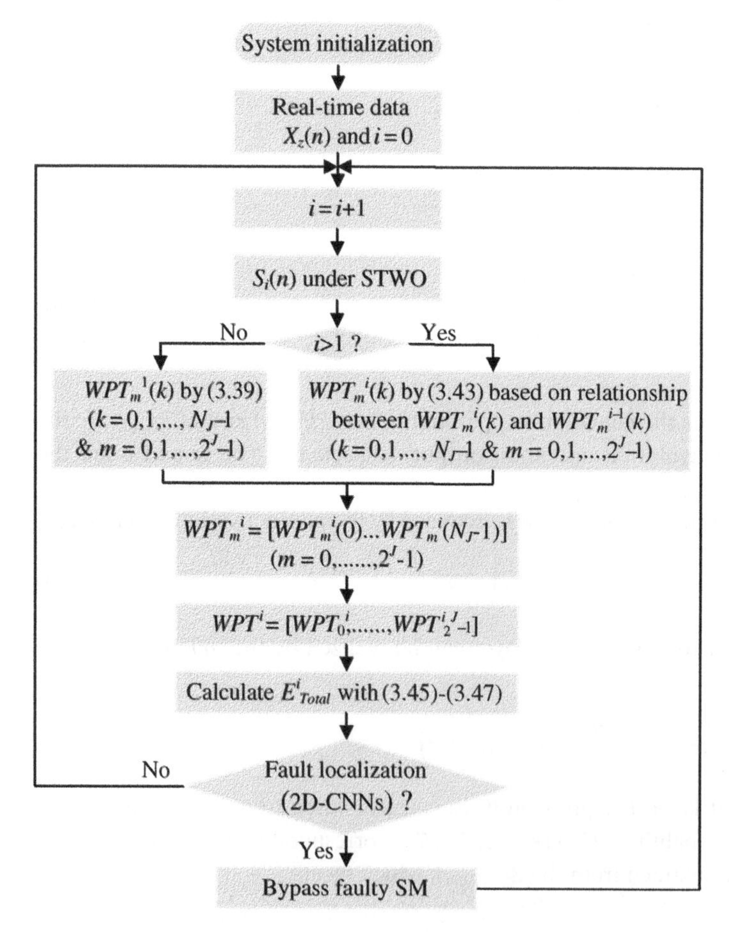

Figure 3.26 STW-FEA based fault localization strategy for MMCs.

- If $i = 1$, the $WPT_m^1 = [WPT_m^1(0), ..., WPT_m^1(N_J-1)]$, $(m = 0, 1, ..., 2^J-1)$ corresponding to FB_m are obtained with equation (3.39). Then, the $WPT^1 = [WPT_0^1, ..., WPT_m^1, ..., WPT_{2^J-1}^1]$, $(m = 0, 1, ..., 2^J-1)$ for $S_1(n)$ is obtained by equation (3.41).
- If $i>1$, the $WPT_m^i = [WPT_m^i(0), ..., WPT_m^i(N_J-1)]$, $(m = 0, 1, ..., 2^J-1)$ corresponding to FB_m are obtained based on relationship between $WPT_m^i(k)$ and $WPT_m^{i-1}(k)$ by equation (3.44). Then, the $WPT^i = [WPT_0^i, ..., WPT_m^i, ..., WPT_{2^J-1}^i]$, $(m = 0, 1, ..., 2^J-1)$ for $S_i(n)$ is obtained by equation (3.41).

The $E^i{}_{Total}$ can be calculated with the equations (3.45)–(3.47) based on the obtained WPT^i, which is sent into the 2D-CNNs to judge the fault. If no fault occurs in the MMC, the $S_{i+1}(n)$ will be captured for the fault localization in the next period. If a fault occurs in the MMC, the faulty SM would be bypassed and the $S_{i+1}(n)$ will be captured for the fault localization in the next period.

The STW-FEA based fault localization strategy may increase computation burden. However, the graphics processing unit (GPU) with higher computing performance has appeared in recent years. The classic Pascal architecture (P40, P100) GPU training 2D-CNNs forward calculation speed is 84 times faster than the classic central processing unit (CPU) E5-2650v2. In theory, the calculation speed of GPU yields 100× speed improvement than that of CPU. The GPU makes the artificial intelligence (AI)-based methods become practical because it uses parallel computing and processing technology, which can greatly alleviate the bottleneck at computing level [25].

3.7.9 Selection of Sliding Interval

Based on Nyquist sampling theorem [26], the spectral range of $S_i(n)$ is $[0, f_s/2]$ and spectrum for FB_0– FB_{2^J-1} of $S_i(n)$ are

$$\begin{cases} FB_0\text{'s spectrum range}: & \left[0, f_s/2^{J+1}\right] \\ \vdots & \vdots \\ FB_m\text{'s spectrum range}: & \left[m \cdot f_s/2^{J+1}, (m+1) \cdot f_s/2^{J+1}\right] \\ \vdots & \vdots \\ FB_{2^J-1}\text{'s spectrum range}: & \left[\left(2^J-1\right) \cdot f_s/2^{J+1}, f_s/2\right] \end{cases} \tag{3.52}$$

The center frequency f_{mc} of the m-th FB is

$$f_{mc} = (2m+1) \cdot 2^{-J-2} \cdot f_s \tag{3.53}$$

Based on relationship between f_{mc} and grid frequency f_g, the optimal sliding interval ΔN_m corresponding to the m-th FB's center frequency can be obtained as

$$\Delta N_m = f_{mc} / f_g \tag{3.54}$$

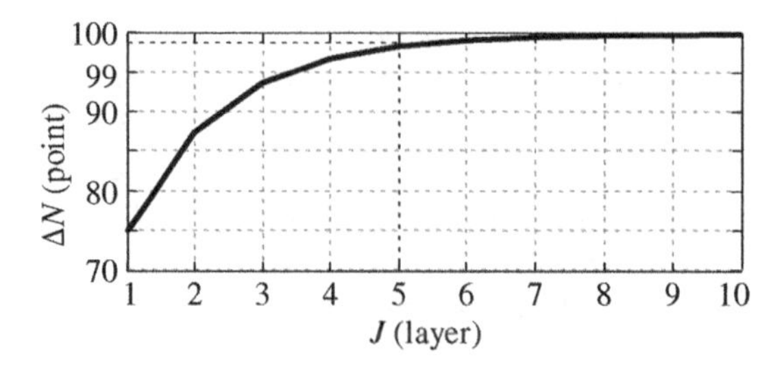

Figure 3.27 ΔN under various J.

The $\Delta N_0 \sim \Delta N_{2^J-1}$ corresponding to FB_0– FB_{2^J-1} are different, where the ΔN_m is larger at higher frequency and smaller at low frequency. According to [27], in order to ensure that all FBs can be effectively decomposed in the frequency domain, ΔN should be set to the maximum value among $\Delta N_0 \sim \Delta N_{2^J-1}$ as

$$\Delta N = \max\left[\Delta N_0, \ldots, \Delta N_{2^J-1}\right] = \left[\left(2^{J+1}-1\right)\cdot f_s\right] / \left(2^{J+2}\cdot f_g\right) \tag{3.55}$$

Figure 3.27 shows the ΔN under various wavelet packet decomposition J, which is derived from simulation in Case Study 3.3. The ΔN increases along with the increase of J. When $J > 5$, the increase of J would not contribute much to the increase of ΔN and the ΔN will eventually stabilize toward 100. Hence, the sliding interval ΔN can be selected as 100 here.

3.7.10 Analysis of Fault Localization Time

To better characterize the fault localization time, the fault ratio D_{fault} of STW is defined as

$$D_{fault} = T_{fault} / T_h \tag{3.56}$$

with

$$\begin{cases} T_{fault} = k \cdot \Delta N \cdot T_s & (k = 0, 1 \ldots) \\ T_h = N \cdot T_s \end{cases} \tag{3.57}$$

where T_{fault} is the fault duration. T_h is the STW duration. Normally, $T_h = 1/f_g$, which is the same to the grid period, e.g. $T_h = 20$ ms for 50 Hz grid.

Figure 3.28a shows the change of $P(F_k)$ under the fault F_k ($k = 1, \ldots, 24$) in phase A of the MMC, which is derived from simulation in Case Study 3.3, where the MMC works in inverter mode. Figure 3.28b shows the change of $P(F_k)$ under the fault F_k ($k = 1, \ldots, 24$) in phase A of the MMC, which is derived from simulation in Case Study 3.3, where the MMC works in rectifier mode. The $P(F_k)$ gradually

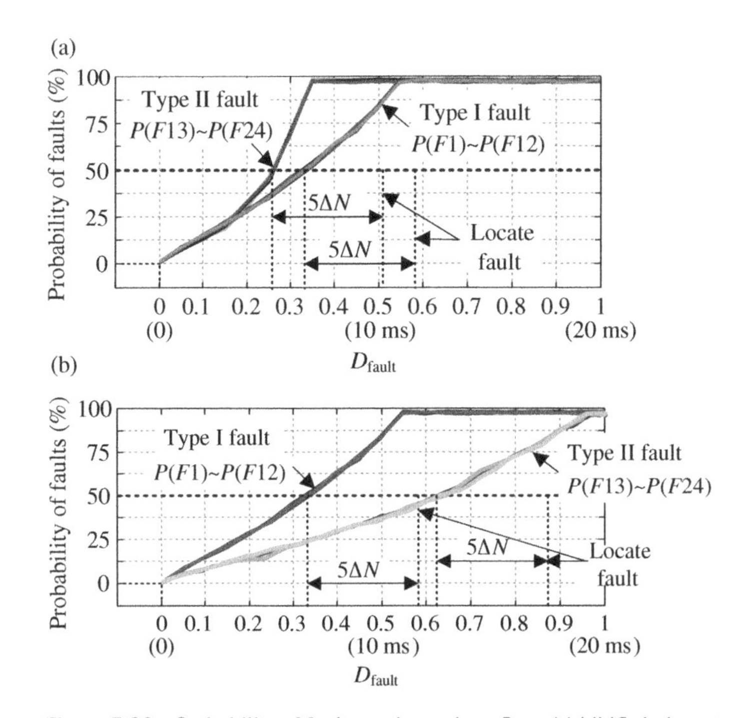

Figure 3.28 Probability of faults under various D_{fault}. (a) MMCs in inverter mode. (b) MMCs in rectifier mode.

increases along with the increase of D_{fault}. The change of $P(F_1)$–$P(F_{12})$ is almost the same to each other.

The change of $P(F_{13})$–$P(F_{24})$ is almost the same to each other. The $P(F_1)$–$P(F_{12})$ increase slower than $P(F_{13})$–$P(F_{24})$ when the MMC works in inverter mode. The $P(F_1)$–$P(F_{12})$ increase faster than $P(F_{13})$–$P(F_{24})$ when the MMC works in rectifier mode. With the fault localization strategy, the fault can be located under $D_{fault} < 1$, which means within one fundamental period of 20 ms.

Case Study 3.3 Analysis of STW-FEA Based Fault Localization Strategy

Objective: In this case study, the performance of the STW-FEA based fault localization strategy for MMCs is studied and shown in Figure 3.27 and Figure 3.28, through professional simulation tool PSCAD/EMTDC. As shown in Figure 3.29, a three-phase MMC system connected to power grid is established.

Parameters: The system parameters are shown in Tables 3.13–3.15.

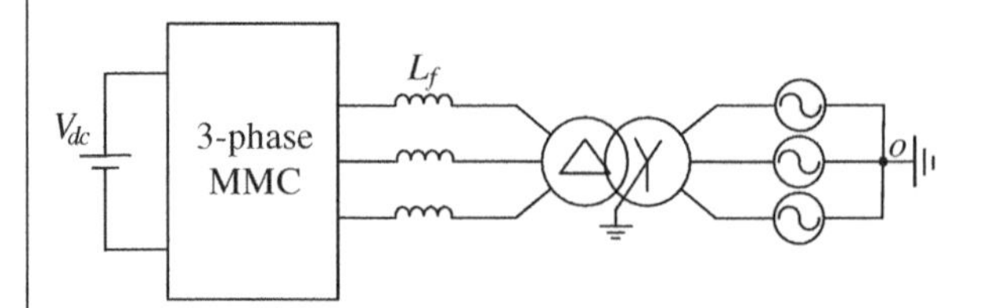

Figure 3.29 Schematic diagram of the simulation system.

Table 3.13 Simulation system parameters.

Parameter	Value
Rated power	6 MW
Grid frequency f_g	50 Hz
DC-link voltage V_{dc}	6 kV
Number of SMs per arm	6
SM capacitance C	12.5 mF
Transformer voltage rating	3 kV/33 kV
Transformer leakage reactance	10%
Arm inductance L_s	3 mH
Carrier frequency	1.5 kHz
Sampling frequency f_s	100 kHz

Table 3.14 Parameters of 2D-CNNs.

Parameter	Value
Weight attenuation parameter λ_s	0.0001
Iterations during model training	2500
Early stopping setting	300
Dropout	0.4

Table 3.15 Parameters of the STW-FEA.

Parameter	Value
L_0	21 points (105 μs)
ΔN	100 points (1 ms)
N	200 nts (20 ms)
J	5 layers

3.8 IF Based Fault Detection Under IGBT Open-Circuit Faults

Isolation forest (IF) shows state-of-the-art performance in the data-mining field, and is widely used in data anomaly detection in industry, network security, and financial transactions due to its linear time complexity and excellent accuracy. This chapter presents an IF based SM switch open-circuit fault localization method for MMCs [28]. Based on the continuous sampling SM capacitor voltages, a number of isolation trees (ITs) are produced to construct the IFs for MMCs. Through the comparison of continuous IFs' outputs, the faulty SM can be effectively localized. The IF based fault localization method only requires SM capacitor voltages in the MMC to construct concise low-data-volume tree models and uses sparsity and difference properties of outlier data to localize fault, accordingly it simplifies calculation complexity. In addition, it does not require the MMC's mathematical models and manual setting of empirical thresholds.

Figure 3.30 shows the IF based fault localization method for the upper arm of phase A, which can localize the fault based on the IF constructed by the capacitor voltages u_{cau1}–u_{caun} in the upper arm of phase A of the MMC.

3.8.1 IT for MMCs

In the MMC, the n SM capacitor voltages u_{cau1}–u_{caun} in the upper arm of phase A are sampled, where the capacitor voltage sampling frequency is f_s and the sampling period is $T_s = 1/f_s$. In each sampling period T_s, an IT is constructed by the sampled n SM capacitor voltages u_{cau1}–u_{caun}. Figure 3.31 shows an IT example. The IT is a nonlinear data structure, which contains a number of levels. Each level is composed of nodes, where the node is separated into two nodes in the next level, as follows.

Level 0: Root Node N_0, which contains n SMs as SM1–SMn.

Level 1: Randomly select a split value u_0, which is between the minimal SM capacitor voltages and the maximal SM capacitor voltages in Node N_0. Afterwards, the Node N_0 is separated into two Nodes N_{1_1} and N_{1_2}. The Node N_{1_1} contains the SMs in the N_0, whose capacitor voltage is less than or equal to u_0. The Node N_{1_2} contains the SMs in the N_0, whose capacitor voltage is more than u_0.

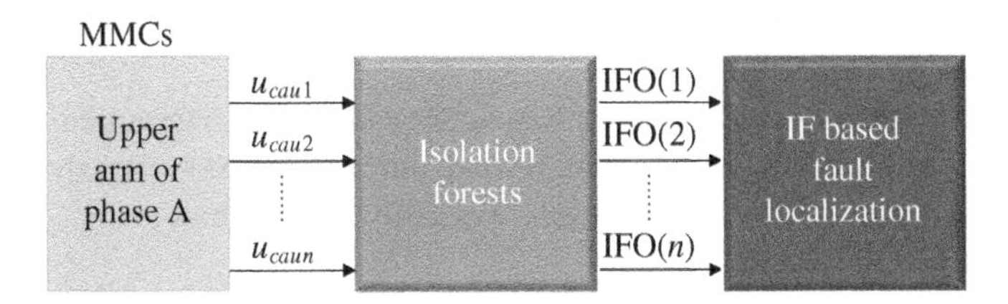

Figure 3.30 IF based fault localization method for upper arm of phase A.

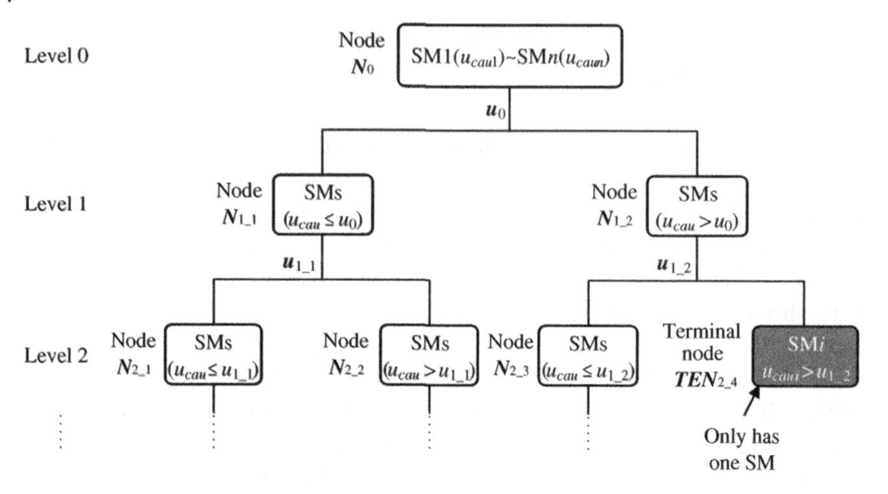

Figure 3.31 An IT example for the upper arm of phase A.

Level 2–Last Level: Referring to Level 1, the Nodes N_{1_1} and N_{1_2} are separated into more nodes, respectively, until each node only contains one SM.

Finally, the IT will have n terminal nodes (TENs) localized at the terminals of these branches, where each TEN contains only one SM. Since the TEN only has one SM, it can no longer be separated. In the IT, the n SMs including SM1–SMn are finally distributed into the n TENs in the light of their corresponding capacitor voltages. In the IT, the TEN containing the SMi ($1 \leq i \leq n$) can be expressed as TEN(SMi). The level of TEN(SMi) in the IT can be expressed as IT[Level(TEN(SMi))].

3.8.2 SM Depth in IT

In the IT, the depth $D(i)$ of the SMi ($1 \leq i \leq n$) can be expressed with the level of the TEN containing the SMi, as

$$D(i) = \mathrm{IT}\left[\mathrm{Level}\left(\mathrm{TEN}\left(\mathrm{SM}i\right)\right)\right] \tag{3.58}$$

According to the IT principle [29], the smaller the $D(i)$, there would be a high probability that the SMi would have a higher anomalous degree; the bigger the $D(i)$, there would be a high probability that the SMi would have a lower anomalous degree, as shown in Table 3.16.

In the MMC, the SM open-circuit fault would cause SM capacitor voltage anomaly. Suppose that the SM anomaly accounts for a small amount, the SM anomaly would result in a high probability that the depth D of faulty SM is small in comparison with that of healthy SM in the IT.

Table 3.16 IT principle about SM anomalous degree and SM depth.

SM depth	Anomalous degree
Small	High (high probability)
Big	Low (high probability)

3.8.3 IF for MMCs

An IF is composed of m_p ($p = 1, 2, 3$ ...) continuous ITs including IT_1–IT_{mp}, which is constructed based on n SM capacitor voltages u_{cau1}–u_{caun} sampled at m_p continuous sampling periods with the sampling interval as T_s.

Figure 3.32 shows the IFs for the MMC with $n = 6$ SMs per arm. In each IF, the j-th ($1 \leq j \leq m_p$) IT has n TENs corresponding to the n SMs (SM1–SMn), respectively, where each SM corresponds with a depth. The $D(i, j)$ represents the depth of the SMi ($1 \leq i \leq n$) in the j-th IT of the IF. Figure 3.32 also shows the anomalous degree for the TENs in the IF, where the node with the color is TEN. Along with the darkening of the color, the anomalous degree of the TEN will have a high probability, and the TEN would have a high probability to contain an anomalous SM.

3.8.4 SM Average Depth in IF

In each IF, the average depth $AD(i)$ of the SMi is

$$AD(i) = \frac{1}{m_p} \sum_{j=1}^{m_p} D(i, j) \tag{3.59}$$

According to IF principle, the $AD(i)$ for the SMi depends on the capacitor voltage anomaly in the SMi, as shown in Table 3.17, as follows.

- If the SMi has no anomaly because its capacitor voltage has no anomaly, there is a high probability that the average depth $AD(i)$ of the SMi in the IF is big.
- If the SMi has anomaly because its capacitor voltage has anomaly, there is a high probability that the average depth $AD(i)$ of the SMi in the IF is small.

3.8.5 IF Output

For each IF, its output IFO is defined as the number (e.g. 1, 2...n) of the SM corresponding to the minimal AD among $AD(1)$–$AD(n)$, as

$$\text{IFO} = \text{No. of SM corresponding to } \min\left[AD(1) - AD(n)\right] \tag{3.60}$$

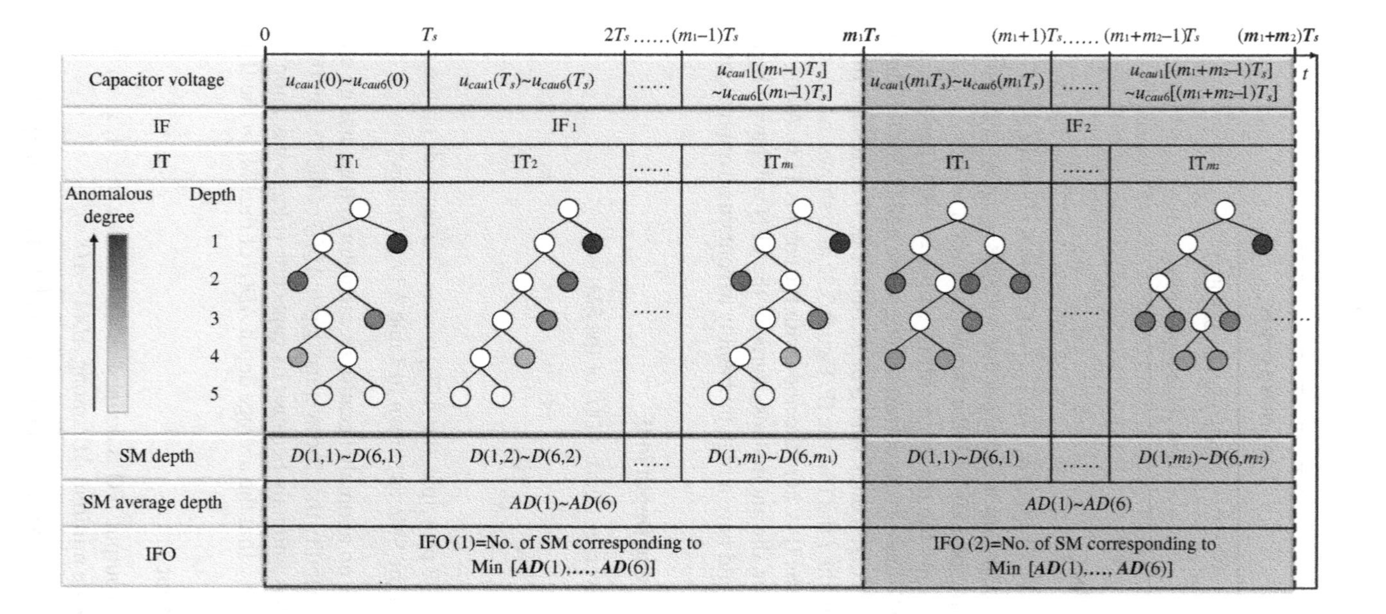

Figure 3.32 IFs and IFOs.

Table 3.17 IF for MMCs under fault.

SM state	SM depth	SM's AD
Open-circuit fault	↓	↓
Health	↑	↑

The switch open-circuit fault in the SMi results in that the $AD(i)$ corresponding to the SMi has a high probability to be smallest. Consequently, the number i corresponding to the faulty SMi would be the output of the IF.

3.8.6 Fault Detection

With the IF principle, the IF_1, IF_2, IF_3... for the MMC are produced over the time, which are not overlapped, as shown in Figure 3.33. In order to achieve fault localization in the MMC, k number of continuous IFs are adopted, whose outputs are stored in the buffer, as shown in Figure 3.33. The buffer can be expressed as

$$\text{Buffer} = \{\text{B}(1), \text{B}(2), \ldots, \text{B}(k)\} \tag{3.61}$$

with

$$\begin{cases} \text{B}(1) = \text{IFO}\{[t] - k + 1\} \\ \quad\vdots \\ \text{B}(k-1) = \text{IFO}\{[t] - 1\} \\ \text{B}(k) = \text{IFO}[t] \end{cases} \tag{3.62}$$

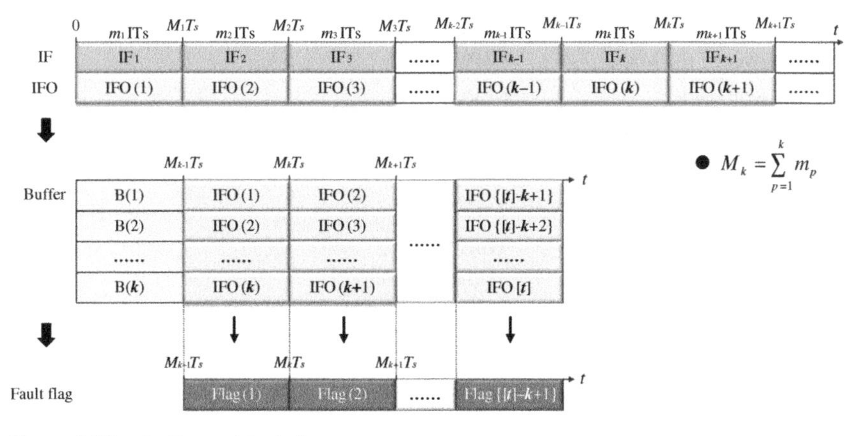

Figure 3.33 Buffer and fault flag.

where B(1), ..., B(k) represent k elements in the buffer. IFO[t] is defined as the latest IFO at time t, IFO{[t]−1} is defined as the sublatest IFO at time t, and so on. The buffer is updated every $m_p T_s$, where the oldest IFO is popped and the newest IFO is inserted.

Based on the k number of values B(1), B(2), ..., B(k) in the buffer, a fault flag is defined as

$$\text{Flag} = \text{TRUE}\{\text{B}(1) = \text{B}(2) = \ldots = \text{B}(k)\} \tag{3.63}$$

The faulty SM in the MMC can be detected based on the flag, as follows.

- When the k values in the buffers are all the same as B(1) = B(2) = ... = B(k) = γ ($1 \le \gamma \le n$), it means that the k numbers of continuous IFOs are the same, the flag will be "1," and the γ-th SM is detected with the fault. And then, the γ-th SM is bypassed from the MMC. Afterwards, the fault localization still continues to work to detect the fault.
- When the k values in the buffers B(1), B(2), ..., B(k) are not the same, it means that the k numbers of continuous IFOs are not the same, the flag will be "0," and the MMC works normally without faults. And then, the fault localization still continues to work to detect the fault.

3.8.7 Selection of m_p

An IF contains m_p ITs. Take the MMC with six SMs per arm as an example, as shown in Table 3.18. Figure 3.34a,b show the average depth AD(1)–AD(6) of a

Table 3.18 Simulation system parameters.

Parameter	Value
Rated power of the MMC	6 MW
DC-link voltage V_{dc}	6 kV
Number of SMs per arm	6
SM capacitance C	12.5 mF
Arm inductance L_s	3 mH
Filter inductance L_f	1 mH
Transformer voltage rating	3 kV/33 kV
Grid voltage	33 kV
Grid frequency f_g	50 Hz
Sampling frequency f_s	100 kHz
E	0.5%

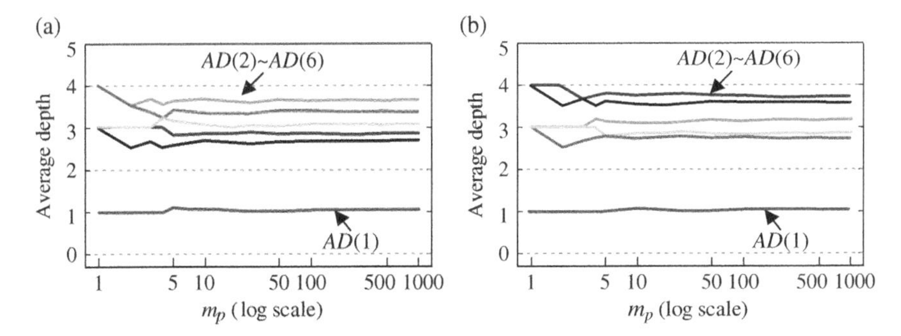

Figure 3.34 *AD* of IT under various m_p. (a) T_1 fault. (b) T_2 fault.

constructed IF with various m_p, where the T_1 open-circuit fault and T_2 open-circuit fault occur in the SM1 of the MMC working in inverter mode, respectively. In the Figure 3.34, the *AD*(1) is the smallest among *AD*(1)–*AD*(6) owing to the fault in SM1.

According to IF algorithm [30], along with the increase of m_p, the *AD* would be stable and nearly constant, as shown in Figure 3.34. As a result, m_p can be selected when the *AD* becomes nearly constant. The selection of m_p can be implemented as follows. A number of latest *AD*(*i*)s are selected and their average value is calculated as $AD(i)_{avg}$. When these latest *AD*(*i*)s are in the range of $[(1-\varepsilon)\cdot AD(i)_{avg}, (1+\varepsilon)\cdot AD(i)_{avg}]$ and lasts for several periods, where ε is error, the *AD*(*i*) can be regarded to be stable, and then the m_p can be selected. For example, the selection of m_p in Figure 3.34 can be implemented, where the m_p can be obtained as 84 when $\varepsilon = 0.5\%$.

3.8.8 Selection of *k*

The buffer has *k* IFOs. Obviously, the bigger the *k*, the longer is the fault localization time, and vice versa. In addition, the *k* is related to the fault localization accuracy, where the fault localization accuracy is increased when the *k* is increased, and vice versa. Figure 3.35 shows the relationship between *k* and fault localization accuracy based on the system shown in Table 3.18, where 250 samples under fault cases and 250 samples under fault-free case are considered. All samples are divided into two types as 0 (health) and 1 (fault), and the accuracy can be expressed as equation (3.64) based on confusion matrix theory [31].

$$\text{Accuracy} = \frac{\text{TP} + \text{TN}}{\text{TP} + \text{FP} + \text{TN} + \text{FN}} \tag{3.64}$$

where TP is true positive defined as the number of samples predicted as 1 and actually is 1. FP is false positive defined as the number of samples predicted as 1

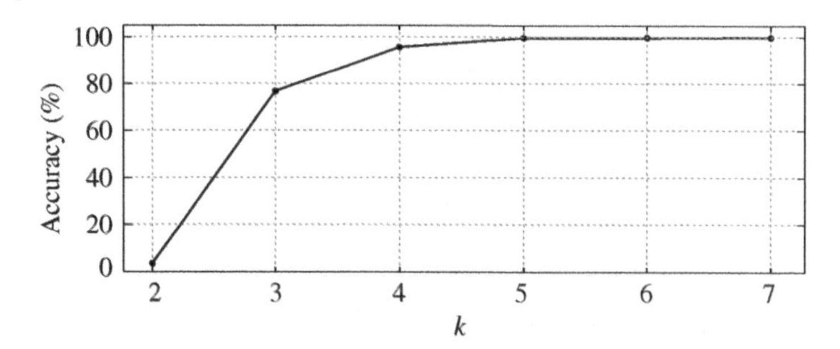

Figure 3.35 Relationship between fault localization accuracy and k.

but actually is 0. FN is false negative defined as the number of samples predicted as 0 but actually is 1. TN is true negative defined as the number of samples predicted as 0 and actually is 0.

Figure 3.35 shows that the accuracy is increased along with the increase of k, where the accuracy reaches 99.2% when $k = 5$. Hence, to ensure the accuracy, the k can be selected as 5 for this MMC system.

Case Study 3.4 Analysis of IF Based Fault Detection Strategy

Objective: In this case study, the working principle and performance of the MMC equipped with the IF based fault detection strategy is studied through professional simulation tool PSCAD/EMTDC.

Parameters: The system parameters are shown in Table 3.18.

Simulation results and analysis:

a) Case 1: T_1 open-circuit fault of MMCs in inverter mode
In this case, the T_1 open-circuit fault of SM1 occurs at 1.2 seconds in the upper arm of phase A, where the MMC works in inverter mode. Figure 3.36 shows the capacitor voltages u_{cau1}–u_{cau6} in the upper arm of phase A. Figure 3.37 shows the IFOs and fault flag in a short time [1.2, 1.2179 seconds]. In the initial stage, the u_{cau1} is normal as u_{cau2}–u_{cau6} and the flag is 0, while u_{cau1} becomes anomaly since 1.2128 seconds. The IFO becomes 1 at 1.2128 seconds and lasts for 5.1 ms. As a result, the flag becomes 1 at 1.2179 seconds and the faulty SM1 is localized at a cost of 17.9 ms.

b) Case 2: T_2 open-circuit fault of MMCs in inverter mode
In this case, the T_2 open-circuit fault of SM1 occurs at 1.2 seconds in the upper arm of phase A, where the MMC works in inverter mode. Figure 3.38 shows the capacitor voltages u_{cau1}–u_{cau6} in the upper arm of

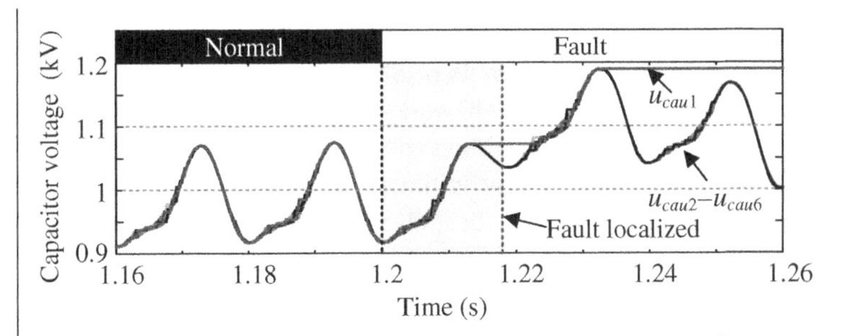

Figure 3.36 Capacitor voltages $u_{cau1}-u_{cau6}$ in the upper arm of phase A.

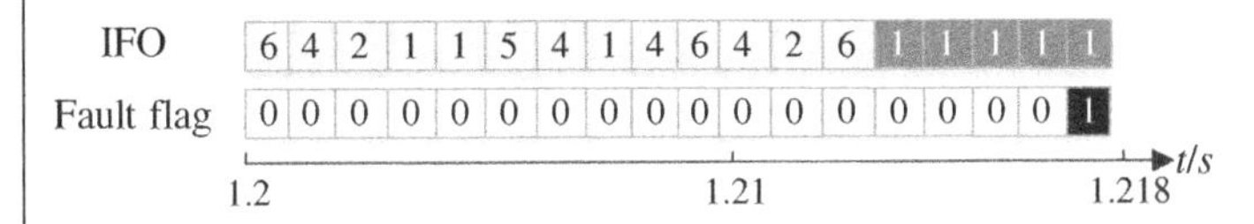

Figure 3.37 IFO and fault flag.

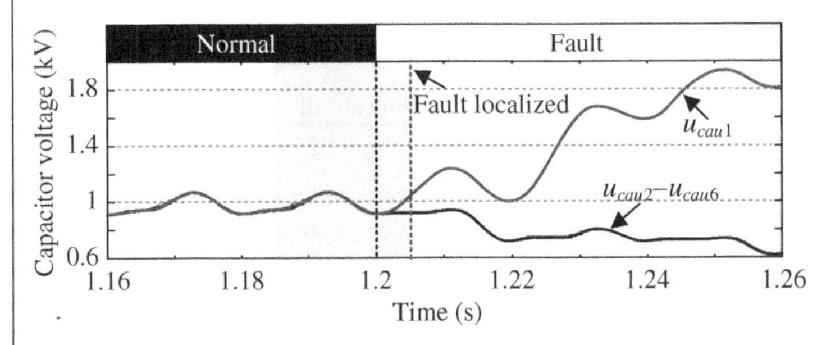

Figure 3.38 Capacitor voltages $u_{cau1}-u_{cau6}$ in the upper arm of phase A.

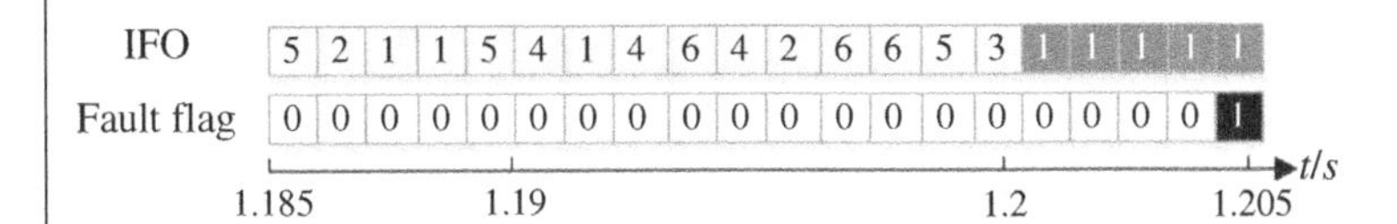

Figure 3.39 IFO and fault flag.

phase A. Figure 3.39 shows the IFOs and fault flag in a short time [1.185, 1.2052 seconds]. In the initial stage, the u_{cau1} is normal as $u_{cau2}-u_{cau6}$ and the flag is 0, while u_{cau1} becomes anomaly after fault. The IFO becomes 1 at 1.2004 seconds and lasts for 4.8 ms. As a result, the flag becomes 1 at 1.2052 seconds and the faulty SM1 is localized at a cost of 5.2 ms.

c) Case 3: T_1 open-circuit fault of MMCs in rectifier mode
In this case, the T_1 open-circuit fault of SM1 occurs at 1.21 seconds in upper arm of phase A, where the MMC works in rectifier mode. Figure 3.40 shows the $u_{cau1}-u_{cau6}$. Figure 3.41 shows the IFOs and fault flag in a short time [1.206, 1.226 seconds]. In the initial stage, the u_{cau1} is normal as $u_{cau2}-u_{cau6}$, and the flag is 0. The u_{cau1} becomes anomaly since 1.221 seconds. The IFO becomes 1 at 1.2212 seconds and lasts for 4.8 ms. As a result, the flag becomes 1 at 1.226 seconds and the faulty SM1 is localized at a cost of 16 ms.

d) Case 4: T_2 open-circuit fault of MMCs in rectifier mode
In this case, the T_2 open-circuit fault of SM1 occurs at 1.21 seconds in upper arm of phase A, where the MMC works in rectifier mode. Figure 3.42

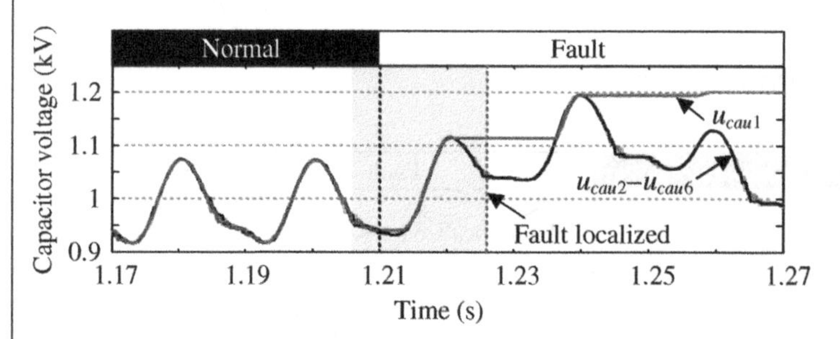

Figure 3.40 Capacitor voltages $u_{cau1}-u_{cau6}$ in the upper arm of phase A.

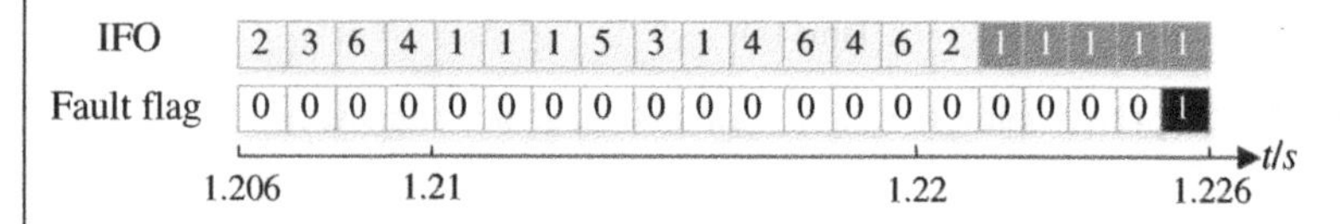

Figure 3.41 IFO and fault flag.

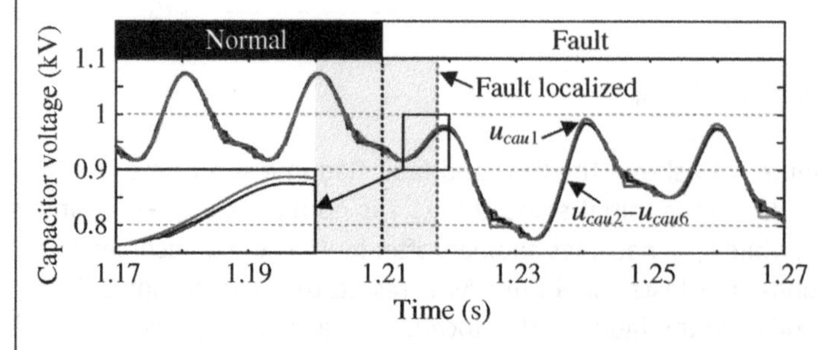

Figure 3.42 Capacitor voltages $u_{cau1}-u_{cau6}$ in the upper arm of phase A.

shows the $u_{cau1}-u_{cau6}$. Figure 3.43 shows the IFOs and fault flag in a short time [1.2, 1.2181 seconds]. In the initial stage, the u_{cau1} is normal as $u_{cau2}-u_{cau6}$ and the flag is 0, while u_{cau1} becomes anomaly since 1.213 seconds. The IFO becomes 1 at 1.2133 seconds and lasts for 4.8 ms. As a result, the flag becomes 1 at 1.2181 seconds and the faulty SM1 is localized at a cost of 8.1 ms.

e) Case 5: T_1 open-circuit fault of MMCs in inverter mode under capacitor parameter inaccuracy
In this case, the performance of the presented method under capacitor parameter inaccuracy is considered, where C_{au2} = 14 mF, C_{au3} = 11 mF, and the other capacitance is 12.5 mF. The T_1 open-circuit fault of SM1 occurs at 1.2 seconds in the upper arm of phase A, where the MMC works in inverter mode. Figure 3.44 shows the capacitor voltages $u_{cau1}-u_{cau6}$ in the upper arm of phase A. Figure 3.45 shows the IFOs and fault flag in a short time [1.2, 1.2183 seconds]. In the initial stage, the u_{cau1} is normal as $u_{cau2}-u_{cau6}$ and the flag is 0, while u_{cau1} becomes anomaly since 1.213 seconds. The IFO becomes 1 at 1.2134 seconds and lasts for 4.9 ms. As a result, the flag becomes 1 at 1.2183 seconds and the faulty SM1 is localized at a cost of 18.3 ms. The inaccuracy of capacitor parameters almost has no effect on fault localization, which shows the robustness of the presented method.

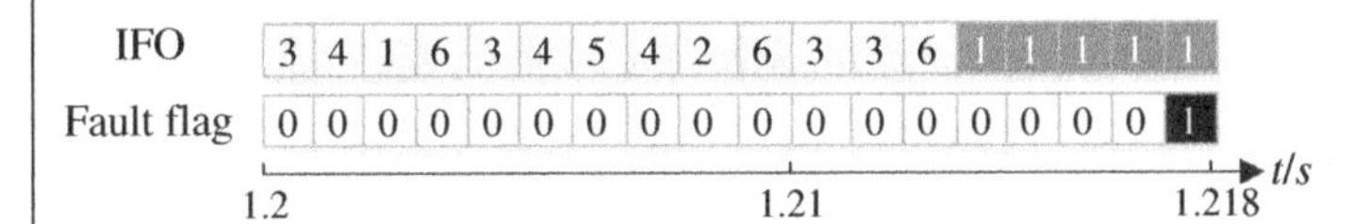

Figure 3.43 IFO and fault flag.

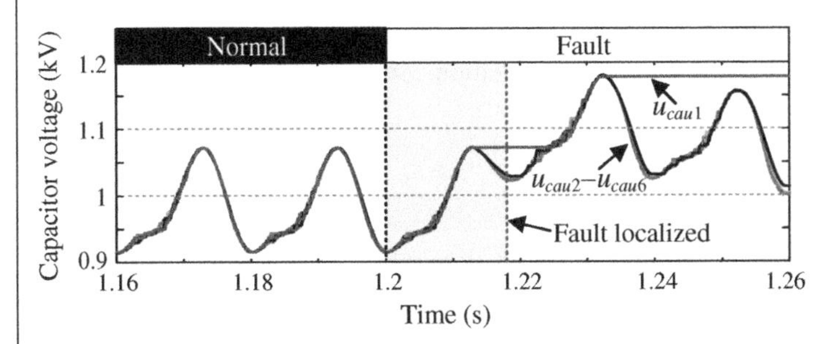

Figure 3.44 Capacitor voltages $u_{cau1}-u_{cau6}$ in the upper arm of phase A.

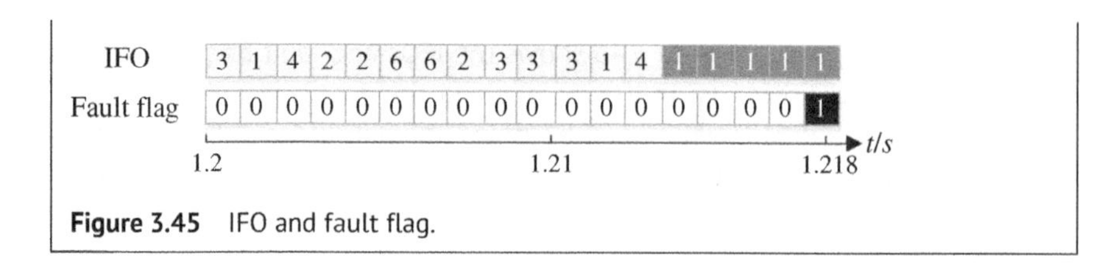

Figure 3.45 IFO and fault flag.

3.9 Summary

The failure of IGBT threatens the reliability and uninterrupted operation of the MMC system. It is required to detect the short-circuit fault and open-circuit fault accurately, locate the faulty IGBTs within a very short time, and protect the devices from destruction, to lay the foundation for further tolerant control of the MMC's continuing operation.

This chapter illustrates recent developments in detection and protection methods of IGBT faults for MMCs. The short-circuit fault and open-circuit fault are two main faults for IGBTs and have different fault features in the MMC system. The short-circuit detection and protection are often integrated in the IGBT drive circuits, while the open-circuit faults remain undetected for a long time. According to the characteristics of IGBT open-circuit in the MMC, the KF based detection method is presented. The localization of the faulty SM is based on the fact that the capacitor voltage is increased under faults. This method is valid in most of the faulty situations, but not available for the SM's lower IGBT open-circuit fault in inverter mode. To overcome this problem, an integrator based localization method is further introduced. Besides, an STW based IGBT open-circuit fault detection method is presented, in which the faulty SMs are located by the 2D CNNs, without modeling the MMC and setting the empirical thresholds. In the last part of this chapter, an IF based SM switch open-circuit fault localization method for MMCs is presented, it uses sparsity and difference properties of outlier data to localize fault, and accordingly it simplifies calculation complexity.

References

1 S. Yang, D. Xiang, A. Bryant, P. Mawby, L. Ran, and P. Tavner, "Condition monitoring for device reliability in power electronic converters: a review," *IEEE Trans. Power Electron.*, vol. 25, no. 11, pp. 2734–2752, Nov. 2010.

2 D. Xiang, L. Ran, P. Tavner, S. Yang, A. Bryant, and P. Mawby, "Condition monitoring power module solder fatigue using inverter harmonic identification," *IEEE Trans. Power Electron.*, vol. 27, no. 1, pp. 235–247, Jan. 2012.

3 S. S. Manohar, A. Sahoo, A. Subramaniam, and S. K. Panda, "Condition monitoring of power electronic converters in power plants-a review," in *2017 20th International Conference on Electrical Machines and Systems (ICEMS)*, Sydney, NSW, Australia, 2017, pp. 1–5.
4 J. Lutz, "IGBT-modules: design for reliability," in *2009 13th European Conference on Power Electronics and Applications*, Barcelona, Spain, 2009, pp. 1–3.
5 B. Lu and S. K. Sharma, "A literature review of IGBT fault diagnostic and protection methods for power inverters," *IEEE Trans. Ind. Appl.*, vol. 45, no. 5, pp. 1770–1777, Sept.-Oct. 2009.
6 U. Choi, F. Blaabjerg, and K. Lee, "Study and handling methods of power IGBT module failures in power electronic converter systems," *IEEE Trans. Power Electron.*, vol. 30, no. 5, pp. 2517–2533, May 2015.
7 B. Li, S. Shi, B. Wang, G. Wang, W. Wang, and D. Xu, "Fault diagnosis and tolerant control of single IGBT open-circuit failure in modular multilevel converters," *IEEE Trans. Power Electron.*, vol. 31, no. 4, pp. 3165–3176, Apr. 2016.
8 R. S. Chokhawala, J. Catt, and L. Kiraly, "A discussion on IGBT short circuit behavior and fault protection schemes," *IEEE Trans. Ind. Appl.*, vol. 31, no. 2, pp. 256–263, Mar./Apr. 1995.
9 Z. Wang, X. Shi, L. M. Tolbert, F. Wang, and B. J. Blalock, "A di/dt feedback-based active gate driver for smart switching and fast overcurrent protection of IGBT modules," *IEEE Trans. Power Electron.*, vol. 29, no. 7, pp. 3720–3732, Jul. 2014.
10 M. S. Kim, B. G. Park, R. Y. Kim, and D. S. Hyun, "A novel fault detection circuit for short-circuit faults of IGBT," in *2011 Twenty-Sixth Annual IEEE Applied Power Electronics Conference and Exposition (APEC)*, Fort Worth, TX, USA, Mar. 2011, pp. 359–363.
11 F. Huang and F. Flett, "IGBT fault protection based on di/dt feedback control," in *Proceedings of IEEE Conference on Power Electronics Specialists*, 2007, pp. 1478–1484.
12 M. A. R. Blanco, A. C. Sanchez, D. Theilliol, L. G. V. Valdes, P. S. Teran, L. H. Gonzalez, and J.A. Alquicira, "A failure-detection strategy for IGBT based on gate-voltage behavior applied to a motor drive system," *IEEE Trans. Ind. Electron.*, vol. 58, no. 5, pp. 1625–1633, May 2011.
13 R. S. Chokhawala and S. Sobhani, "Switching voltage transient protection schemes for high-current IGBT modules," *IEEE Trans. Ind. Appl.*, vol. 33, no. 6, pp. 1601–1610, Nov./Dec. 1997.
14 J. Wang, H. Ma, and Z. Bai, "A submodule fault ride-through strategy for modular multilevel converters with nearest level modulation," *IEEE Trans. Power Electron.*, vol. 33, no. 2, pp. 1597–1608, Feb. 2018.
15 C. Liu *et al.*, "Fault localization strategy for modular multilevel converters under submodule lower switch open-circuit fault," *IEEE Trans. Power Electron.*, vol. 35, no. 5, pp. 5190–5204, May 2020.

16 B. Tamhane and S. Kurode, "Estimation of states of seeker system of a missile using sliding mode observer and kalman filter approaches-A comparative study," in *Proceedings of the 1st International and 16th National Conference on Machines and Mechanisms (iNaCoMM2013)*, IIT Roorkee, India, Dec. 2013, pp. 276–282.

17 M. R. Khan, G. Mulder, and J. Van Mierlo, "An online framework for state of charge determination of battery systems using combined system identification approach," *J. Power Sources*, vol. 246, pp. 629–641, Jan. 2014.

18 J. Kanieski, R. Cardoso, H. Pinheiro, and H. Gründling, "Kalman filter based control system for power quality conditioning devices," *IEEE Trans. Ind. Electron.*, vol. 60, no. 11, pp. 5214–5227, Nov. 2013.

19 M. D. Islam, R. Razzaghi, and B. Bahrani, "Arm-sensorless sub-module voltage estimation and balancing of modular multilevel converters," *IEEE Trans. Power Delivery*, vol. 35, no. 2, pp. 957–967, Apr. 2020.

20 F. Deng, Z. Chen, M. R. Khan, and R. Zhu, "Fault detection and localization method for modular multilevel converters," *IEEE Trans. Power Electron.*, vol. 30, no. 5, pp. 2721–2732, May 2015.

21 F. Deng, M. Jin, C. Liu, M. Liserre, and W. Chen, "Switch open-circuit fault localization strategy for MMCs using sliding-time window based features extraction algorithm," *IEEE Trans. Ind. Electron.*, vol. 68, no. 10, pp. 10193–10206, Oct. 2021.

22 S. Chen, R. Zhang, H. Su, J. Tian and J. Xia, "SAR and multispectral image fusion using generalized IHS transform based on à trous wavelet and EMD decompositions," *IEEE Sens. J.*, vol. 10, no. 3, pp. 737–745, Mar. 2010.

23 A. Y. Panov, S. V. Kuznetsov, and S. V. Ivanov, "Forming a common information space for mechanical engineering cluster product life cycle based on CALS technologies principles," in *Proceedings of ICIEAM*, St. Petersburg, 2017, pp. 1–4.

24 M. Martin, B. Sciolla, M. Sdika, P. Quétin, and P. Delachartre, "Segmentation of neonates cerebral ventricles with 2D CNN in 3D US data: suitable training-set size and data augmentation strategies," in *Proceedings of IEEE IUS*, Glasgow, United Kingdom, 2019, pp. 2122–2125.

25 G. Chen, H. Meng, Y. Liang, and K. Huang, "GPU-accelerated real-time stereo estimation with binary neural network," *IEEE Trans. Parallel Distrib. Syst.*, vol. 31, no. 12, pp. 2896–2907, Dec. 2020.

26 V. V. Zamaruiev, "The use of Kotelnikov-Nyquist-Shannon sampling theorem for designing of digital control system for a power converter," in *Proceedings of 2017 IEEE First UKRCON, Kiev*, 2017, pp. 522–527.

27 T. Lee and H. Shen, "Efficient local statistical analysis via integral histograms with discrete wavelet transform," *IEEE Trans. Visual Comput. Graphics*, vol. 19, no. 12, pp. 2693–2702, Dec. 2013.

28 F. Deng, Y. Chen, J. Dou, C. Liu, Z. Chen, and F. Blaabjerg, "Isolation forest based submodule open-circuit fault localization method for modular multilevel converters," *IEEE Trans. Ind. Electron.*, vol. 70, no. 3, pp. 3090–3102, Mar. 2023.

29 F. T. Liu *et al.*, "Isolation-based anomaly detection," *ACM Trans. Knowl. Discovery Data*, vol. 6, no. 1, pp. 3-1–3-39, Mar. 2012.

30 C. C. Aggarwal, "Isolation forests," in *Outlier Analysis*, 2nd ed. New York, NY, USA: Springer Nature, 2017, pp. 161–164.

31 M. Ohsaki, P. Wang, K. Matsuda, S. Katagiri, H. Watanabe, and A. Ralescu, "Confusion-matrix-based kernel logistic regression for imbalanced data classification," *IEEE Trans. Knowl. Data Eng.*, vol. 29, no. 9, pp. 1806–1819, Sept. 2017.

4

Condition Monitoring and Control of MMCs Under Capacitor Faults

4.1 Introduction

Reliability is one of the important issues for the modular multilevel converter (MMC). The MMC contains a considerable number of capacitors, where the capacitor is one type of reliability-critical component and each capacitor could be considered as a potential fault point [1]. Owing to the chemical process, aging effects, etc., the capacitors would deteriorate and their characteristics would change, which is also called aging effect. The aging effect takes effect along with the operation of capacitors, which eventually causes the failure of the capacitor.

Capacitor failure is due to a combined effect of thermal, electrical, mechanical, and environmental stress and has several root causes and failure modes. The primary failure mode is wear-out failure mechanism due to the evaporation of the electrolyte solution and its loss through the seal, which would result in a decrease in capacitance and an increase in equivalent series resistance (ESR).

When one of the capacitors fails in an MMC, the normal operation of the MMC may be disrupted, and the entire system may fail. Therefore, an effective capacitor condition monitoring method is essential for reliable, efficient, and safe operation of the MMC, which enables the indication of future failure occurrences and preventive maintenance. It is widely applied in reliable or safety-critical power applications as well.

This chapter deals with the state-of-the-art and recent advancements in the capacitor monitoring method and corresponding tolerant control in the MMC. The capacitor equivalent circuit and capacitor parameter characteristics in the MMC are presented in Sections 4.2 and 4.3, respectively. Based on the model established in Section 4.2 and the parameter features discussed in Section 4.3, a brief description of capacitor aging is presented in Section 4.4. The drop of capacitance and the

Modular Multilevel Converters: Control, Fault Detection, and Protection, First Edition.
Fujin Deng, Chengkai Liu, and Zhe Chen.

Published 2023 by John Wiley & Sons, Inc.

rise of ESR are the prominent characteristics to monitor the lapsed capacitor [2–4]. Normally, the capacitor is needed to be replaced with new one when its capacitance drops below the corresponding threshold value or its ESR is over the corresponding threshold value [5]. Most capacitor monitoring methods are implemented based on effective detection of fault characteristics, such as the reference submodules (RSMs) based capacitor condition monitoring method. Since ESR rises and capacitance drops along with the aging process of capacitors, consequently, the capacitance and ESR monitoring methods and their detailed strategies are illustrated in Sections 4.5 and 4.6, respectively. In Section 4.7, the capacitor lifetime monitoring is introduced. The capacitor should be replaced when its capacitance or ESR reaches the lapsed criteria. However, before the replacement of the capacitor, the capacitance declines in the submodule (SM) and the capacitance difference between SMs would cause the circulating current between the upper arm and the lower arm, as well as the unbalanced SM power losses in the same phase unit [6, 7]. Section 4.8 discusses arm current optimal control to suppress circulating current caused by capacitor parameters faults. In Section 4.9, the SM power losses optimal control under capacitor parameters faults is presented to balance the SM power losses distribution in the same arm of the MMC. At last, the summary of this chapter is discussed in Section 4.10.

4.2 Capacitor Equivalent Circuit in MMCs

Typically, the aluminum electrolytic capacitors are the preferred choice for MMCs in some applications, such as micro power grid and motor drives, due to their large capacitance per volume and low cost per capacitance. In this section, based on aluminum electrolytic capacitor structure, its simplified model is introduced and key parameters are discussed.

Aluminum electrolytic capacitor consists of two aluminum electrodes (foils) separated by a porous strip of paper soaked in electrolyte solution, as shown in Figure 4.1a. An oxide layer (Al_2O_3) attaches electrochemically to the surface of the anode and cathode electrodes, serving as the dielectric material of the capacitor [8]. In addition, by etching the surface of the aluminum foil, the effective area of the foil is enlarged. Usually, as shown in Figure 4.1b, electrodes are wound into a cylindrical shape to minimize the volume, enabling combining high capacitance with small size.

The electrolytic capacitors can be modeled as a series combination of a capacitor in parallel with a resistor, a resistor, and an inductor, as shown in Figure 4.2a [9, 10]. The total capacitance C_{AK} is the ideal anode-cathode capacitance, which equals to $\varepsilon S/d$ and is independent of the capacitor current frequency f. ε is the dielectric constant. S is the effective electrode surface area. d is the thickness of

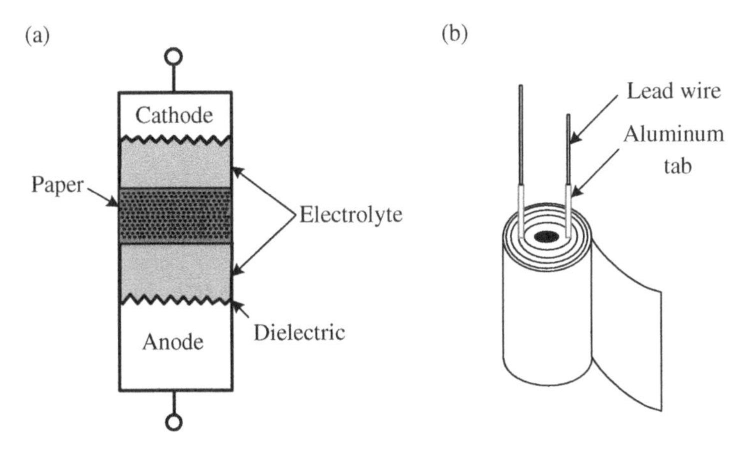

Figure 4.1 Aluminum electrolytic capacitors. (a) Schematic of dielectric structure. (b) Construction.

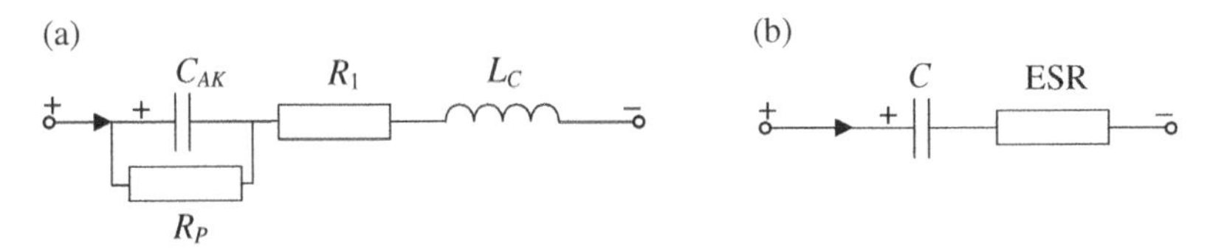

Figure 4.2 (a) Capacitor equivalent circuit. (b) Simplified capacitor equivalent circuit in the MMC.

the dielectric material of the capacitor (Al_2O_3). R_P is the parallel resistance due to leakage current. R_1 in the model represents the sum of the resistances of the electrolyte, separator paper, oxide layers, foils, leads, and connectors, which decreases with the increase of frequency f. L_C is the series equivalent inductance of connections and windings, which is mainly determined by wound structure of the capacitor and independent of frequency. The L_C of the capacitor is normally on the order of nH [10], and the main frequencies of the capacitor current in the MMC are low. The combined effect results in that the inductive reactance of the capacitor in the MMC is very small, which is usually neglected. As a result, the capacitor in the MMC can be simplified as a capacitance in series with an ESR, as shown in Figure 4.2b, which can be expressed as

$$\begin{cases} C = C_{AK} \cdot \left(1 + \dfrac{1}{4\pi^2 \cdot R_p^2 \cdot C_{AK}^2 \cdot f^2}\right) \\ \mathrm{ESR} = R_1 + \dfrac{R_p}{1 + 4\pi^2 \cdot R_p^2 \cdot C_{AK}^2 \cdot f^2} \end{cases} \tag{4.1}$$

In the MMC, the capacitance is normally on the order of mF [11–13], which makes the parallel resistance R_P on the order of MΩ [10]. Accordingly, the capacitance C and ESR can be considered to be the same as C_{AK} and R_1, respectively, which means that C is independent of frequency and ESR decreases with the rise of frequency.

4.3 Capacitor Parameter Characteristics in MMCs

Aluminum electrolytic capacitors can be modeled as the series combination of one capacitance and one ESR. This section describes the characteristics of the capacitor current, impedance, and voltage in the MMC based on the capacitor model established in Section 4.2.

4.3.1 Capacitor Current Characteristics

Figure 4.3 shows the upper arm of phase A in the MMC, where both ESRs R_{au1}–R_{aun} and capacitances C_{au1}–C_{aun} are considered.

Supposing that the second-order harmonic circulating current is suppressed, the upper arm current i_{au} is

$$i_{au} = \frac{i_{dc}}{3} + \frac{1}{2} I_m \sin \omega t \tag{4.2}$$

and the reference y_{au} for the upper arm of phase A is

$$y_{au} = m \sin\left(\omega t + \phi\right) \tag{4.3}$$

where ϕ is the phase angle. Suppose that the capacitor voltages u_{cau1}–u_{caun} shown in Figure 4.3 are kept the same with the voltage-balancing control [14], the equivalent reference (ER) y_{er_aui} for the i-th SM ($i = 1, 2, ..., n$) in the arm of the MMC is mainly related to C_{au1}–C_{aun}, as shown in equation (4.4) [15], because the capacitor's ESR is much smaller than its capacitive reactance at lower frequency [10] and the voltage of the ESR is much smaller than that of the capacitive reactance.

$$y_{er\_aui} = \left(\frac{C_{aui}}{\frac{1}{n}\sum_{i=1}^{n} C_{aui}} - 1 \right) + \frac{C_{aui}}{\frac{1}{n}\sum_{i=1}^{n} C_{aui}} y_{au} \tag{4.4}$$

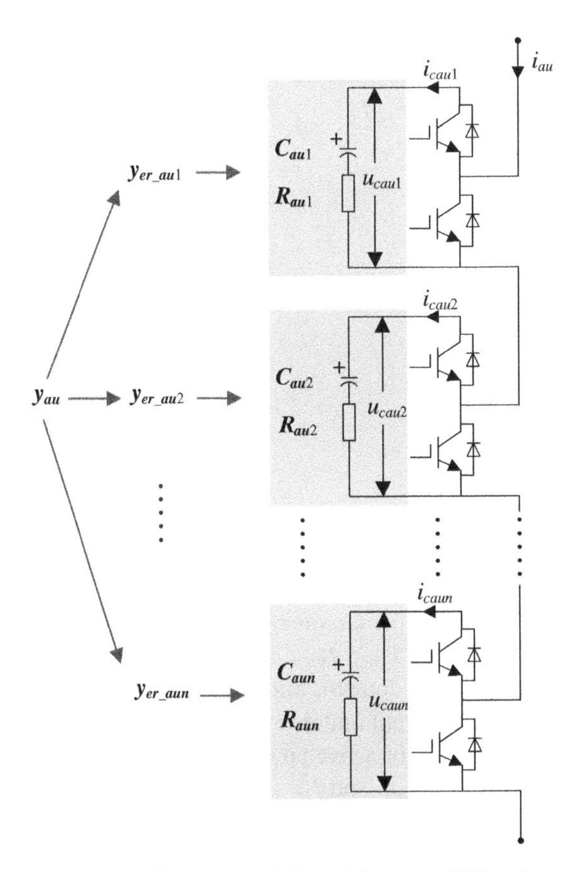

Figure 4.3 Upper arm of phase A based on ESR and capacitance.

According to equations (4.2)–(4.4), the capacitor current i_{caui} of the i-th SM can be expressed as

$$i_{caui} = i_{au}\frac{1 + y_{er\_aui}}{2} = \underbrace{\frac{i_{dc}}{6}\left(\frac{C_{aui}}{\sum_{i=1}^{n} C_{aui}}\right) + \frac{nmC_{aui}I_m}{8\sum_{i=1}^{n} C_{aui}}\cos\phi}_{\text{DC component}} \underbrace{+ I_{1m\_i}\sin(\omega t + \beta_{1\_i})}_{\text{Fundamental component } i_{caui\_1f}} \underbrace{+ I_{2m\_i}\sin(2\omega t + \beta_{2\_i})}_{\text{Second-order harmonic component } i_{caui\_2f}} \tag{4.5}$$

with

$$\begin{cases} I_{1m\_i} = \dfrac{nC_{aui}}{2\sum\limits_{i=1}^{n} C_{aui}} \sqrt{\left(\dfrac{I_m}{2}\right)^2 + \left(\dfrac{mi_{dc}}{3}\right)^2 + \dfrac{mi_{dc} I_m}{3} \cos\phi} \\ \beta_{1\_i} = \arctan\left(\dfrac{2mi_{dc} \sin\phi}{3I_m + 2mI_{dc} \cos\phi}\right) \\ I_{2m\_i} = -\dfrac{1}{8} \dfrac{nmI_m C_{aui}}{\sum\limits_{i=1}^{n} C_{aui}} \\ \beta_{2\_i} = \dfrac{\pi}{2} + \phi \end{cases} \tag{4.6}$$

where I_{1m_i} and β_{1_i} are the amplitude and angle of the fundamental component i_{caui_1f} of the capacitor current in the i-th SM. I_{2m_i} and β_{2_i} are the amplitude and angle of the second-order harmonic component i_{caui_2f} of the capacitor current in the i-th SM. In the steady state, the DC component in the capacitor current is zero. Figure 4.4 shows the amplitudes I_{1m_1}, I_{2m_1}, I_{3m_1}, I_{4m_1}, and I_{5m_1} of the fundamental component and the second-, third-, fourth-, and fifth-order harmonic component, respectively, in the capacitor current under various active power. The I_{1m_1} and I_{2m_1} drop along with the decrease of active power, and the I_{3m_1}, I_{4m_1} and I_{5m_1} are quite small compared to the I_{1m_1} and I_{2m_1} and can be neglected. Actually, the capacitor current mainly consists of the fundamental component and the second-order harmonic component, which can be expressed as

$$i_{caui} = i_{caui\_1f} + i_{caui\_2f} \tag{4.7}$$

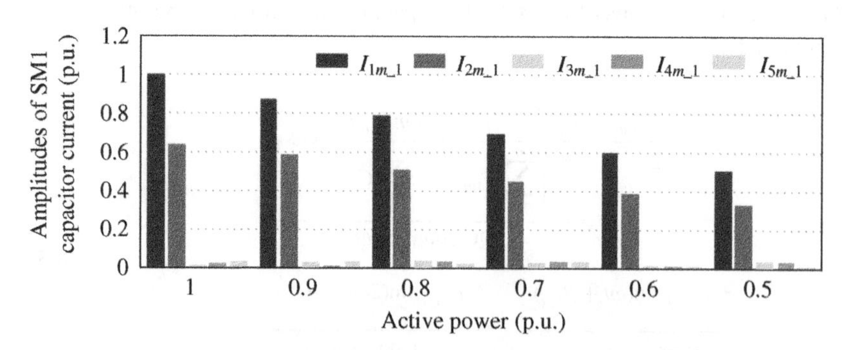

Figure 4.4 Amplitudes of fundamental component and second-, third-, fourth-, and fifth-order harmonic component in capacitor current under various active power.

4.3.2 Capacitor Impedance Characteristics

Since the capacitor current mainly contains the fundamental component i_{caui_1f} and the second-order harmonic component i_{caui_2f}, the capacitor's ESR is mainly expressed by the fundamental component R_{aui_1f} and the second-order component R_{aui_2f} [16]. Figure 4.5 shows the capacitor impedance characteristics in the MMC considering both ESR and C, which can be described as

$$\begin{cases} R_{aui\_1f} = \tan\delta / \omega C_{aui} \\ R_{aui\_2f} = \tan\delta / \omega_2 C_{aui} = R_{aui\_1f}/2 \end{cases} \tag{4.8}$$

$$\begin{cases} Z_{aui\_1f} = R_{aui\_1f} + 1/j\omega C_{aui} \\ Z_{aui\_2f} = R_{aui\_2f} + 1/j\omega_2 C_{aui} = Z_{aui\_1f}/2 \end{cases} \tag{4.9}$$

where Z_{aui_1f} and Z_{aui_2f} are the capacitor impedance corresponding to i_{caui_1f} and i_{caui_2f}. δ is the capacitor loss angle. tan δ is the capacitor dissipation factor, which is nearly a constant at low frequency [2, 10] such as fundamental frequency and double-line frequency. $\omega_2 = 2\omega$.

4.3.3 Capacitor Voltage Characteristics

In Figure 4.5, the capacitor voltage ripple u_{caui_1f} and u_{caui_2f} caused by i_{caui_1f} and i_{caui_2f} can be obtained as

$$\begin{cases} u_{caui\_1f} = i_{caui\_1f} \cdot Z_{aui\_1f} = u_{caui\_c1f} + u_{caui\_r1f} \\ u_{caui\_2f} = i_{caui\_2f} \cdot Z_{aui\_2f} = u_{caui\_c2f} + u_{caui\_r2f} \end{cases} \tag{4.10}$$

with

$$\begin{cases} u_{caui\_r1f} = i_{caui\_1f} \cdot R_{aui\_1f} \\ u_{caui\_c1f} = i_{caui\_1f} / j\omega C_{aui} \end{cases} \tag{4.11}$$

$$\begin{cases} u_{caui\_r2f} = i_{caui\_2f} \cdot R_{aui\_2f} \\ u_{caui\_c2f} = i_{caui\_2f} / j\omega_2 C_{aui} \end{cases} \tag{4.12}$$

where u_{caui_c1f} and u_{caui_r1f} are voltage ripples caused by C_{aui}, i_{caui_1f} and R_{aui_1f}, i_{caui_1f}, respectively. u_{caui_c2f} and u_{caui_r2f} are voltage ripples caused by C_{aui}, i_{caui_2f} and R_{aui_2f}, i_{caui_2f}, respectively.

According to equations (4.5)–(4.12), the capacitor voltage ripple Δu_{caui} in the i-th SM can be expressed as

$$\Delta u_{caui} = \underbrace{U_{1m\_i}\sin\left(\omega t + \alpha_{1\_i}\right)}_{\substack{\text{Fundamental} \\ \text{component } u_{caui\_1f}}} + \underbrace{U_{2m\_i}\sin\left(2\omega t + \alpha_{2\_i}\right)}_{\substack{\text{Second-order harmonic} \\ \text{component } u_{caui\_2f}}} \tag{4.13}$$

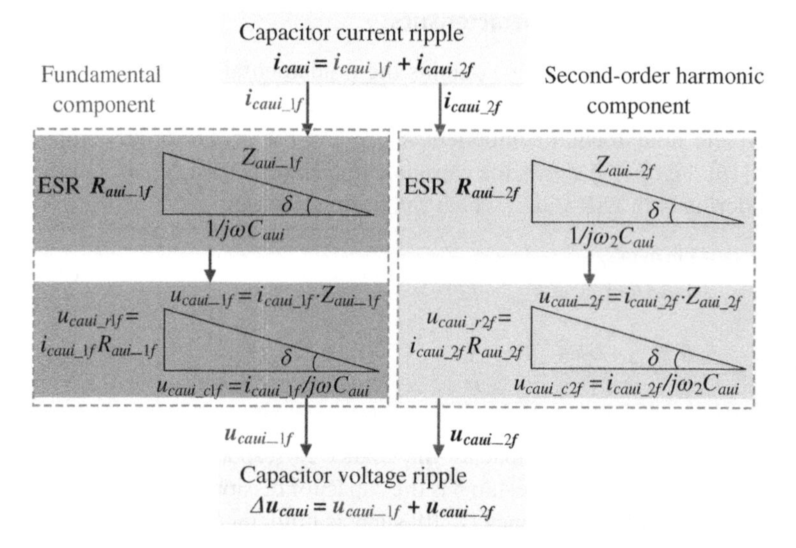

Figure 4.5 Capacitor impedance and voltage characteristics in the MMC.

with

$$\begin{cases} U_{1m\_i} = I_{1m\_i}\sqrt{R_{aui\_1f}^2 + \left(\dfrac{1}{\omega C_{aui}}\right)^2} \\ \alpha_{1\_i} = \beta_{1\_i} - \arctan\left(\dfrac{1}{\omega C_{aui} R_{aui\_1f}}\right) \\ U_{2m\_i} = \dfrac{1}{2} I_{2m\_i}\sqrt{R_{aui\_1f}^2 + \left(\dfrac{1}{\omega C_{aui}}\right)^2} \\ \alpha_{2\_i} = \beta_{2\_i} - \arctan\left(\dfrac{1}{\omega C_{aui} R_{aui\_1f}}\right) \end{cases} \tag{4.14}$$

where U_{1m_i} and α_{1_i} are the amplitude and angle of the fundamental component u_{caui_1f} in capacitor voltage. U_{2m_i} and α_{2_i} are the amplitude and angle of the second-order harmonic component u_{caui_2f} in capacitor voltage.

The capacitor voltage ripple in the i-th SM is composed of the fundamental component u_{caui_1f} and the second-order component u_{caui_2f}. Owing to this, the arm capacitor voltages are kept the same by voltage-balancing control [14] and the amplitude U_{1m_1}–U_{1m_n} in arm SM capacitors is kept the same as U_{1m}; the U_{2m_1}–U_{2m_n} in arm SM capacitors would be the same as U_{2m} as well.

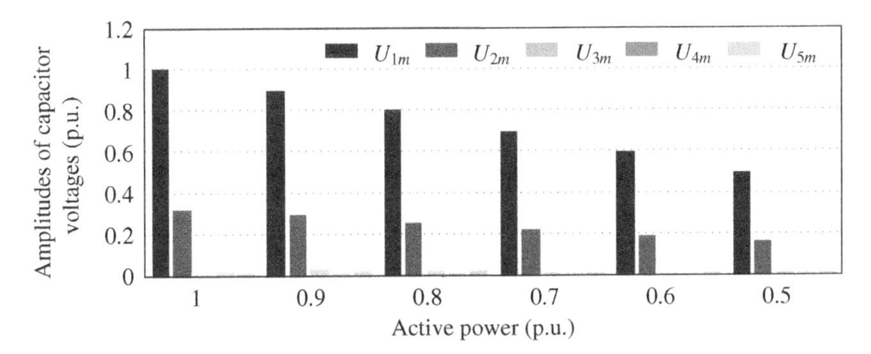

Figure 4.6 Amplitudes of fundamental component and second-, third-, fourth-, and fifth-order harmonic component in capacitor voltage under various active power.

Figure 4.6 shows the amplitudes of the fundamental component and the second-, third-, fourth-, and fifth-order harmonic components in the capacitor voltage under various active power. The U_{1m} and U_{2m} decline along with the decrease of active power; the U_{3m}, U_{4m}, and U_{5m} are quite small compared with U_{1m} and U_{2m} and can be neglected.

Figure 4.7a shows the amplitudes U_{1m_c} and U_{2m_c} of u_{cau1_c1f} and u_{cau1_c2f}, respectively, under different output active power of the MMC. Figure 4.7b shows the amplitudes U_{1m_r} and U_{2m_r} of u_{cau1_r1f} and u_{cau1_r2f}, respectively, under different output active power of the MMC. It is vividly shown that U_{1m_c}, U_{2m_c}, U_{1m_r}, and U_{2m_r} decline along with the decrease of the active power of the MMC. Besides, U_{1m_r} and U_{2m_r} are quite small compared with U_{1m_c} and U_{2m_c} and can be neglected. In actual capacitor parameter analysis of the MMC, u_{cau1_c1f} along with u_{cau1_c2f} are widely accepted and considered as the ripple voltage of the capacitor. In addition, the U_{1m_c} and U_{1m_r} are much bigger than the U_{2m_c} and U_{2m_r}, respectively.

4.4 Capacitor Aging

Aluminum electrolytic capacitors are the desirable choice for power converter applications such as the MMC due to large capacitance and low cost. Although electrolytic capacitors have many attractive features in various applications, they have undesirable properties such as finite lifetime and high failure rate due to wear-out degradation failure. As the aluminum electrolytic capacitors age, the capacitance would drop and ESR would rise, which can further reflect the health status of electrolytic capacitors. It has been reported that capacitors are one of the weakest components in power electronic devices [6], as shown in Figure 4.8, where 30% of the power electronic device failures were due to electrolytic

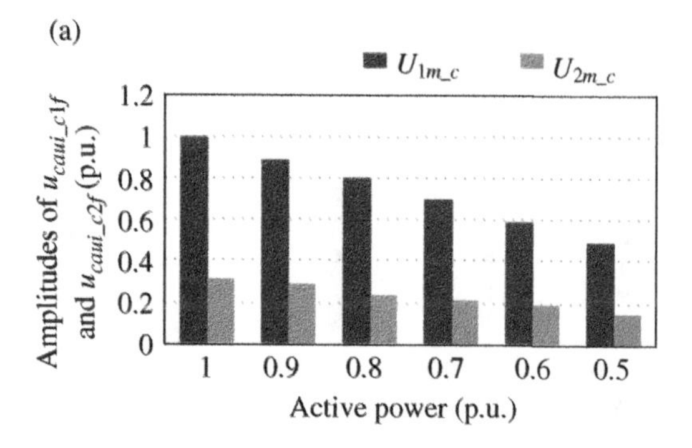

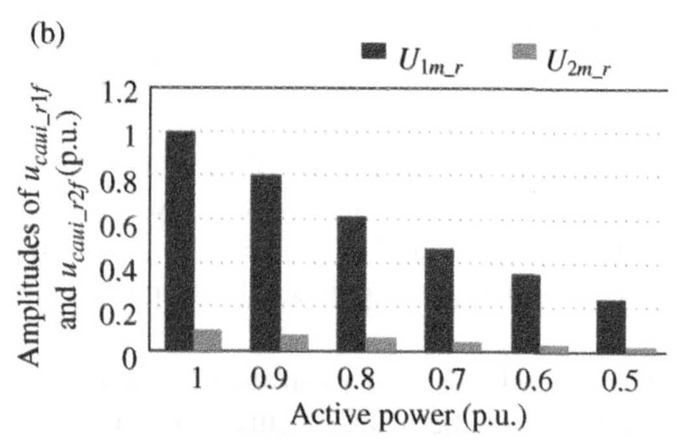

Figure 4.7 (a) U_{1m_c} and U_{2m_c}. (b) U_{1m_r} and U_{2m_r}.

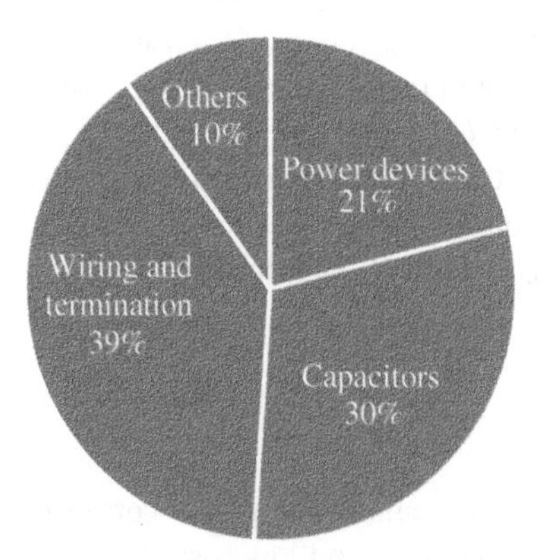

Figure 4.8 Failure rates of different power electronic devices.

capacitor breakdown. This section mainly discusses its degradation mechanism and end-of-life criterion.

Electrolytic capacitor degradation mechanism is a combined effect of thermal, electrical, mechanical, and environmental issues. The primary wear-out failure mechanism is the evaporation of the electrolyte solution and its loss through the end seal. The evaporation of electrolyte is accelerated with temperature rise during operation due to ripple currents, ambient-temperature increase, overvoltage, etc. As the electrolyte solution dries up, the effective contact area between the electrodes decreases because the amount of electrolyte decreases. This results in a decrease in capacitance and increase in ESR, which further increases the temperature. This further accelerates the degradation process and drastically changes the capacitor characteristics, namely capacitor aging, which eventually drifts out of specification and is considered to have failed.

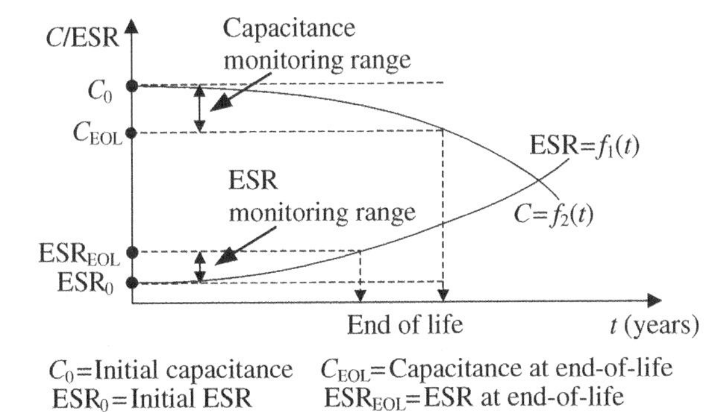

Figure 4.9 Capacitor deterioration curve.

Figure 4.9 shows the deterioration curve of capacitors [3], where the capacitance decreases and ESR increases at the same time during operation due to aging effect. Once the capacitance reaches the value C_{EOL} or the ESR reaches the value $\mathrm{ESR}_{\mathrm{EOL}}$, the capacitor can be considered to have been lapsed and must be replaced for the normal operation of power electronic applications including the MMCs. The range between the initial value of capacitance/ESR and the aged value is the condition monitoring range, as shown in Figure 4.9.

Based on deterioration curve in Figure 4.9, an end-of-life or threshold criterion is needed before further deciding the health condition of the capacitors. The criteria of the end-of-life of the aluminum electrolytic capacitors is considered based on the following two aspects [7, 17]:

- The capacitor degradation rate becomes considerably faster (e.g. dC/dt, $d\mathrm{ESR}/dt$) after the capacitance or ESR reaches the specified threshold criteria.
- The power electronic conversion systems may not function appropriately once the capacitance drops or the ESR rises to a specified level.

Normally, capacitance and ESR are two typical indicators of the degradation of capacitors, based on which threshold criterion of lapsed capacitor is implemented. For electrolytic capacitors, the widely accepted end-of-life criterion is 20% capacitance reduction or double of the ESR.

4.5 Capacitance Monitoring

Based on the end-of-life criterion in Section 4.4, the electrolytic capacitors are needed to be replaced with a brand new one once its capacitance drops below 80% of the rated value or its ESR rises over two times of the rated value to avoid operating under deteriorated condition. Accordingly, the condition monitoring scheme should be implemented until the capacitor reach the end-of-life criterion. Capacitance drop is one of the prominent features in capacitor aging, which can indicate capacitor's health condition. This section presents several capacitor monitoring schemes including capacitor voltage and current based strategy, arm average capacitance based strategy, RSM-based strategy, and sorting-based monitoring strategy. Temperature effect of capacitors is discussed at last to illustrate the influence of temperature to capacitance.

4.5.1 Capacitor Voltage and Current Based Monitoring Strategy

Capacitor voltage and current based monitoring strategy is widely used in capacitor monitoring of power electronic converters. For the MMC, the capacitance can be monitored based on the capacitor voltage, arm current, and switching function. Equation (4.15) shows the estimation of the capacitance C_{aui} in the i-th SM of the MMC in upper arm of phase A, which can be directly estimated based on the arm current i_{au}, capacitor voltage u_{caui}, and the switching function S_{aui}.

$$C_{aui} = \frac{1}{u_{caui}} \int S_{aui} \cdot i_{au} dt \tag{4.15}$$

Typically, equation (4.15) is a most widely used formula to estimate capacitance and based on which many voltage-ripple based methods are introduced [15–19]. Normally, the error of the capacitance estimation method is below 1% [18].

4.5.2 Arm Average Capacitance Based Monitoring Method

4.5.2.1 Equivalent Arm Structure

Arm average capacitance based monitoring strategy is a capacitor monitoring method with simple algorithm for the MMC. It reveals the relationship between the average capacitance and the capacitance in each SM of the arm, which can be used for capacitance estimation. Figure 4.10 shows the n series-connected SMs in the upper arm of phase A, where the capacitances C_{au1}–C_{aun} in each SM are uncertain. With the voltage-balancing control, all the capacitor voltages in the upper arm of phase A can be kept the same as

$$u_{cau1} = u_{cau2} = \cdots\cdots = u_{caun} = u_{cau} \tag{4.16}$$

With the PD-SPWM strategy and voltage-balancing control, the synthesized multilevel arm voltage u_{au} for the upper arm of phase A can be expressed as

$$u_{au} = n \cdot u_{cau} \cdot \frac{1 + y_{au}}{2} \tag{4.17}$$

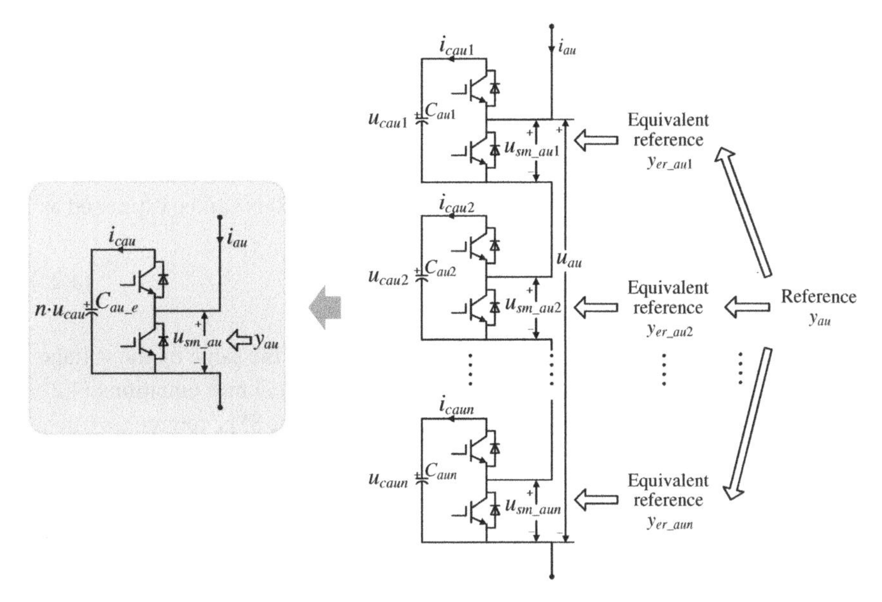

Figure 4.10 Equivalent structure of n series-connected SMs in the upper arm of phase A.

The n series-connected SMs in the upper arm of phase A can be equivalent to one SM, as shown in Figure 4.10, where the voltage relationship can be expressed as

$$\frac{dnu_{cau}}{dt} = \frac{i_{cau}}{C_{au\_e}} = \frac{i_{au}}{C_{au\_e}} \cdot \frac{1 + y_{au}}{2} \tag{4.18}$$

where C_{au_e} is the equivalent arm capacitor.

4.5.2.2 Capacitor Monitoring Method

The voltage and current in the i-th SM ($i = 1, 2 \dots n$) can be expressed as

$$\begin{cases} u_{sm\_aui} = u_{caui} \cdot \dfrac{1 + y_{er\_aui}}{2} \\ \dfrac{du_{caui}}{dt} = \dfrac{i_{caui}}{C_{aui}} = \dfrac{i_{au}}{C_{aui}} \cdot \dfrac{1 + y_{er\_aui}}{2} \end{cases} \tag{4.19}$$

where y_{er_aui} is the equivalent reference for the i-th SM. In addition, the output voltage and capacitor current of each SM can be described as

$$\begin{cases} u_{sm\_aui} = S_{aui} \cdot u_{caui} \\ i_{caui} = S_{aui} \cdot i_{au} \end{cases} \tag{4.20}$$

With the Fourier transform, neglecting the harmonics, the i-th SM's switching function S_{aui} can be expressed using Fourier series as

$$S_{aui} \approx \frac{y_{er\_aui} + 1}{2} \tag{4.21}$$

The total output voltage u_{au} of the n series-connected SMs can be expressed as

$$u_{au} = \sum_{i=1}^{n} u_{sm\_aui} = \sum_{i=1}^{n} \left(S_{aui} \cdot u_{caui} \right) \tag{4.22}$$

Suppose that the capacitor voltage in each arm is kept the same by the voltage-balancing control, substituting equations (4.16) and (4.17) into equations (4.21) and (4.22), the relationship between arm reference and the SMs' references can be obtained as

$$y_{au} = \frac{1}{n} \sum_{i=1}^{n} y_{er\_aui} \tag{4.23}$$

Combining equations (4.16), (4.18), and (4.19), there is

$$\frac{1 + y_{er\_au1}}{C_{au1}} = \cdots\cdots = \frac{1 + y_{er\_aun}}{C_{aun}} = \frac{1 + y_{au}}{nC_{au\_e}} \tag{4.24}$$

and

$$\begin{cases} y_{er\_au1} = \dfrac{C_{au1}}{C_{aui}} \left(1 + y_{er\_aui}\right) - 1 = \dfrac{C_{au1}}{nC_{au\_e}} \left(1 + y_{au}\right) - 1 \\ y_{er\_au2} = \dfrac{C_{au2}}{C_{aui}} \left(1 + y_{er\_aui}\right) - 1 = \dfrac{C_{au2}}{nC_{au\_e}} \left(1 + y_{au}\right) - 1 \\ \qquad \cdots\cdots \\ y_{er\_aun} = \dfrac{C_{aun}}{C_{aui}} \left(1 + y_{er\_aui}\right) - 1 = \dfrac{C_{aun}}{nC_{au\_e}} \left(1 + y_{au}\right) - 1 \end{cases} \tag{4.25}$$

Substituting equation (4.25) into equation (4.24), there is

$$\begin{cases} y_{er\_aui} = y_{aui} + \overline{y}_{aui} \\ y_{aui} = y_{au} \dfrac{C_{aui}}{C_{ave}} \\ \overline{y}_{aui} = \dfrac{C_{aui}}{C_{ave}} - 1 \end{cases} \tag{4.26}$$

with

$$C_{ave} = nC_{au\_e} = \frac{1}{n}\sum_{i=1}^{n} C_{aui} \tag{4.27}$$

where $\tilde{y}_{aui}$ and $\overline{y}_{aui}$ are the AC component and DC component in y_{er_aui}, respectively. C_{ave} is the arm average capacitance.

The capacitance C_{aui} in the i-th SM can be estimated based on equation (4.26) as

$$C_{aui} = C_{ave}\left(1 + \overline{y}_{aui}\right) \tag{4.28}$$

The capacitance C_{aui} in the i-th SM can be estimated with C_{ave} and $\overline{y}_{aui}$. The DC component $\overline{y}_{aui}$ in y_{er_aui} can be easily obtained as shown in Figure 4.11, the switching function S_{aui} of the i-th SM can be used to calculate the ER y_{er_aui} for the i-th SM by equation (4.21). The y_{er_aui} is integrated and passed through a notch filter tuned at the fundamental frequency and then divided by the integration period Δt for $\overline{y}_{aui}$.

Figure 4.12 shows the arm average capacitance based capacitor monitoring method, which is based on two parts. One part is the arm average capacitance C_{ave}. The other part is the DC component in the ER for each SM in the arm. According to equations (4.18) and (4.27), the C_{ave} can be calculated based on capacitor voltage, arm current, and arm reference y_{au}. The DC components $\overline{y}_{au1} \sim \overline{y}_{aun}$ of y_{er_au1}–y_{er_aun} in the arm can be obtained from Figure 4.11 based on the switching function S_{au1}–S_{aun} for each SM in the arm. Finally, the C_{au1}–C_{aun} in the arm can be obtained based on equation (4.28). The capacitor monitoring for the other arms of the MMC can be realized with the same method as that for the upper arm of phase A, as shown in Figure 4.12, which is not repeated here.

In order to reduce the effect of measurement noises in the actual MMC system and improve the capacitor monitoring accuracy, several sets capacitances are estimated such as ten sets. Removing the maximum and the minimum among the estimated capacitances, the average values of the rest capacitances are calculated as the capacitance monitoring results.

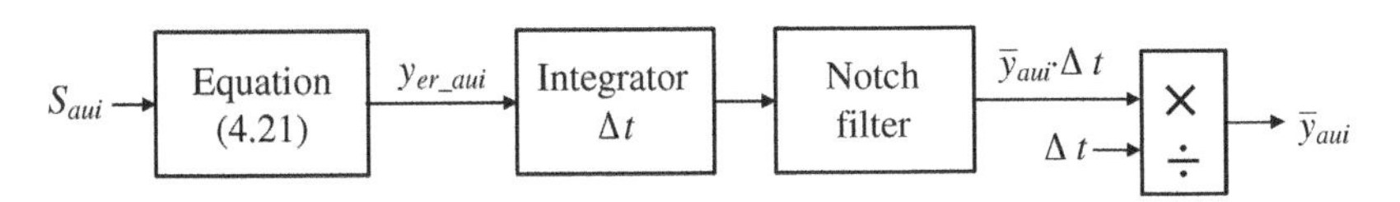

Figure 4.11 Calculation for $\overline{y}_{aui}$.

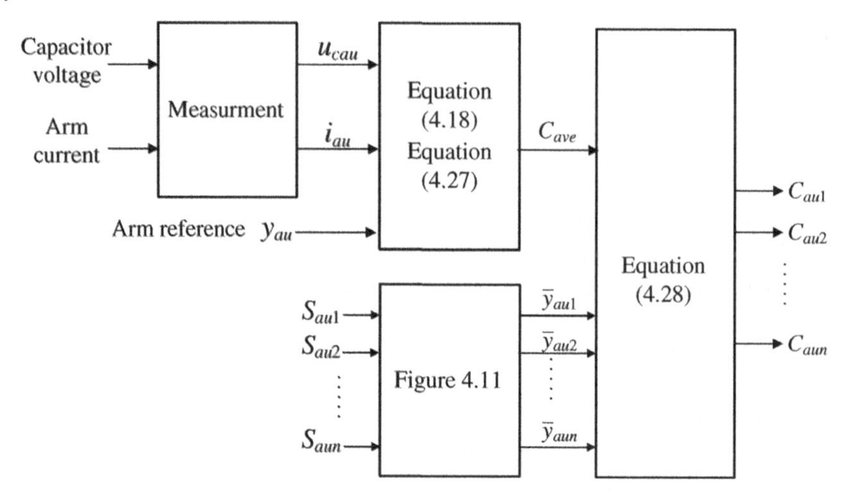

Figure 4.12 Arm average capacitance based capacitor monitoring method for the upper arm of phase A.

Case Study 4.1 Analysis of Arm Average Capacitance Based Monitoring Strategy

Objective: In this case study, the working principle and the performance of the arm average capacitance based capacitor condition monitoring strategy are studied through professional time-domain simulation software PSCAD/EMTDC. A three-phase MMC system equipped with the DC voltage V_{dc} and the RL load are simulated for this study, which is shown in Figure 4.13. The circulating current is eliminated with the circulating current suppression control introduced in [20].

Parameters: The simulation system parameters are shown in Tables 4.1 and 4.2.

Simulation results and analysis:

Figure 4.14 shows the performance of the MMC with the arm average capacitance based capacitor monitoring method, where the capacitances in the upper arm of phase A are as listed in Table 4.2. Figures 4.14a,b show the MMC voltage u_{ab}, u_{bc}, u_{ca} and current i_a, i_b, i_c, respectively. The upper and lower arm currents i_{au} and i_{al} are shown in Figure 4.14c, where the circulating current is eliminated. The capacitor voltage is kept balanced, as shown in Figure 4.14d. Figure 4.14e shows the reference for the upper arm of phase A, which is a sinusoidal wave. The estimated capacitances C_{au1}–C_{au6} are shown in Figure 4.14f. With the comparison of the estimated and actual capacitance, the estimated error can be calculated, which is very small and less than 0.17%, as shown in Figure 4.14g.

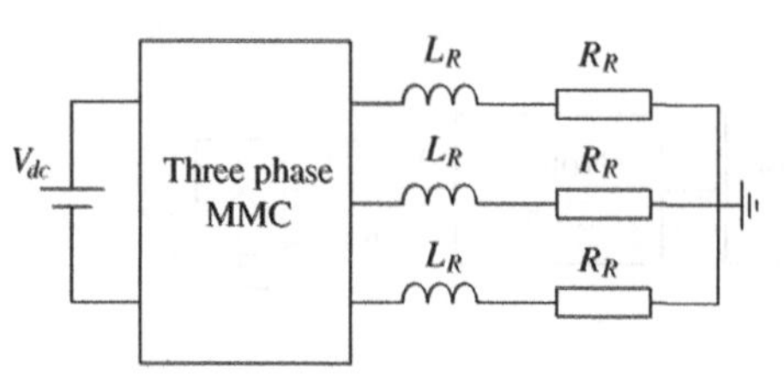

Figure 4.13 Block diagram of the simulation system.

Table 4.1 Simulation system parameters.

Parameters	Value
DC-link voltage V_{dc} (kV)	6
Load frequency (Hz)	50
Number of SMs per arm n	6
Nominal SM capacitance C (mF)	9
Inductance L_s (mH)	5
Load inductance L_R (mH)	3
Load resistance R_R (Ω)	2.5
Carrier frequency fs (kHz)	3

Table 4.2 Capacitance used in upper arm of phase A.

Capacitance (mF)	C_{au1}	C_{au2}	C_{au3}	C_{au4}	C_{au5}	C_{au6}
	8	8.2	8.4	8.6	8.8	9

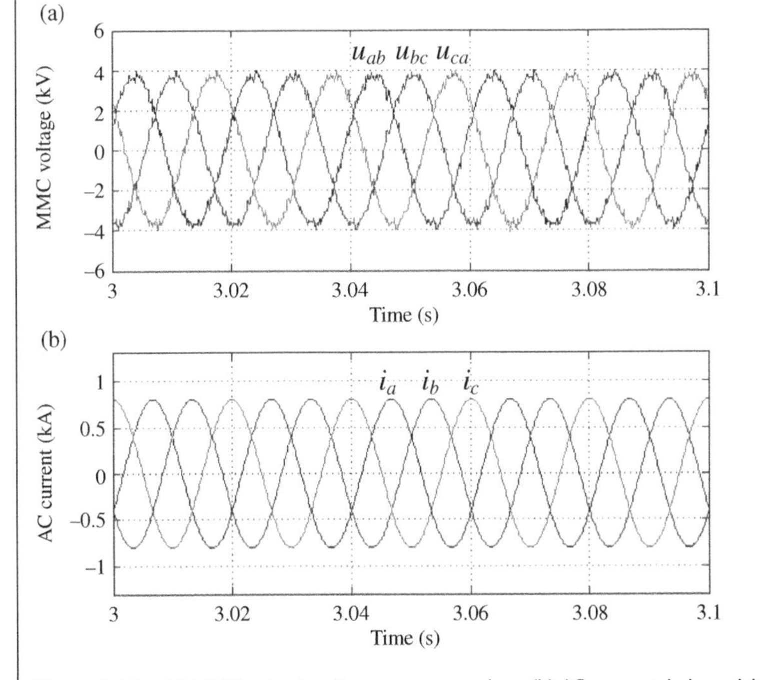

Figure 4.14 (a) MMC output voltage u_{ab}, u_{bc}, and u_{ca}. (b) AC current i_a, i_b, and i_c. (c) Current i_{au}, i_{al}, and i_{diff_a}. (d) Capacitor voltage. (e) Reference for upper arm of phase A. (f) Estimated capacitance $C_{au1}-C_{au6}$. (g) Capacitance estimation error.

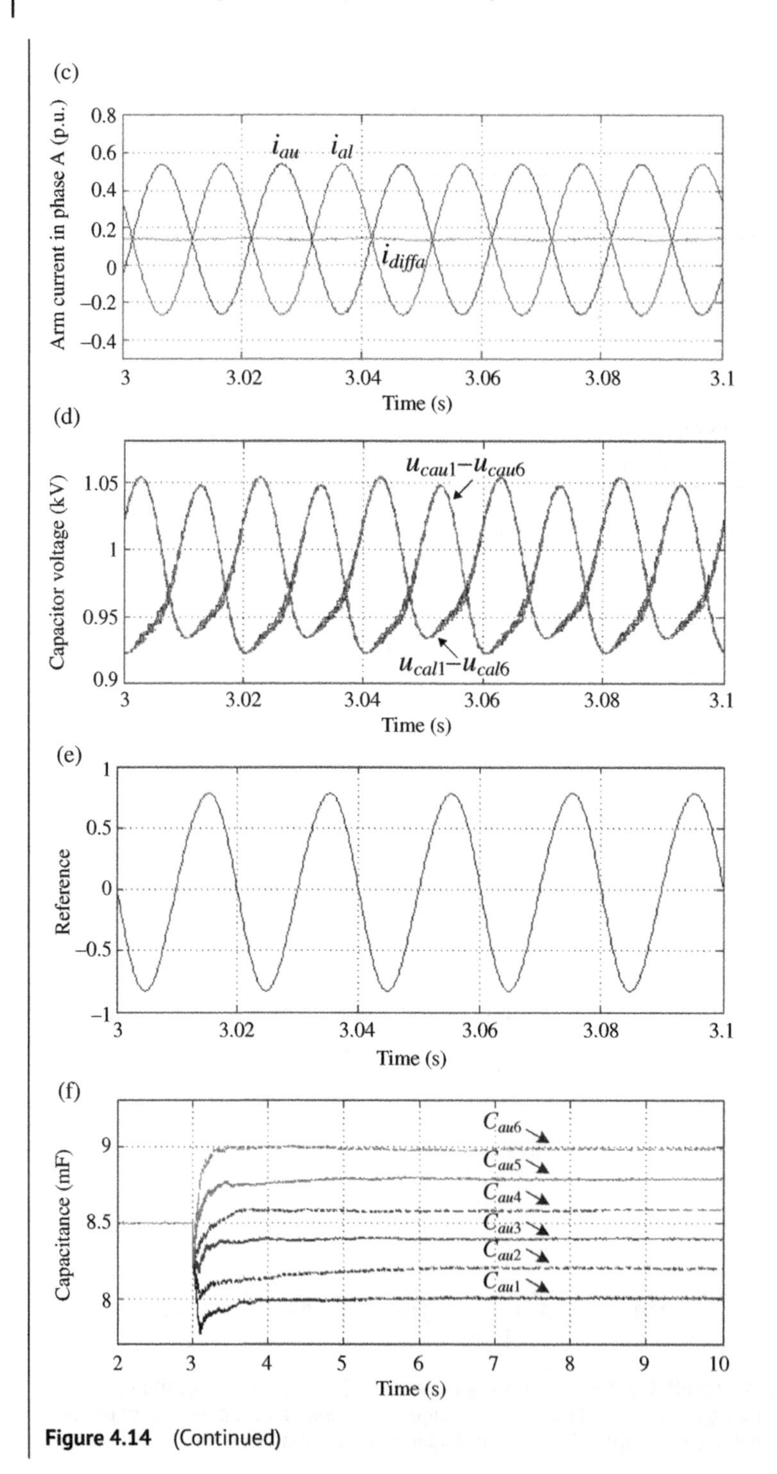

Figure 4.14 (Continued)

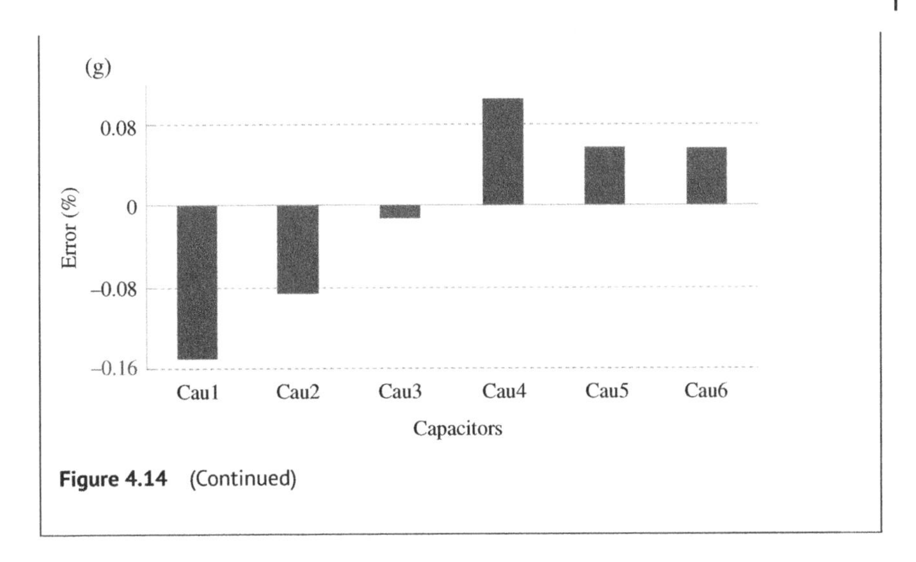

Figure 4.14 (Continued)

4.5.3 Reference SM based Monitoring Method

RSM-based capacitor monitoring method is another simplified capacitor monitoring method for the MMC, where some SMs are selected as the RSM and the capacitances in the other SMs of the same arm can be estimated based on the capacitance in the RSM. The RSM-based capacitor monitoring method does not rely on the information of all capacitor's current for the complicated integral computation, which effectively simplifies the capacitor monitoring algorithm and saves computing resources.

4.5.3.1 Principle of the RSM-Based Capacitor Monitoring Strategy

Figure 4.15 shows the RSM-based capacitor monitoring method. Suppose that the capacitance in the RSM has been estimated, in order to monitor the capacitance C_{aui} in the SM$_i$, $i \in (1, 2, \ldots n)$, the switching function S_{rsm} for the RSM must be the same to S_{aui} for the SM$_i$, as $S_{rsm} = S_{aui}$. In this situation, the RSM capacitor voltage u_{rsm} and the SM$_i$ capacitor voltage u_{caui} can be expressed as

$$\begin{cases} u_{rsm} = \dfrac{1}{C_{rsm}} \int S_{rsm} \cdot i_{au} dt \\ u_{caui} = \dfrac{1}{C_{aui}} \int S_{rsm} \cdot i_{au} dt \end{cases} \tag{4.29}$$

According to equation (4.29), the relationship between the capacitance C_{rsm} and C_{aui} can be obtained as

$$C_{aui} = C_{rsm} \frac{\Delta u_{rsm}}{\Delta u_{caui}} \tag{4.30}$$

where Δu_{rsm} and Δu_{caui} are the peak-to-peak values of the voltages u_{rsm} and u_{caui}, respectively, as shown in Figure 4.15. According to equation (4.30), the capacitance C_{aui} can be estimated in one fundamental period based on the measured RSM capacitor voltage ripple Δu_{rsm}, the measured SM$_i$ capacitor voltage ripple Δu_{caui}, and the estimated RSM capacitance C_{rsm}, as shown in Figure 4.15.

4.5.3.2 Capacitor Monitoring-Based Voltage-Balancing Control

To realize the RSM-based capacitor monitoring, Figure 4.16a shows a capacitor monitoring-based voltage-balancing control (CM-VBC). The CM-VBC not only balances the capacitor voltages in the arm to provide a stable and controllable operation of the MMC but also ensures that the S_{rsm} for the RSM follows the S_{aui} for the monitoring SM$_i$, so as to implement the RSM-based capacitor monitoring.

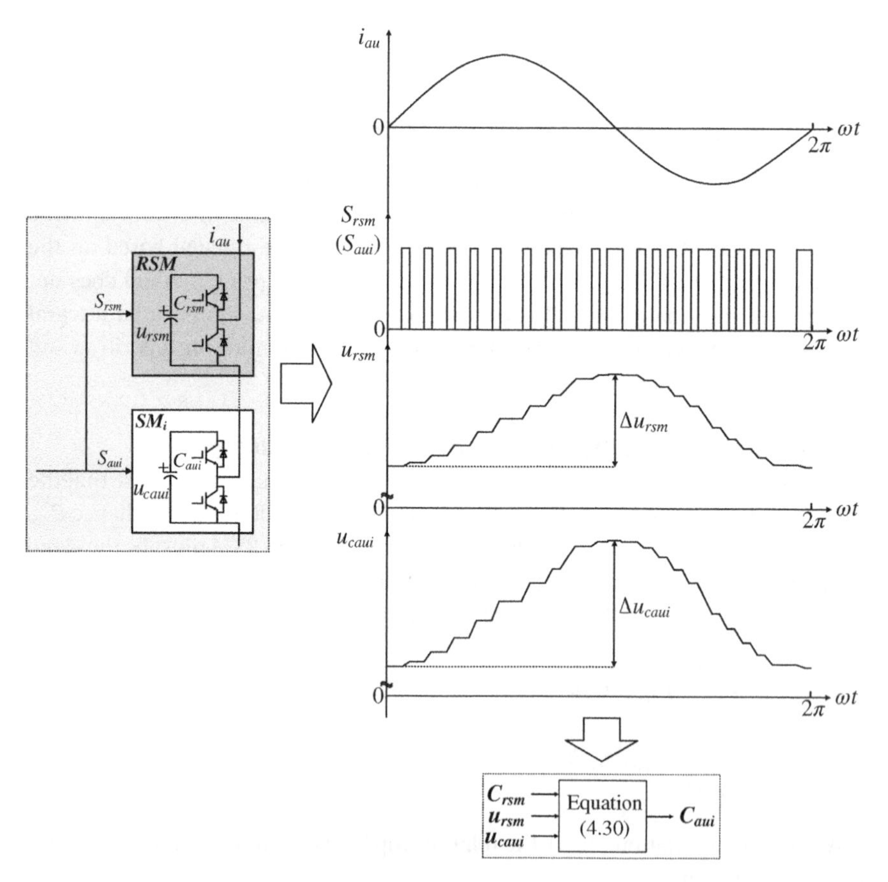

Figure 4.15 RSM-based capacitor monitoring method.

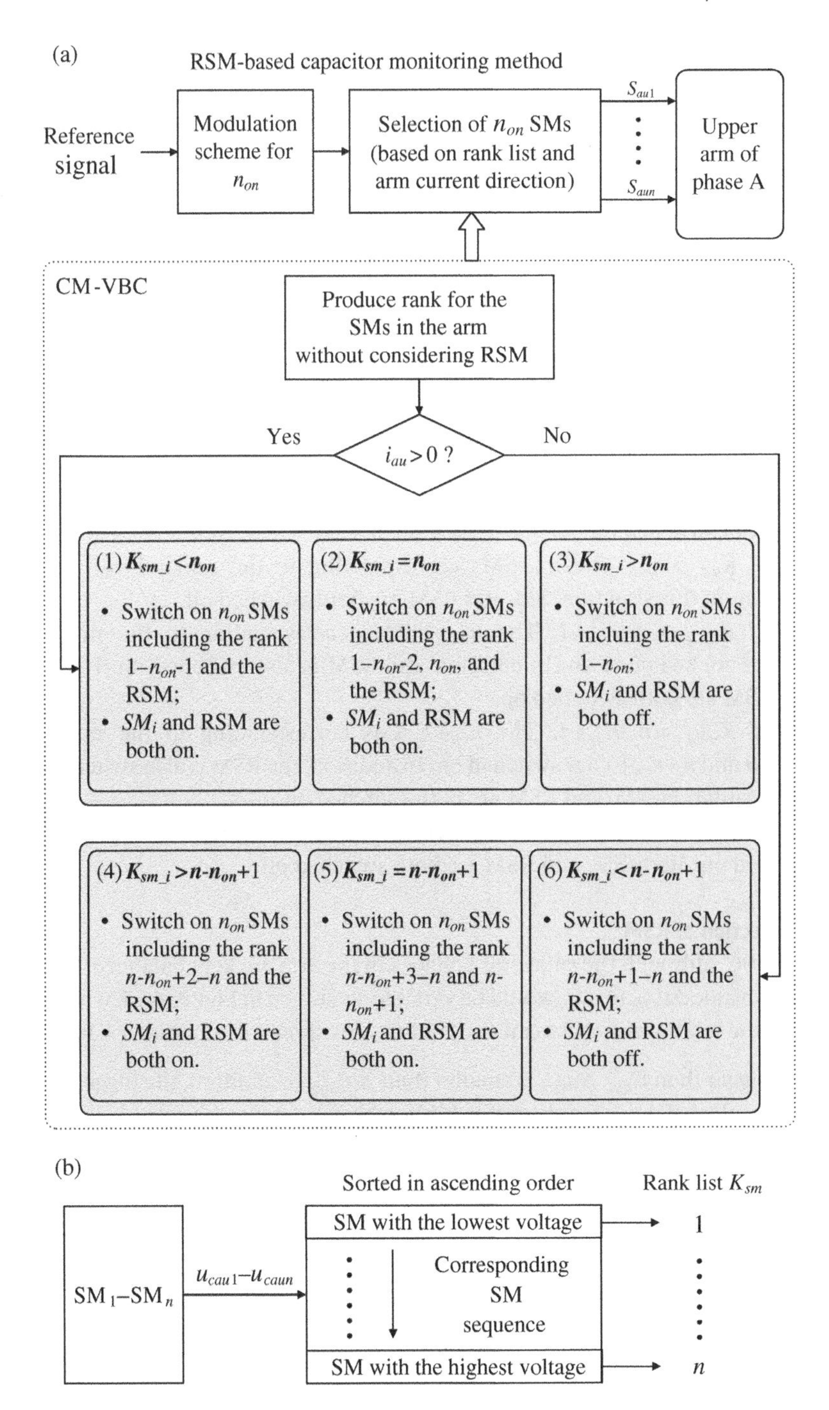

Figure 4.16 (a) CM-VBC for RSM-based capacitor monitoring. (b) Rank list production.

In Figure 4.16a, under the modulation schemes, through the comparison between the reference signal and the carriers, the number n_{on} of the on-state SMs can be obtained. The capacitor voltages u_{cau1}–u_{caun} in the arm are measured by voltage sensors. The n–1 SMs in the arm except RSM are sorted based on their capacitor voltages in the ascending order, so as to produce the rank K_{sm} for the corresponding SM, where the K_{sm} for the SM with the lowest voltage is 1 and the K_{sm} for the SM with the highest voltage is n, as shown in Figure 4.16b. Suppose that the SM_i corresponding to the rank K_{sm_i} is to be monitored, the selection of n_{on} on-state SMs will be decided based on the arm current i_{au} and the ranks of the n–1 SMs as follows:

- $i_{au}>0$ *and* $K_{sm\_i}<n_{on}$: The $n_{on}-1$ SMs corresponding to the ranks 1–$n_{on}-1$ are switched on. In addition, the RSM is also switched on. In this situation, SM_i and RSM are both switched on.
- $i_{au}>0$ *and* $K_{sm\_i}=n_{on}$: The $n_{on}-1$ SMs corresponding to the ranks 1–$n_{on}-2$ and n_{on} are switched on. In addition, the RSM is also switched on. Here, SM_i and RSM are both switched on.
- $i_{au}>0$ *and* $K_{sm\_i}>n_{on}$: The n_{on} SMs corresponding to the ranks 1–n_{on} are switched on. In this situation, SM_i and RSM are both switched off.
- $i_{au}<0$ *and* $K_{sm\_i}>n-n_{on}+1$: The $n_{on}-1$ SMs corresponding to the ranks $n-n_{on}+2$–n are switched on. In addition, the RSM is also switched on. Here, SM_i and RSM are both switched on.
- $i_{au}<0$ *and* $K_{sm\_i}=n-n_{on}+1$: The $n_{on}-1$ SMs corresponding to the ranks $n-n_{on}+3$–n and $n-n_{on}+1$ are switched on. In addition, the RSM is also switched on. In this situation, SM_i and RSM are both switched on.
- $i_{au}<0$ *and* $K_{sm\_i}<n-n_{on}+1$: The n_{on} SMs corresponding to the ranks $n-n_{on}+1$–n are switched on. Here, SM_i and RSM are both switched off.

4.5.3.3 Selection of RSM

In the CM-VBC, although the capacitor voltages in the arm are kept balanced, the peak-to-peak value Δu_{rsm} in u_{rsm} would be variable, as shown in Figure 4.15, which depends on the relationship between C_{rsm} and C_{aui}, as shown in equation (4.30).

- If C_{rsm} is bigger than C_{aui}, Δu_{rsm} is smaller than Δu_{caui}. In addition, the bigger of C_{rsm}, the smaller of Δu_{rsm}, as shown in Table 4.3.

Table 4.3 Voltage ripple in RSM.

RSM's capacitance C_{rsm}	RSM's voltage ripple Δu_{rsm}
$>C_{aui}$	$<\Delta u_{caui}$
$<C_{aui}$	$>\Delta u_{caui}$

- If C_{rsm} is smaller than C_{aui}, Δu_{rsm} is bigger than Δu_{caui}. The smaller of C_{rsm}, the bigger of Δu_{rsm}, as shown in Table 4.3.

The RSM should be selected with the SM, whose capacitance is the biggest one among C_{au1}–C_{aun}. It ensures that the Δu_{rsm} is always not bigger than the peak-to-peak value of the other SMs' capacitor voltages, which reduces the capacitor voltage ripple in the RSM.

Suppose that the electrolytic capacitor is used in the MMC and the capacitor needs to be replaced when its capacitance drops to 80% of the rated value, the RSM's Δu_{rsm} would be reduced by 0.2 p.u. at most. On the other hand, the capacitor voltages in the arm are kept balanced with the CM-VBC. Although the peak-to-peak value Δu_{rsm} of the RSM's capacitor voltage is less than that of the other SMs' capacitor voltages during the capacitor monitoring period, the capacitor voltage's peak-to-peak value is far less than the capacitor voltage, and therefore the impact of the variable RSM's Δu_{rsm} on the MMC performance during the capacitor monitoring period can be omitted.

4.5.3.4 Capacitor Monitoring Strategy

Figure 4.17 shows the capacitor monitoring strategy, which estimates the SM's capacitances in the sequence from SM_1 to SM_n round. First, the SM_1 is selected as the RSM, and the RSM capacitance is estimated based on the conventional method as equation (4.15). Based on the RSM, the capacitance in the SM_2 is estimated with the RSM-based monitoring method as shown in Figure 4.18a.

In Figure 4.18a, the selection signal (SS) is initially "0" and the conventional capacitor voltage-balancing control (C-VBC) is enabled, where u_{cau1}–u_{caun} are kept the same with each other and the MMC works in normal operation mode. The principle of C-VBC is shown in Figure 4.18b. In details, under the modulation schemes, through the comparison between the reference signal and the carriers, the number n_{on} of the on-state SMs can be obtained. The switching function for each SM can be decided based on n_{on}, the rank list K_{sm}, and the arm current. If the arm current is positive, the n_{on} SMs with the lowest voltage corresponding to the rank 1–n_{on} will be switched on. If the arm current is negative, the n_{on} SMs with the highest voltage corresponding to the rank n–n_{on}+1–n will be switched on, which ensures the capacitor voltage balancing in the arm.

When the capacitor in the SM_2 is demanded to be monitored, the SS is switched to "1" and the RSM-based CM-VBC is enabled for one fundamental period and the MMC enters monitoring mode. Here, the capacitance of the SM_2 can be estimated after a fundamental period based on equation (4.30). Afterward, the *SS* is switched back to "0" again to keep the u_{cau1}–u_{caun} the same with each other, which waits for the next command to monitor the capacitance in the other SMs.

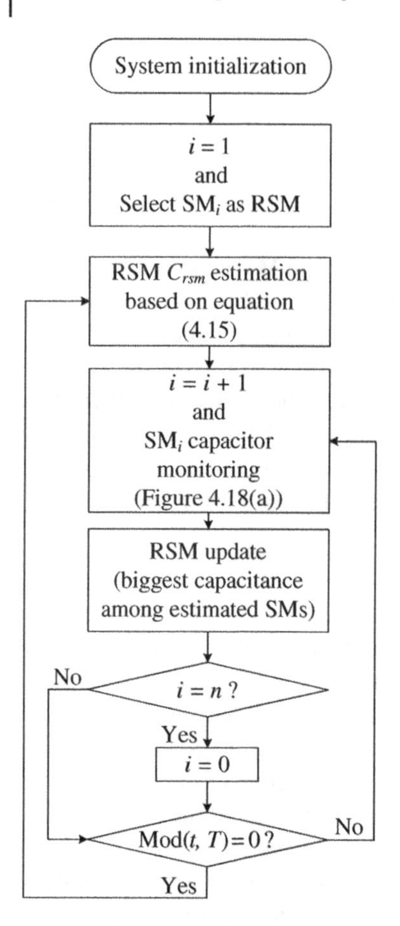

Figure 4.17 Flowchart of the capacitor monitoring strategy for the MMC.

After the capacitance in SM_2 is estimated, the RSM will be updated. Among the estimated SMs, the SM with the biggest capacitance is selected as the RSM. Afterward, the other SMs within the same arm will be monitored one by one with the same method, which is not repeated here.

Although the capacitor voltage ripple of the RSM is a little smaller than those of the other SMs in the CM-VBC, the capacitor voltage ripple is far less than the capacitor voltage, which makes that the capacitor monitoring has little effect on the performance of the MMC. In addition, due to the chemical process and aging effect, the speed of the capacitance drop is quite slow. Therefore, the SM capacitors in the arm can be monitored in the sequence from SM1 to SM_n round with some short interval such as one fundamental period.

In the presented method, the capacitance of the RSM should be measured and the estimation accuracy of the SM's capacitances relies on that of the RSM's capacitance.

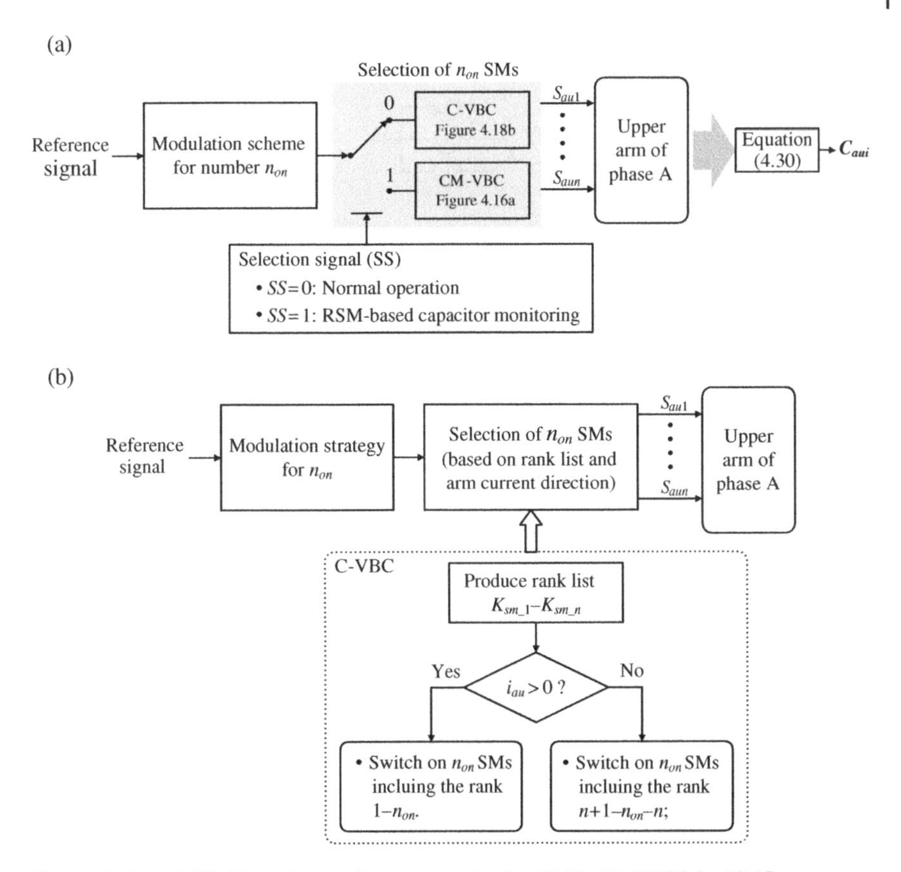

Figure 4.18 (a) RSM-based capacitor monitoring for MMCs. (b) C-VBC for MMCs.

According to [15, 19], the RSM's capacitance can be estimated by equation (4.15) with very high accuracy and the error is below 1%, which guarantee the capacitance estimation of the presented method. On the other hand, in order to obtain high accurate estimation precision, the capacitance in the selected RSM would be estimated by equation (4.15) at an interval of time T, as shown in Figure 4.17.

Case Study 4.2 Analysis of the RSM-Based Monitoring Strategy

Objective: To verify the RSM-based capacitance monitoring strategy, a three-phase MMC system, as shown in Figure 4.19, is simulated with the time-domain simulation tool PSCAD/EMTDC. The three-phase MMC system is linked to a three-phase AC system and works in the inverter mode. The system parameters are shown in Tables 4.4 and 4.5.

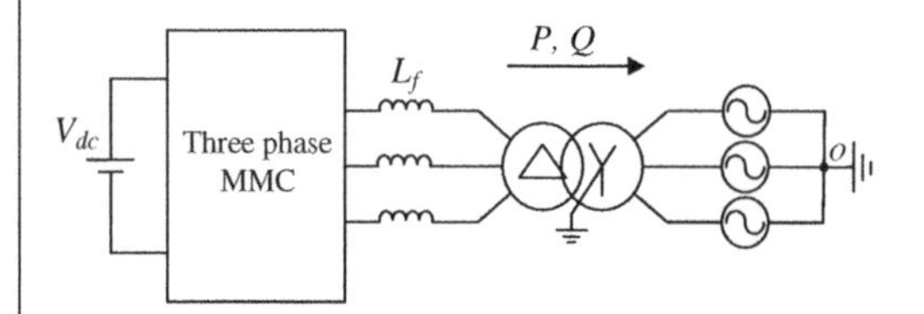

Figure 4.19 Block diagram of the simulation system.

Table 4.4 Simulation system parameters.

Parameters	Value
DC-link voltage V_{dc} (kV)	100
Grid line-to-line voltage (kV)	220
Grid frequency (Hz)	50
Transformer voltage rating	50 kV/220 kV
Transformer leakage reactance	10%
Number of SMs per arm n	100
Nominal SM capacitance C (mF)	15
Inductance L_s (mH)	10
Load inductance L_f (mH)	2

Table 4.5 Capacitances used in upper arm of phase A.

Capacitance (mF)	C_{au1}	C_{au2}	C_{au3}	C_{au4}	C_{au5}	C_{au6}
	15	13.5	12	10.5	9	7.5

Parameters: In the simulation, the active power P and the reactive power Q are 100 MW and 0 MVar, respectively. The capacitances C_{au1}–C_{au6}, which are presented in Table 4.5, drop gradually. The other system parameters are shown in Table 4.4.

Simulation results and analysis:
Figure 4.20 shows the performance of the MMC with the RSM-based capacitor monitoring strategy, where the average switching frequency is about 2.5 kHz. Figure 4.20a shows the arm current i_{au} and i_{al}. Figure 4.20b shows capacitor voltage u_{cau1}–u_{cau10}. Before 1.8 seconds, the C-VBC is used with "*SS*=0." With the conventional capacitor monitoring method, the C_{au1} is estimated as 15 mF, as shown in Figure 4.20c. And then, the SM_1 is selected as the RSM. Between

1.8 and 1.82 seconds, the C_{au2} is monitored with "SS=1", where the switching function for SM_1 is the same to that for SM_2. The different C_{au1} and C_{au2} result in the different capacitor voltages u_{cau1} and u_{cau2}, as shown in Figure 4.20d, where Δu_{cau1} and Δu_{cau2} are 0.126 and 0.139, respectively. According to equation (4.29), the capacitor C_{au2} is estimated as 13.5 mF as shown in Figure 4.20c. And then, the MMC goes back to the C-VBC with "SS = 0", where all capacitor voltages are kept the same again between 1.82 and 1.84 seconds. Similarly, the C_{au3}, C_{au4}, C_{au5}, and C_{au6} are monitored with "SS = 1" one after another. In Figure 4.20e, C_{au3} is monitored as 12 mF with Δu_{cau1} = 0.111 and Δu_{cau3} = 0.139. In Figure 4.20f, C_{au4} is monitored as 10.5 mF with Δu_{cau1} = 0.098 and Δu_{cau4} = 0.139. In Figure 4.20g, C_{au5} is monitored as 9 mF with Δu_{cau1} = 0.084 and Δu_{cau5} = 0.139. In Figure 4.20h, C_{au6} is monitored as 7.5 mF with Δu_{cau1} = 0.07 and Δu_{cau6} = 0.139. Owing to this, C_{au1} is the biggest among $C_{au1}-C_{au6}$, the SM_1 is always selected as the RSM, and Δu_{cau1} is less than $\Delta u_{cau2}-\Delta u_{cau6}$, as shown in Figure 4.20b.

In Figure 4.21, the THDs of i_{au} are almost the same with each other. It shows that the variable RSM capacitor voltage ripple has little impact on the MMC performance in the RSM-based capacitor monitoring strategy.

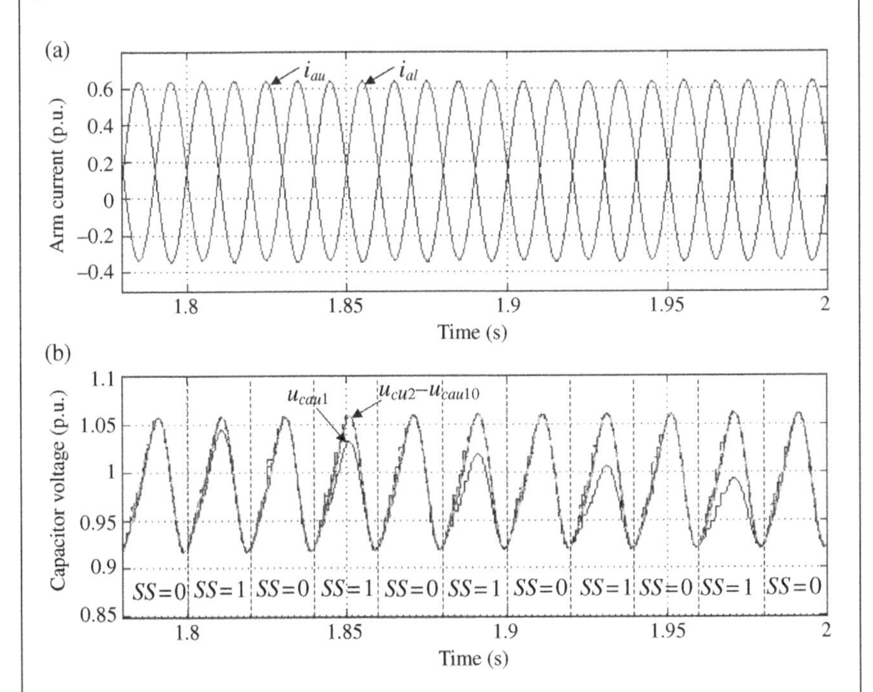

Figure 4.20 (a) i_{au} and i_{al}. (b) Upper arm capacitor voltage $u_{cau1}-u_{cau10}$. (c) Calculated capacitance $C_{au1}-C_{au6}$. (d) u_{cau1} and u_{cau2}. (e) u_{cau1} and u_{cau3}. (f) u_{cau1} and u_{cau4}. (g) u_{cau1} and u_{cau5}. (h) u_{cau1} and u_{cau6}.

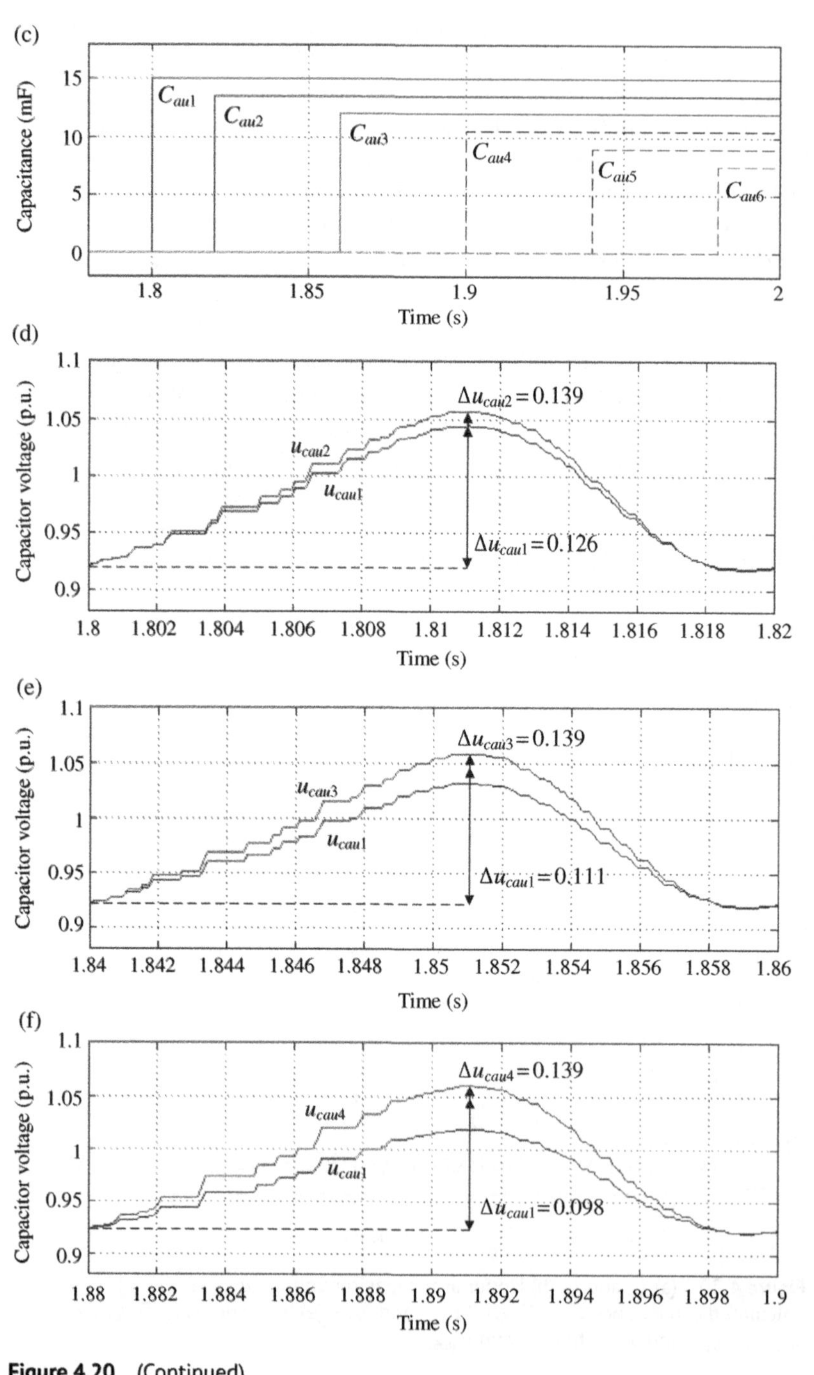

Figure 4.20 (Continued)

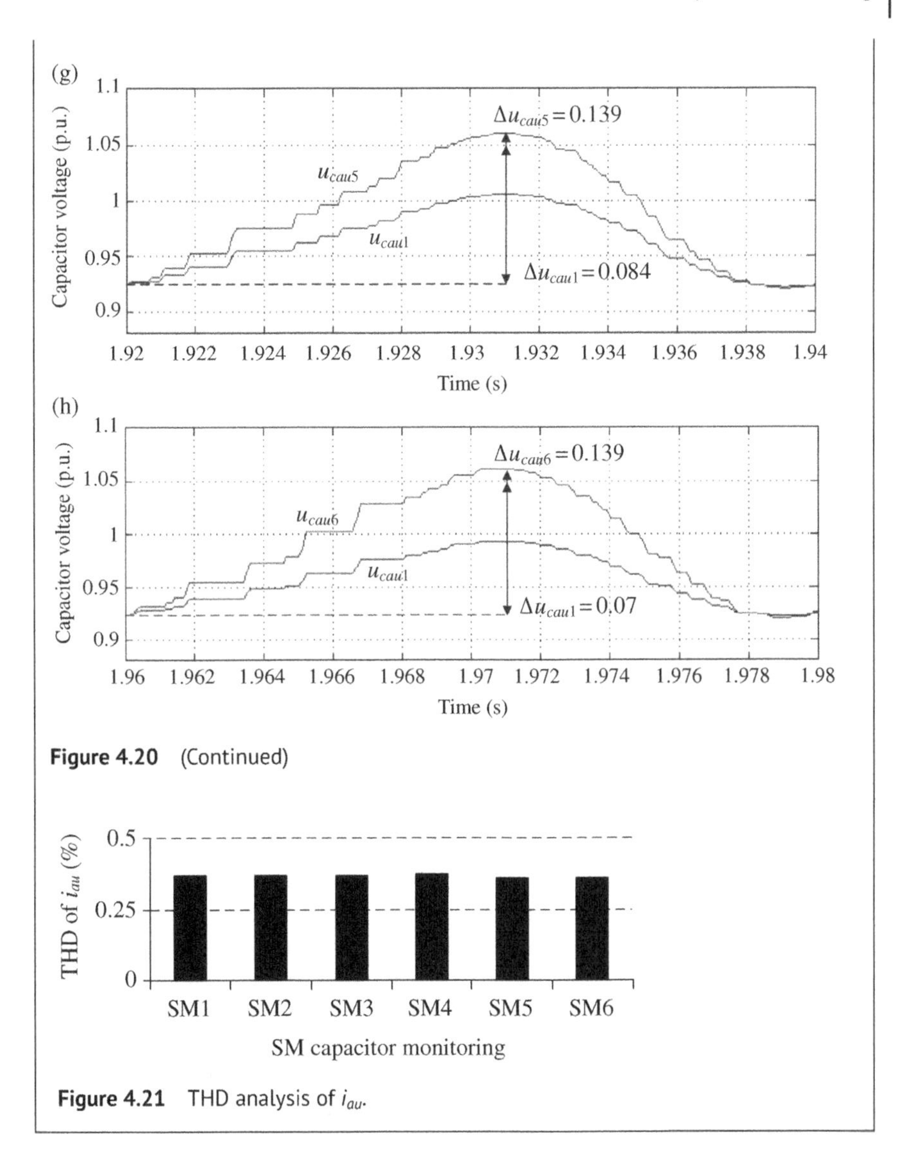

Figure 4.20 (Continued)

Figure 4.21 THD analysis of i_{au}.

4.5.4 Sorting-Based Monitoring Strategy

In the sorting-based monitoring strategy for capacitor estimation in the MMC, the SM capacitances in the arm are indirectly sorted based on the relationship among capacitance and current, and only the capacitor with smallest capacitance in the arm is estimated. This strategy not only realizes capacitance monitoring in the MMC but also simplifies monitoring algorithm through studying the characteristics of capacitance, voltage, and current of capacitors in the MMC.

Table 4.6 Capacitance relationship of SMs.

I_{1m_i}	C_{aui}
↑	↑
↓	↓

Since R_{aui_1f} is much smaller than $1/\omega C_{aui}$ [9], equation (4.14) can be rewritten as

$$U_{1m} = \frac{I_{1m\_i}}{\omega C_{aui}} \tag{4.31}$$

With voltage-balancing control (VBC) [14], the voltage amplitudes U_{1m_1}–U_{1m_n} in arm SM capacitors are kept the same as U_{1m}, equation (4.31) reveals that the I_{1m_i} of the i-th SM is proportional to the C_{aui}, where I_{1m_i} increases along with the increase of C_{aui} and I_{1m_i} declines along with the decrease of C_{aui}, as shown in Table 4.6.

Figure 4.22a shows a sorting-based monitoring strategy for capacitor's capacitance in the MMC. In the sorting-based monitoring strategy, I_{1m_i} is calculated with the characteristic variables calculation (CVC) block based on the switch function S_{aui} and arm current i_{au}, as shown in Figure 4.22b. At first, i_{caui} is obtained based on equation (4.20). And then fundamental component i_{caui_1f} is extracted from i_{caui} by band-pass filter (BPF) tuned at fundamental frequency, which is used to calculate its amplitude I_{1m_i}.

According to the obtained I_{1m_1}–I_{1m_n}, the corresponding SM_1–SM_n are sorted in descending order to find the SM with smallest capacitance among the n SMs within the arm, because the SM with smallest I_{1m} has the smallest capacitance. Afterward, the capacitance of the SM with the smallest I_{1m} will be calculated based on the capacitance estimation method [19], which derives the capacitance based on the relationship between the capacitor voltage variation and capacitor current integration. Normally, the accuracy of the capacitance estimation method is below 1% [19].

In order to reduce the effect of measurement noises in the actual MMC system and improve the capacitor monitoring accuracy, several sets capacitances are estimated such as ten sets. Removing the maximum and the minimum among the estimated capacitances, the average values of the rest capacitances are calculated as the capacitance monitoring results.

Once the capacitance monitoring result, which is the smallest capacitance in the SMs of the arm, is below the threshold value C_{limit}, the corresponding capacitor would be considered to be replaced with the new one. For electrolytic capacitors, the threshold value C_{limit} is normally 80% of the capacitance rated value. The sorting-based capacitor monitoring strategy is implemented periodically with some interval, which can effectively monitor capacitances in the MMC.

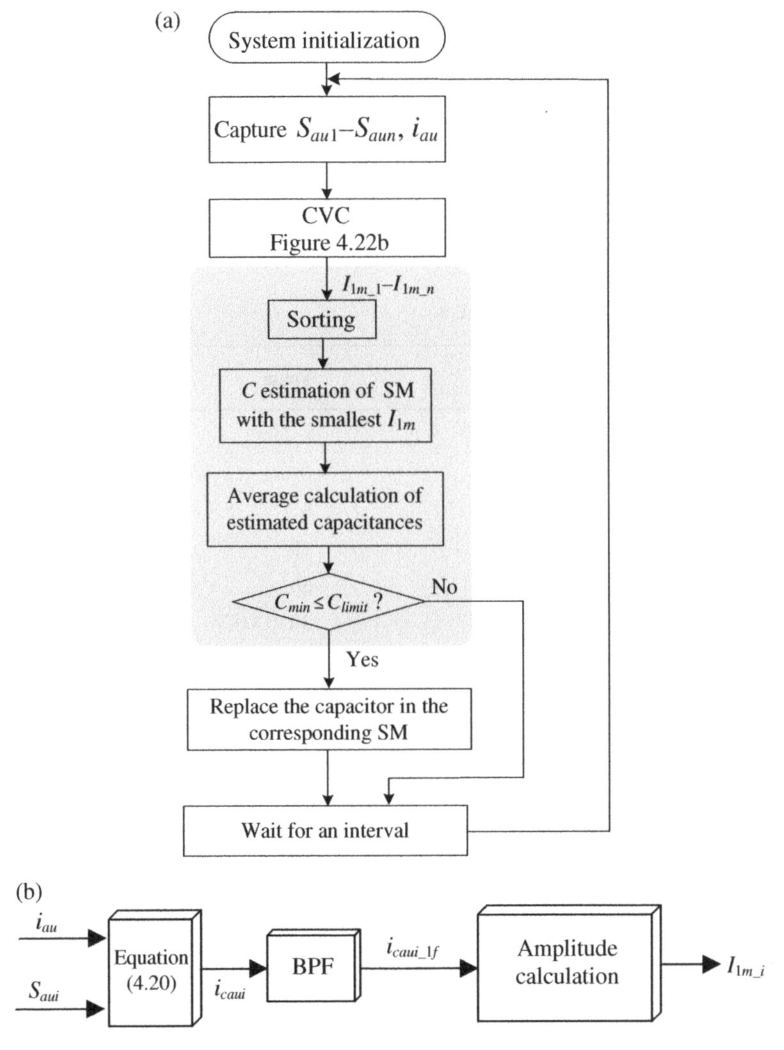

Figure 4.22 Sorting-based capacitor monitoring strategy. (a) Flowchart of the monitoring algorithm. (b) Characteristic variables calculation for I_{1m_1}–I_{1m_n}.

Case Study 4.3 Analysis of Sorting-Based Capacitance Monitoring Strategy

Objective: In this case study, the working principle and performance of the MMC equipped with the sorting-based capacitance monitoring strategy are studied through professional simulation tool PSCAD/EMTDC. As shown in Figure 4.23, a three-phase MMC system connected to power grid is established and the capacitances of SMs in the upper arm of phase A need to be

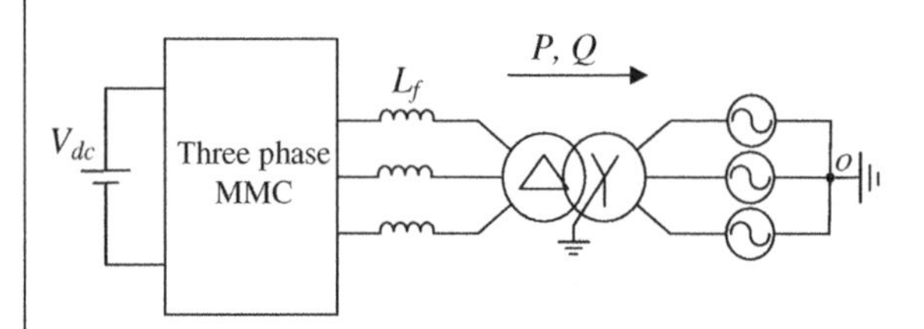

Figure 4.23 Schematic diagram of the simulation system.

Table 4.7 Simulation system parameters.

Parameters	Value
Active power (MW)	6
Reactive power (MVar)	0
DC-link voltage V_{dc} (kV)	6
Grid line-to-line voltage (kV)	33
Grid frequency (Hz)	50
Transformer rating voltage	3 kV/33 kV
Transformer leakage reactance	10%
Number of SMs per arm n	6
Inductance L_s (mH)	1.5
Inductance L_f (mH)	1

monitored. The MMC system works in inverter mode, and the power flow is from DC side to AC side.

Parameters: The system parameters are shown in Table 4.7. The nominal SM capacitance C is 13.2 mF. The capacitances of six SMs in the upper arm of phase A are presented in Table 4.8, which is shown in unitary value.

Simulation results and analysis:

Figure 4.24 shows the performance of the MMC. Figure 4.24a,b show the output voltages u_{ab}, u_{bc}, u_{ca} and currents i_a, i_b, i_c of the MMC. Figure 4.24c shows the upper arm current i_{au} and the lower arm current i_{al} in phase A. Figure 4.24d shows the upper arm capacitor voltages u_{cau1}–u_{cau6} in phase A. The MMCs equipped with the sorting-based capacitance monitoring scheme can function appropriately, the output power quality of which is desirable.

Figure 4.25 shows the performance of the capacitance monitoring strategy for the MMC. Figure 4.25a shows that $I_{1m\_1} > I_{1m\_2} > I_{1m\_3} > I_{1m\_4} > I_{1m\_5} > I_{1m\_6}$, which is consistent with the relationship among C_{au1}–C_{au6}, as shown in

Table 4.8 Capacitances in the upper arm of phase A.

SM	1	2	3	4	5	6
C_{aui} (p.u.)	1	0.96	0.92	0.88	0.84	0.8

Table 4.8. As I_{1m_6} is the smallest one among them, the C_{au6} in the SM6 is decided to be monitored. At 1.04 seconds, the C_{au6} is estimated as 10.49 mF and the error is approximately 0.66%, as shown in Figure 4.25b.

In order to reduce the effect of measurement noises in the actual MMC system and improve the capacitor monitoring accuracy, several sets capacitances are estimated such as ten sets. Removing the maximum and the minimum among the estimated capacitances, the average values of the rest capacitances are calculated as the capacitance monitoring results.

From this case study, it can be found that the sorting-based capacitance monitoring method is effective, which can realize capacitor condition monitoring successfully with considerably little estimation error. The sorting-based monitoring strategy does not have to monitor the capacitances of all capacitors in the arm, which simplifies the monitoring algorithm.

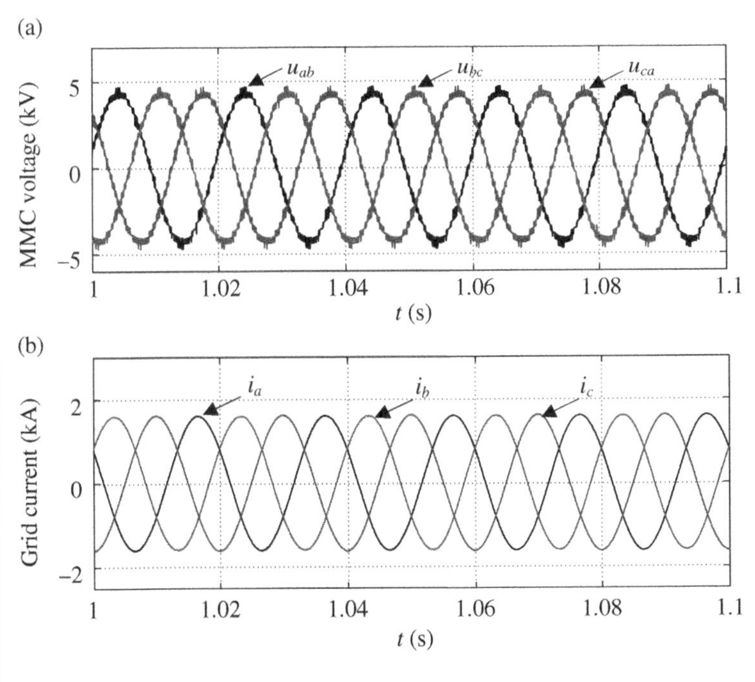

Figure 4.24 (a) u_{ab}, u_{bc} and u_{ca}. (b) i_a, i_b and i_c. (c) i_{au} and i_{al}. (d) u_{cau1}−u_{cau6}.

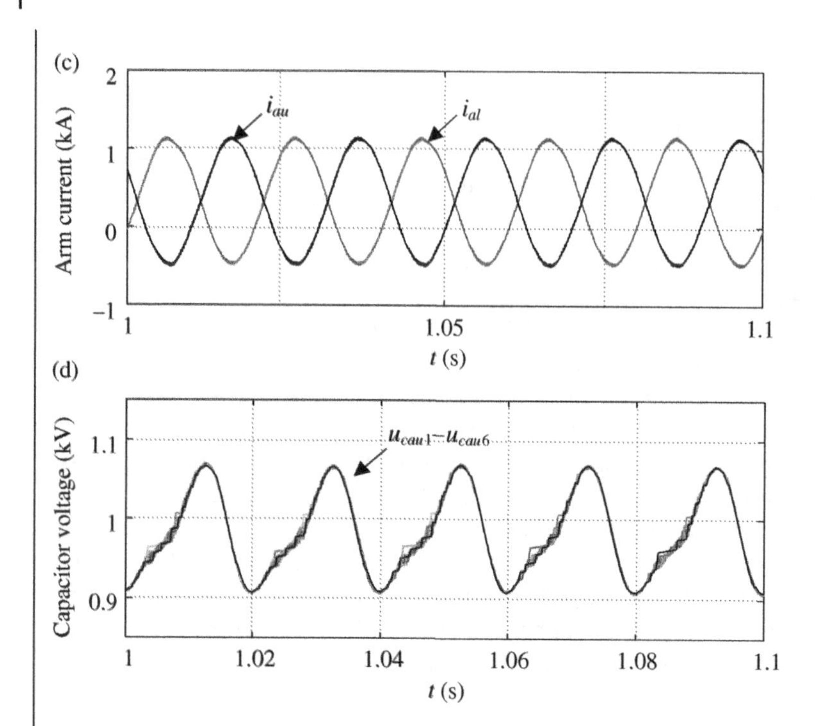

Figure 4.24 (Continued)

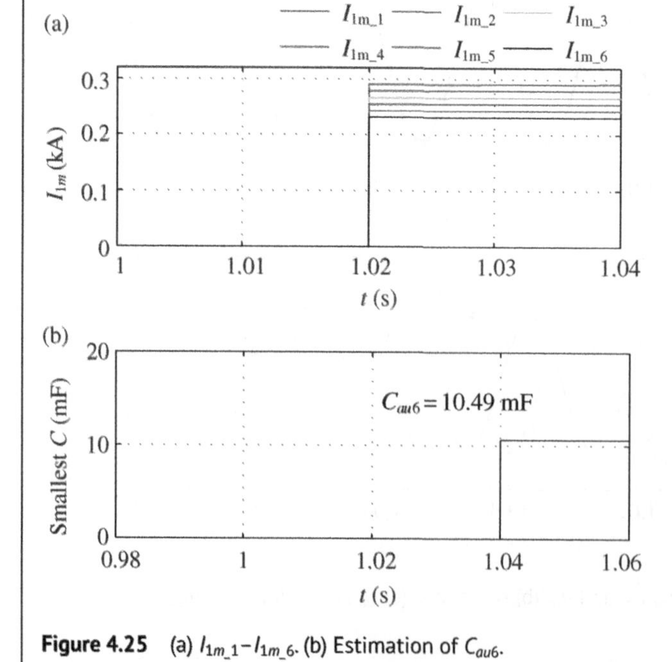

Figure 4.25 (a) I_{1m_1}–I_{1m_6}. (b) Estimation of C_{au6}.

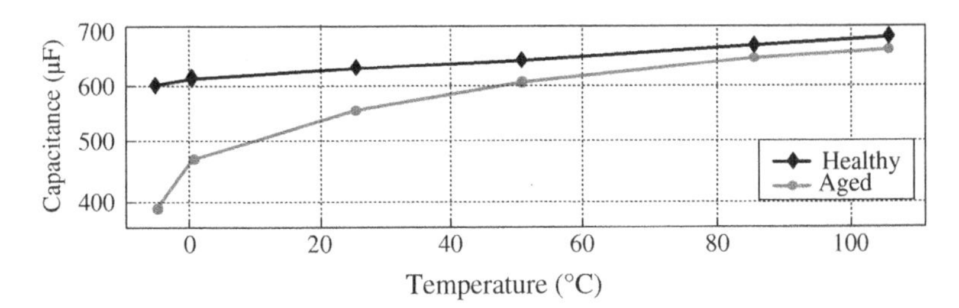

Figure 4.26 Measured relationship between C and temperature for a healthy capacitor and an aged capacitor where testing frequency is 360 Hz.

4.5.5 Temperature Effect of Capacitance

Aluminum electrolytic capacitors are known to be sensitive to temperature variation [10], where the temperature dependence is mainly due to the electrolyte solution [22]. Along with the temperature rising, the effective cross-sectional area of contact increases due to thermal expansion of the electrolyte and aluminum electrode, which accordingly results in an increase in capacitance C.

Figure 4.26 shows the relationship between the capacitance and the temperature for a healthy capacitor and an aged capacitor, where the testing frequency is 360 Hz. The C increases with an increase in temperature for both healthy and aged samples. However, due to the evaporation of electrolyte, the C of the aged capacitor is lower than the C of the healthy capacitor for any given temperature. Thus, it is desirable to consider the temperature variation in the capacitance monitoring process because the variation range of C with temperature is comparable to the variation of C caused by capacitor degradation.

4.6 ESR Monitoring

According to the capacitor deterioration curve in Figure 4.9, the increase of ESR is normally more pronounced in the deteriorated capacitor in comparison with the reduction of capacitance, which makes the ESR another important characteristic to monitor the health state of the electrolytic capacitor and detect the deteriorated capacitor. This section lists two ESR monitoring methods including direct ESR monitoring method and sorting-based monitoring method. Temperature effect of capacitors is discussed at last to illustrate the influence of temperature to ESR.

4.6.1 Direct ESR Monitoring Strategy

The health status of capacitors in the MMCs can be obtained by direct ESR monitoring strategy. Equation (4.32) shows the estimation of ESR, which can be directly estimated based on the modulus of the capacitor voltage $|u_c|$, modulus of the capacitor current $|i_c|$, phase angle α_{cv} of the capacitor voltage, and the phase angle β_{ci} of the capacitor current.

$$\mathrm{ESR} = \frac{|u_c|}{|i_c|}\cos(\alpha_{cv} - \beta_{ci}) \tag{4.32}$$

Generally, the capacitor voltage and current should be captured at high frequency when monitoring electrolytic capacitors' ESR. The accuracy of ESR estimation at low frequency is limited because the capacitor impedance at low frequency is almost dominated by its capacitance and its impedance phase angle is almost 90°. The ESR declines steadily with the increase of frequency and becomes approximately constant after crossing over the capacitive reactance of the capacitor. In order to obtain a higher level of accuracy, its value should be estimated at a higher range of frequencies, such as 1 kHz.

The main drawback of the direct ESR monitoring strategy is that the voltage component and current component of capacitor at 1 kHz should be extracted by BPF, which complicates monitoring method. On the other hand, as shown in Figures 4.4 and 4.6, capacitor voltage and current are mainly composed of fundamental and the second-order harmonic component, and components at high frequency is quite small compared with those at low frequency, which makes ESR monitoring difficult to conduct.

4.6.2 Sorting-Based ESR Monitoring Strategy

Together with the sorting-based capacitance monitoring strategy discussed in Section 4.5, sorting-based monitoring strategy can also be used to monitor ESR of the electrolytic capacitors in the MMC. This part introduces a sorting-based ESR monitoring strategy, by which the ESRs of SM capacitors in the arm are indirectly sorted based on the relationship among capacitor's ESR, current, and energy, and only the capacitor with biggest ESR in the arm is monitored. This strategy can effectively monitor capacitor health status without bringing about heavy computation burden.

The capacitor energy variety E_{aui} of the i-th ($i = 1, 2, \ldots, n$) SM within a fundamental period T_f can be obtained based on equations (4.5)–(4.9) and (4.13) and [16], as

$$E_{aui} = \int_0^{T_f} \left(\frac{V_{dc}}{n} + \Delta u_{caui} \right) i_{caui} dt = E_{aui\_r1f} + E_{aui\_r2f} \tag{4.33}$$

with

$$\begin{cases} E_{aui\_r1f} = \int_0^{T_f} u_{caui\_1f} i_{caui\_1f} dt = \dfrac{2\pi U_{1m}^2 / \omega}{R_{aui\_1f} + 1/\left(\omega C_{aui}\right)^2 / R_{aui\_1f}} \\ E_{aui\_r2f} = \int_0^{T_f} u_{caui\_2f} i_{caui\_2f} dt = \dfrac{4\pi U_{2m}^2 / \omega_2}{R_{aui\_2f} + 1/\left(\omega_2 C_{aui}\right)^2 / R_{aui\_2f}} \end{cases} \tag{4.34}$$

where E_{aui} represents the energy consumption of ESR in the capacitor C_{aui}; E_{aui_r1f} and E_{aui_r2f} are the energy consumption caused by R_{aui_1f} and R_{aui_2f}, respectively; Δu_{caui} is the ripple component of the SM$_i$'s ($i = 1, 2, \dots n$) capacitor voltage u_{caui}.

Figure 4.27 shows the capacitor energy variety E_{au1_r1f} and E_{au1_r2f}, respectively, under various power of the MMC, where E_{au1_r1f} and E_{au1_r2f} decline with the decrease of power. In addition, E_{au1_r1f} is much bigger than E_{au1_r2f}, which means that R_{aui_1f} results in the main energy consumption in the capacitor and is the main cause of accelerating the aging process in comparison with R_{aui_2f}.

Substituting equations (4.8) and (4.34) into (4.33), the variation E_{aui} can be rewritten as

$$E_{aui} = \frac{2\pi}{\omega}\left(U_{1m}^2 + 2U_{2m}^2\right)\frac{1}{R_{aui\_1f} + 1/\left(\omega C_{aui}\right)^2 / R_{aui\_1f}} \tag{4.35}$$

The E_{aui} in the SM of the arm is determined by corresponding R_{aui_1f} and C_{aui}, as shown in Figure 4.28, as follows:

1) E_{aui} *and* C_{aui}: E_{aui} increases along with the increase of C_{aui} and vice versa, as shown in Table 4.9.
2) E_{aui} *and* R_{aui_1f}: The SM capacitance C_{aui} in the MMC is normally on the order of mF, which results in that $R_{aui\_1f} < 1/(\omega_1 C_{aui})$ [2]. Hence, E_{aui} increases along with the increase of R_{aui_1f} and vice versa, as shown in Table 4.10.

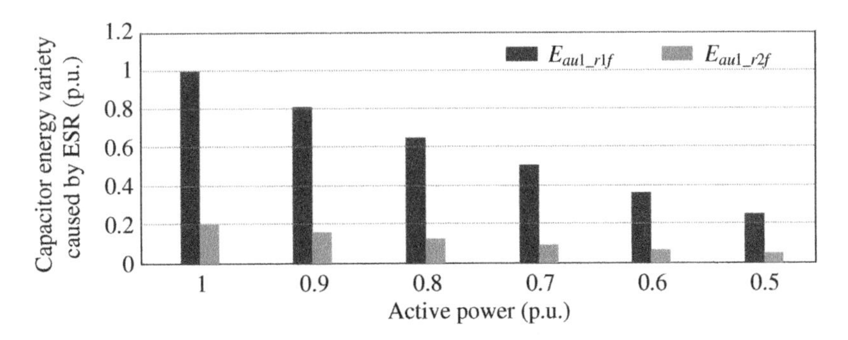

Figure 4.27 E_{aui_r1f} and E_{aui_r2f} under various active power.

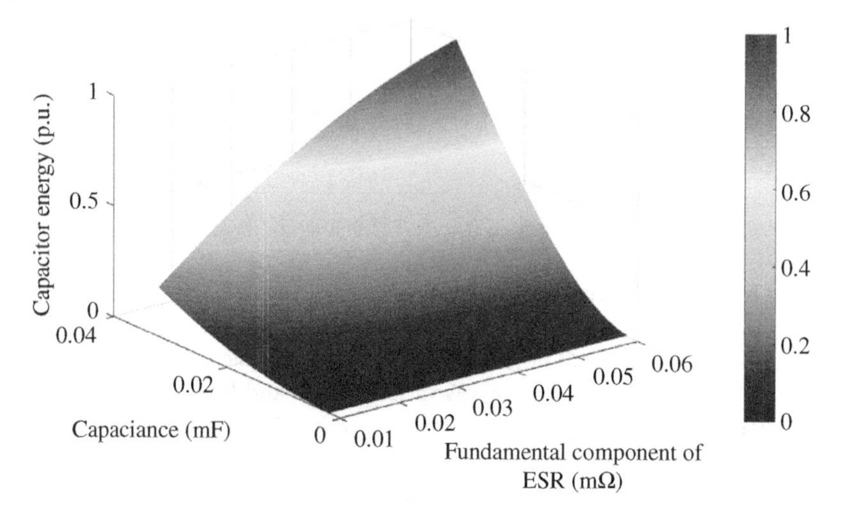

Figure 4.28 Relationship between SM E_{aui} and R_{aui_1f}, C_{aui} in the arm of the MMC.

Table 4.9 Relationship between E_{aui} and C_{aui}.

C_{aui}	E_{aui}
↑	↑
↓	↓

Table 4.10 Relationship between E_{aui} and R_{aui_1f}.

R_{aui_1f}	E_{aui}
↑	↑
↓	↓

The variation K_{aui} is defined as

$$K_{aui} = \frac{E_{aui}}{I_{1m\_i}^2} = \frac{2\pi R_{aui\_1f}}{\omega}\left(1 + \frac{2U_{2m}^2}{U_{1m}^2}\right) \tag{4.36}$$

The K_{aui} of the i-th SM is proportional to R_{aui_1f}, where K_{aui} increases along with the increase of R_{aui_1f} and K_{aui} reduces along with the decrease of R_{aui_1f}, as shown in Table 4.11.

Table 4.11 ESR relationship of SMs.

K_{aui}	R_{aui_1f}
↑	↑
↓	↓

Figure 4.29a shows a sorting-based monitoring strategy for capacitor's ESR in the MMC. Because ESR corresponding to fundamental frequency is nearly double the ESR corresponding to double-line frequency and is the main energy consumption, only the ESR corresponding to fundamental frequency is estimated here so as to simplify the computation.

In the capacitor monitoring strategy, capacitor voltages u_{cau1}–u_{caun} and currents i_{cau1}–i_{caun} are captured at first. Then, the K_{aui} are calculated with the CVC block based on the capacitor voltage u_{caui}, switch function S_{aui}, and arm current i_{au}, as shown in Figure 4.29b. In the CVC block, the capacitor current can be obtained based on equation (4.20). Capacitor energy variety E_{aui} can be calculated through the integral of capacitor ripple voltage and current based on equation (4.33). And then, the fundamental component of capacitor current i_{caui_1f} can be obtained by a BPF tuned at the fundamental frequency, which is used to calculate its amplitude I_{1m_i}, as shown in Figure 4.29b. The K_{aui} is obtained based on equation (4.36). Afterward, according to the obtained K_{au1}–K_{aun}, the corresponding SM_1–SM_n are sorted in ascending order, so as to find the SM with biggest K_{au} among the n SMs in the arm, because the SM with the biggest K_{au} has the biggest ESR. Afterward, the ESR corresponding to the fundamental frequency in the SM with biggest K_{au} will be calculated based on the ESR estimation method in [21], which utilizes the average capacitor power divided by the square of capacitor current at the fundamental frequency to obtain the ESR. To date, the accuracy of the ESR estimation methods is about 4%–10% [9, 23].

In addition, in order to reduce the effect of measurement noises in the actual MMC system and improve the capacitor monitoring accuracy, several sets ESRs are estimated such as ten sets. Removing the maximum and the minimum among the estimated ESRs, the average values of the rest ESRs are calculated as the ESR monitoring results. Finally, once the ESR monitoring result, which is the biggest ESR in the SMs of the arm, is over the threshold value ESR_{limit}, the corresponding capacitor would be considered to be replaced with the new one. For electrolytic capacitors, the threshold value ESR_{limit} is normally two times of the ESR rated value [16]. The presented capacitor monitoring strategy is implemented periodically with some interval, which can effectively monitor capacitor's ESR in the MMC.

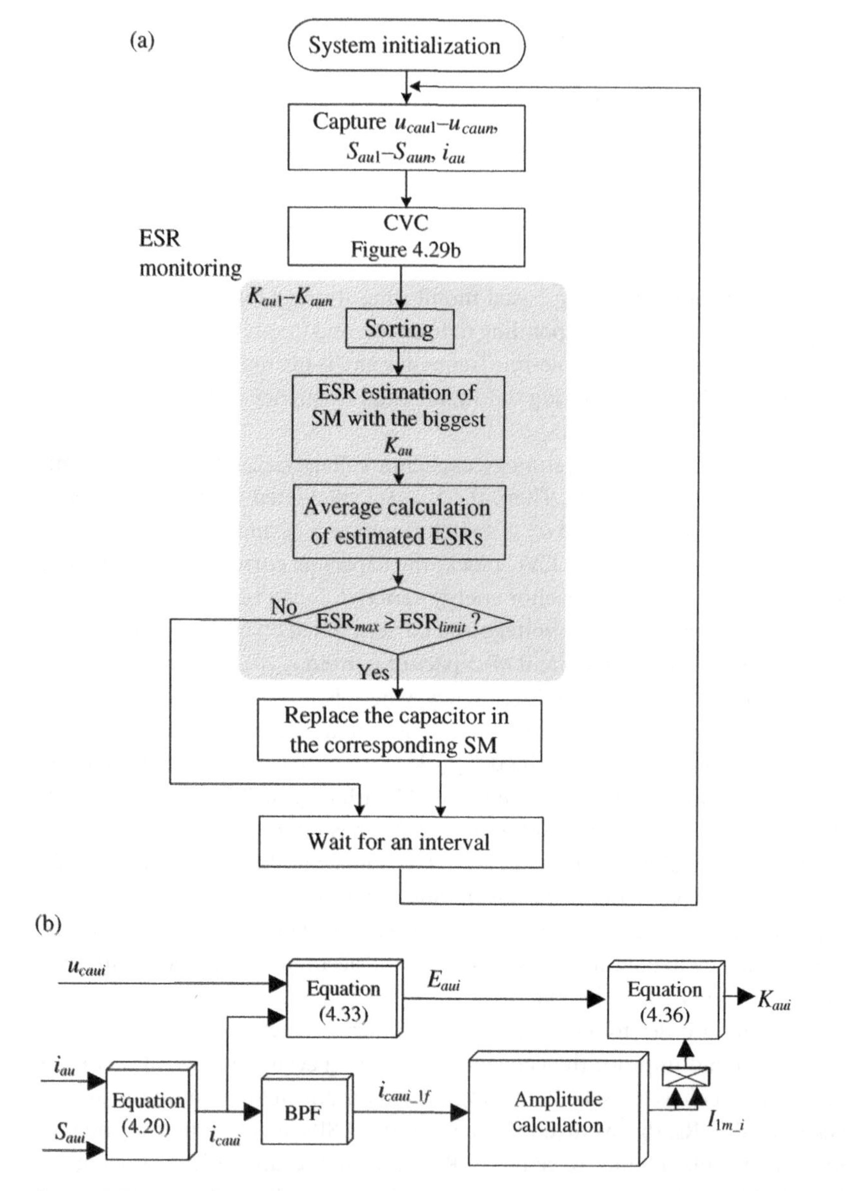

Figure 4.29 Capacitor monitoring strategy. (a) Flowchart of ESR monitoring algorithm. (b) Characteristic variables calculation for K_{au1}–K_{aun}.

Case Study 4.4 Analysis of Sorting-Based ESR Monitoring Strategy

Objective: In this case study, the working principle and performance of the MMC equipped with the sorting-based ESR monitoring strategy are studied through professional simulation tool PSCAD/EMTDC. As shown in Figure 4.23, a three-phase MMC system connected to power grid is established and the capacitances of SMs in the upper arm of phase A need to be monitored. The MMC system works in inverter mode, and the power flow is from DC side to AC side.

Parameters: The system parameters are shown in Table 4.7, and the nominal SM ESR has been chosen as 25.2 mΩ. The ESRs of six SMs in the upper arm of phase A are presented in Table 4.12, which is shown in unitary value.

Simulation results and analysis:
Figure 4.30 shows the performance of the MMC. Figure 4.30a,b show the output voltages u_{ab}, u_{bc}, u_{ca} and currents i_a, i_b, i_c of the MMC. Figure 4.30c shows the upper arm current i_{au} and the lower arm current i_{al} in phase A. Figure 4.30d shows the upper arm capacitor voltages $u_{cau1}-u_{cau6}$ in phase A. The MMCs equipped with the sorting-based ESR monitoring strategy can function appropriately, and the output power quality is desirable.

Table 4.12 Capacitances in the upper arm of phase A.

SM	1	2	3	4	5	6
R_{aui_1f} (p.u.)	1	1.2	1.4	1.6	1.8	2

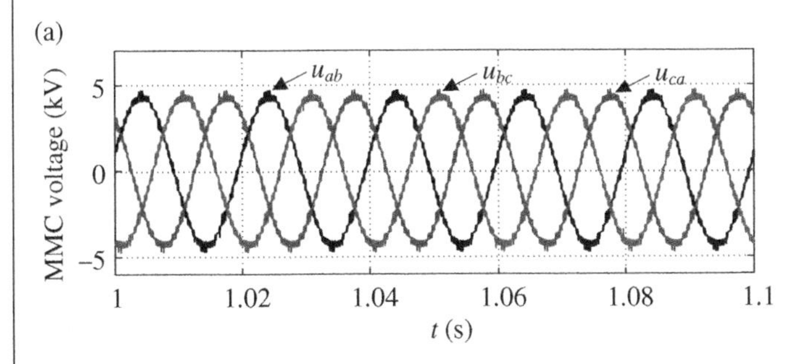

Figure 4.30 (a) u_{ab}, u_{bc}, and u_{ca}. (b) i_a, i_b, and i_c. (c) i_{au} and i_{al}. (d) $u_{cau1}-u_{cau6}$.

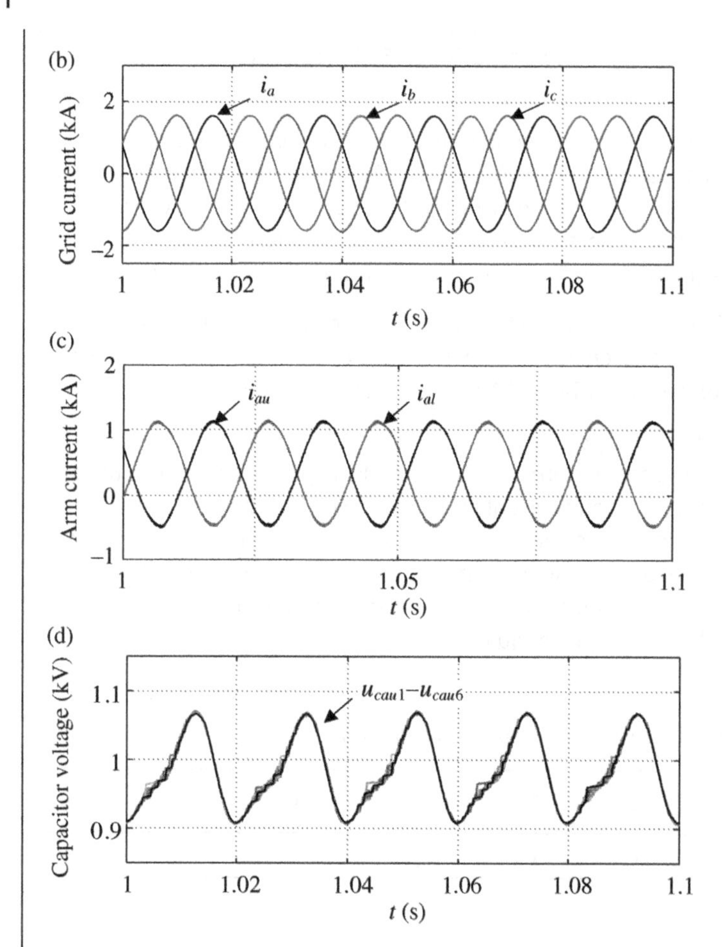

Figure 4.30 (Continued)

Figure 4.31 shows the performance of the ESR monitoring for the MMC. Figure 4.31a shows that $K_{au1} < K_{au2} < K_{au3} < K_{au4} < K_{au5} < K_{au6}$, which is consistent with the relationship among R_{au1_1f}–R_{au6_1f}, as shown in Table 4.12. As K_{au6} is the biggest one among them, the ESR in the SM6 is decided to be monitored. At 1.04 seconds, R_{au6_1f} in the SM6 is estimated as 52.15 mΩ and the error is approximately 3.47%, as shown in Figure 4.31b.

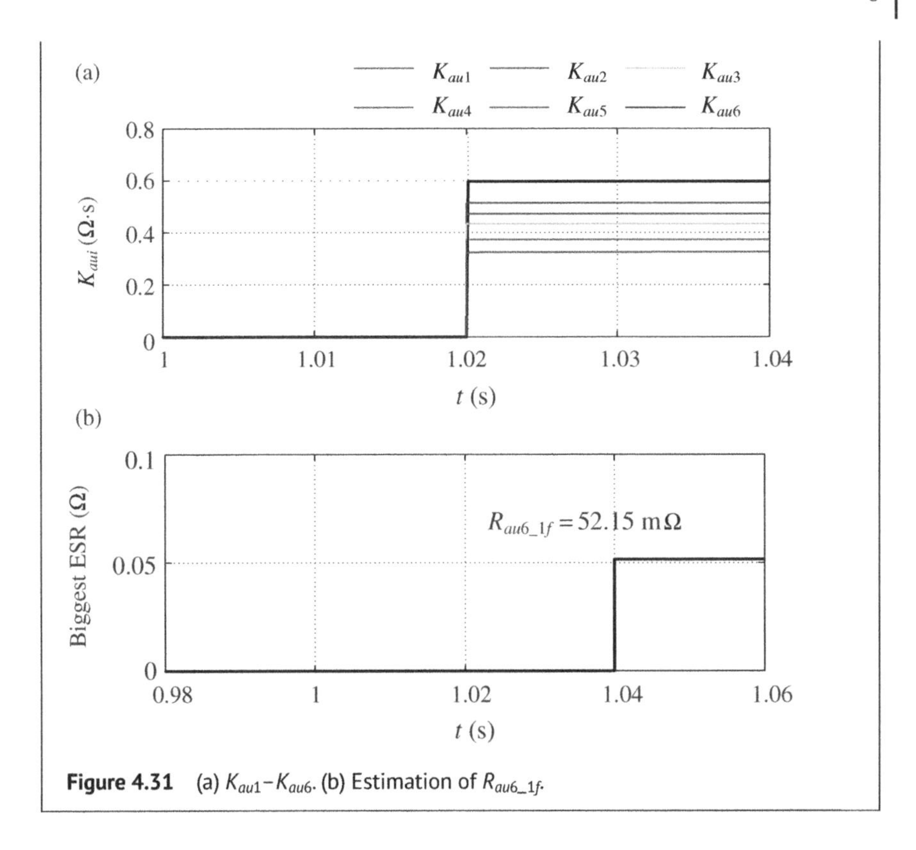

Figure 4.31 (a) K_{au1}–K_{au6}. (b) Estimation of R_{au6_1f}.

4.6.3 Temperature Effect of ESR

Temperature dependence of aluminum electrolytic capacitors is an important issue for ESR monitoring, which is mainly due to the electrolyte solution. The ESR of electrolytic capacitors represents the sum of the resistance of all components such as electrolyte, oxide layers, foils, leads, and connectors. As the capacitor temperature increases, the resistivity of the electrolyte and oxide layer decrease due to increased polarization and carrier mobility, and the change in the dielectric constant in the temperature range of interest can be ignored. On the other hand, an increase in temperature also increases the effective cross-sectional area of contact due to thermal expansion of the electrolyte and aluminum electrode. This combined effect results in a decrease of ESR along with the rising of temperature.

Figure 4.32 shows the values of ESR for a healthy and aged capacitor as a function of temperature, where the frequency is fixed at 360 Hz. The ESR decreases with a rising of the temperature for both healthy and aged capacitors. However, the evaporation of electrolyte in capacitor aging causes the increase of the ESR for any given temperature, which may cancel out the decrease of ESR when

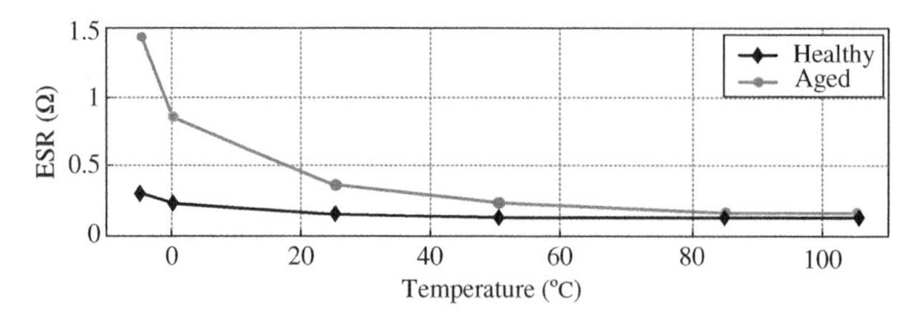

Figure 4.32 Relationship between the ESR and temperature for the healthy capacitor and the aged capacitor where testing frequency is 360 Hz.

temperature increases. Therefore, it is desirable to take ambient temperature variation into consideration in ESR monitoring process because the range of ESR variation due to temperature increase could be comparable to the variation caused by the capacitor degradation.

4.7 Capacitor Lifetime Monitoring

As a key element for the MMC, capacitor's lifetime is very important for the converter and the system reliability. The failure of capacitor should be detected as soon as possible when it occurs because the failed capacitor would loss its function and it may cause serious consequences. To prevent capacitor failure, it is necessary to monitor the capacitor lifetime.

Generally, a capacitor lifetime analysis is based on Arrhenius law. Sensors such as voltage sensor, current sensor, and temperature sensor are equipped to collect the required data of capacitor. The capacitor lifetime can be calculated as [24]

$$L_c = L_o \times M_V \times 2^{\left(\frac{T_{core,\max} - T_{core}}{10}\right)} \tag{4.37}$$

where L_c is capacitor lifetime (hours), L_o is maximum capacitor lifetime (hours), T_{core} is core temperature(°C), $T_{core,\max}$ is maximum core temperature(°C), and M_V is voltage multiplier. In addition, L_0, $T_{core,\max}$, and M_V can be obtained from the provided manufacturer datasheets. However, electrical model and heat transfer model of capacitor should be applied to calculate core temperature, because the core temperature (T_{core}) could not be directly measured. The core temperature (T_{core}) can be calculated with equations (4.38)–(4.40) as

$$T_{core} = T_{ambient} + \left(\Delta T \times M_I\right) \tag{4.38}$$

$$\Delta T = P_{\text{ESR}} \times T_{resistance} \tag{4.39}$$

$$P_{\text{ESR}} = I^2 \times \text{ESR} \tag{4.40}$$

where $T_{ambient}$ is the ambient temperature (°C), ΔT is the temperature rise (°C), M_I is the ripple current multiplier, P_{ESR} is the power loss of ESR (W), $T_{resistance}$ is the thermal resistance (°C/W), I is the capacitor current (A), and ESR is the ESR (mΩ). The parameters M_I and $T_{resistance}$ can be obtained from the manufacturer datasheets.

To sum up, capacitor lifetime monitoring is an effective method to prevent capacitor failure and ensure the normal operation of the MMC. According to the capacitor parameters obtained from the manufacturer datasheets and the measurement of capacitor current, voltage, and ambient temperature, capacitor lifetime can be calculated.

4.8 Arm Current Optimal Control Under Capacitor Aging

As capacitors age, the capacitances in the upper and lower arms would decrease and be different from each other due to different operation, which would cause a fundamental component in the circulating current within one-phase unit. The circulating current would increase the capacitors' voltage ripple and may disturb the normal operation of the MMC. This section mainly analyzes the mechanism of the circulating current by establishing equivalent circuit of MMCs and presents an effective arm current optimization scheme to suppress the circulating current caused by capacitance difference.

4.8.1 Equivalent Circuit of MMCs

The electrolyte of capacitor reduces with its working time, which makes the capacitance of capacitor gradually decreases [3, 16]. Due to the cascade structure of the MMC, there are large numbers of capacitors in the MMC, whose capacitances could be different because of various operation conditions. Referring to Section 4.5.2, the n SMs within the same arm are equivalent to one SM, as shown in Figure 4.33.

According to Section 4.5.2, the relationship between the equivalent SM's capacitance C_{au_e} and the i-th SM's capacitance ($i = 1, 2, \ldots, n$) C_{aui} of the upper arm of phase A can be expressed as

$$C_{au\_e} = \frac{1}{n^2} \sum_{i=1}^{n} C_{aui} \tag{4.41}$$

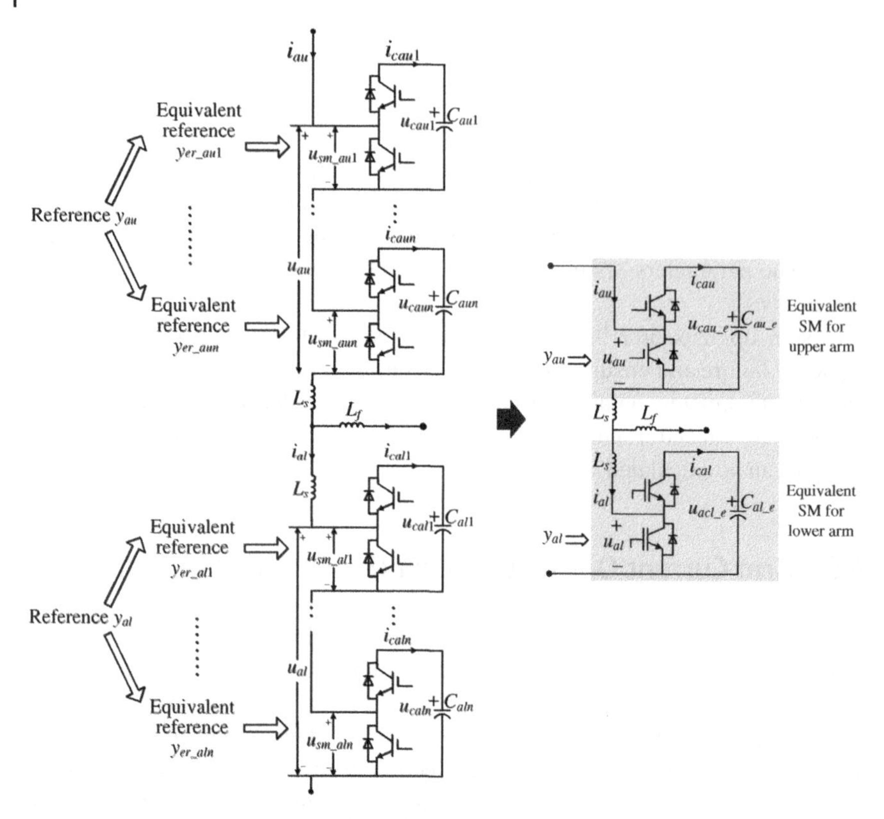

Figure 4.33 Equivalent arm circuit of phase A.

The relationship between the equivalent SM's capacitance C_{al_e} and the i-th SM's capacitance (i = 1, 2, ..., n) C_{ali} of the lower arm of phase A can be expressed as

$$C_{al\_e} = \frac{1}{n^2}\sum_{i=1}^{n} C_{ali} \tag{4.42}$$

The equivalent circuit and analysis of the other four arms are the same as that of the upper arm and lower arm in phase A, which are not repeated here. When the capacitors deteriorate, the capacitances in the upper and lower arm would decrease and be different from each other due to different operations. According to equations (4.41) and (4.42), the capacitances of the equivalent SMs in the upper and lower arm would be unequal, which would affect the normal operation of the MMC.

4.8.2 Arm Current Characteristics

The capacitance decline in the SM and the capacitance difference between SMs would affect the normal performance of the MMC to some extent. A fundamental component in the circulating current would be caused with the rise of the difference between the upper and the lower arm capacitances. Each SM in the MMC contains a capacitor, and capacitance change of each capacitor would affect the MMC operation, which bring about heavy computation burden. However, according to equivalent arm circuit principle in Section 4.8.1, the SMs within an arm can be equivalent to one SM and the difference between the upper and lower arm equivalent capacitance C_{au_e} and C_{al_e} decides the fundamental component in the circulating current. Accordingly, to show the influence caused by the difference between C_{au} and C_{al} clearly, the situation that only capacitors within phase A are deteriorated is considered in this section.

According to Kirchhoff voltage principle, the relationship between u_{au} and u_{al} can be described as

$$V_{dc} = u_{au} + u_{al} + 2L_s \frac{di_{diff\_a}}{dt} \tag{4.43}$$

with

$$i_{diff\_a} = \frac{i_{au} + i_{al}}{2} \tag{4.44}$$

Suppose that the second-order harmonic component of the circulating current has been suppressed, the arm currents can be described as

$$i_{au} = \bar{i}_{au} + \tilde{i}_{au} = \bar{i}_{au} + I_{aum} \sin(\omega t + \alpha_r) \tag{4.45}$$

$$i_{al} = \bar{i}_{al} + \tilde{i}_{al} = \bar{i}_{al} + I_{alm} \sin(\omega t + \beta_r) \tag{4.46}$$

where $\bar{i}_{au}$ and $\tilde{i}_{au}$ are the DC and AC components of i_{au}, respectively. I_{aum} and α_r are the amplitude and phase angle of $\tilde{i}_{au}$. $\bar{i}_{al}$ and $\tilde{i}_{al}$ are the DC and AC components of i_{al}, respectively. I_{alm} and β_r are the amplitude and phase angle of $\tilde{i}_{al}$.

Suppose that the references y_a for phase A is

$$y_a = m \cdot \sin(\omega t + \phi) \tag{4.47}$$

and the references for the upper arm and the lower arm are $-y_a$ and y_a, respectively, based on the equivalent circuit shown in Figure 4.33, combining equations (4.18), (4.19), and (4.45)–(4.47), the fundamental components u_{au_1f} and u_{al_1f} of the upper and lower arm voltages can be obtained as

$$u_{au\_1f} = -\frac{m\bar{i}_{au}}{4\omega C_{au\_e}}\cos(\omega t+\phi) - \frac{8I_{mau}+m^2I_{mau}}{32\omega C_{au\_e}}\cos(\omega t+\alpha_r) \tag{4.48}$$

$$u_{al\_1f} = \frac{m\bar{i}_{al}}{4\omega C_{al\_e}}\cos(\omega t+\phi) - \frac{8I_{mal}+m^2I_{mal}}{32\omega C_{al\_e}}\cos(\omega t+\beta_r) \tag{4.49}$$

Substituting equations (4.48) and (4.49) into equation (4.43), it can be obtained

$$\frac{di_{1f\_a}}{dt} = -\frac{u_{au\_1f}+u_{al\_1f}}{2L_s} \tag{4.50}$$

where i_{1f_a} is the fundamental component of the circulating current in the phase A.

Since the C_{au_e} is not equal to C_{al_e}, the sum u_{au_1f} and u_{al_1f} is not zero and would increase with the increase of the difference between C_{au_e} and C_{al_e}. Therefore, a fundamental component in the circulating current would be caused and the amplitude of the i_{1f_a} increases with the increase of the difference between C_{au_e} and C_{al_e}.

4.8.3 Arm Current Optimal Control

A fundamental component i_{1f_j} in the circulating current of phase j would emerge with the increase of the difference between C_{au_e} and C_{al_e}, which would cause the corresponding voltage $u_{diff\_1f\_a}$ in phase A, as shown in Figure 4.34. The relationship for i_{1f_j} and $u_{diff\_1f\_j}$ in phase j is

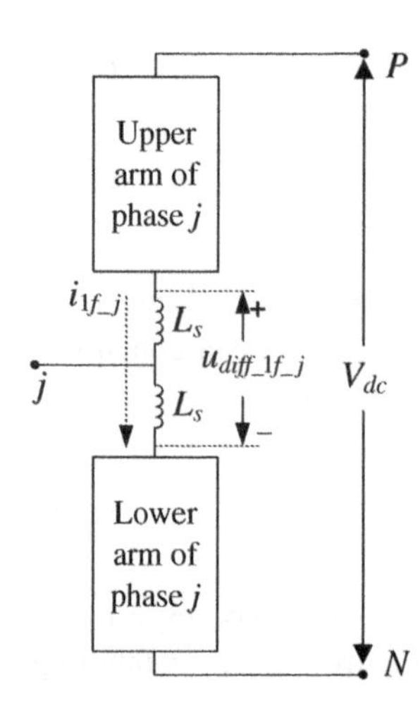

Figure 4.34 Relationship of i_{1f_j} and $u_{diff\_1f\_j}$ in phase j.

$$u_{diff\_1f\_j} = 2L_s\frac{di_{1f\_j}}{dt} \tag{4.51}$$

According to the method from single-phase system to dq transferring [25], the behavior of phase j in the synchronous rotating reference frame can be expressed as

$$\begin{bmatrix} u_{diff\_1f\_j\_d} \\ u_{diff\_1f\_j\_q} \end{bmatrix} = 2L_s\frac{d}{dt}\begin{bmatrix} i_{1f\_j\_d} \\ i_{1f\_j\_q} \end{bmatrix} + 2L_s\begin{bmatrix} 0 & \omega \\ -\omega & 0 \end{bmatrix}\cdot\begin{bmatrix} i_{1f\_j\_d} \\ i_{1f\_j\_q} \end{bmatrix} \tag{4.52}$$

where $u_{diff\_1f\_j\_d}$, $u_{diff\_1f\_j\_q}$ and $i_{1f\_j\_d}$, $i_{1f\_j\_q}$ are the dq components of the voltage $u_{diff\_1f\_j}$ and the current i_{1f_j} in the rotating reference frame, respectively.

Figure 4.35 shows an arm current optimal control scheme for the MMC, which can effectively suppress the fundamental component i_{1f_j} in the circulating current caused by the capacitance difference. Based on equation (4.44), the circulating current i_{diff_j} can be obtained. And then the fundamental component of the circulating current can be obtained based on the BPF, which is tuned at the fundamental frequency. i_{1f_j} is transformed into dq domain by a single-phase Park transform, where one product by 2sin(ωt) and the other one by 2cos(ωt). Then, a notch filter tuned at twice the fundamental frequency. Finally, the dq component $i_{1f\_j\_d}$ and $i_{1f\_j\_q}$ of the i_{1f_j} can be obtained. In Figure 4.35a, the fundamental phase angle θ_e and angular frequency ω are obtained by the phase locked loop. With the PI controller, the fundamental component i_{1f_j} in phase j can be eliminated when the $i_{1f\_a\_d}$ and $i_{1f\_a\_q}$ are regulated to zero by the voltage reference u_{1f_j} in phase j. Figure 4.35b shows the control of the MMC with the arm current optimal control. e_d, e_q and i_d, i_q are dq-axis components of grid voltage e_a, e_b, e_c and current i_a, i_b, i_c, respectively. According to the control objective such as active power, reactive power, and DC-link voltage control, the current reference i_{d_ref} and i_{q_ref} can be obtained [26]. The vector control method is used here for grid current control [26] and produce the three-phase reference voltage u_{jm_ref}. To eliminate the fundamental component in the arm current, the references for the upper and lower arms of phase j are $u_{ju\_ref} = -u_{jm\_ref} + u_{1f\_j}$ and $u_{jl\_ref} = u_{jm\_ref} + u_{1f\_j}$, respectively.

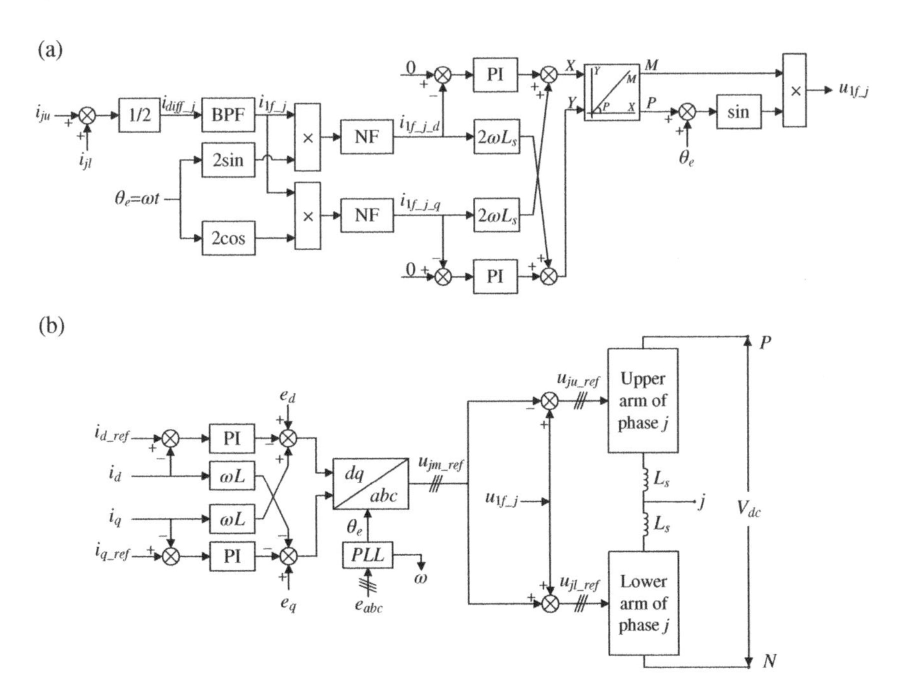

Figure 4.35 Block diagram of the arm current optimal control for MMCs. (a) Single-phase control in phase j. (b) MMCs control.

Case Study 4.5 Analysis of Arm Current Optimal Control Under Capacitor Aging

Objective: In this case study, the working principle and the performance of the arm current optimal control under capacitor aging are studied through professional time-domain simulation software PSCAD/EMTDC. A three-phase MMC system, as shown in Figure 4.36, equipped with the DC voltage V_{dc} and connected to power grid through filter inductance L_f is simulated for this study.

Parameters: The main parameters of the simulation system are shown in Table 4.13. In addition, to show the influence caused by the difference between C_{au} and C_{al} clearly, the situation that the SM capacitance in the upper arm of phase A and the SM capacitance in the lower arm of phase B drop by 20%.

Simulation results and analysis: Figure 4.37 shows the performance of the optimal control, where the optimal control is enabled since 2 seconds. Figure 4.37a shows the three-phase AC-grid current i_a, i_b, i_c. Figure 4.37b shows the upper arm current i_{au} and the lower arm current i_{al} in phase A. Figure 4.37c shows the upper arm current i_{bu} and the lower arm current i_{bl} in phase B. Figure 4.37d shows the upper arm current i_{cu} and the lower arm current i_{cl} in phase C. Figure 4.37e shows the circulating current i_{diff_a}, i_{diff_b}, i_{diff_c} in the three phases. Figure 4.37f shows the DC-link current of the MMC. Obviously,

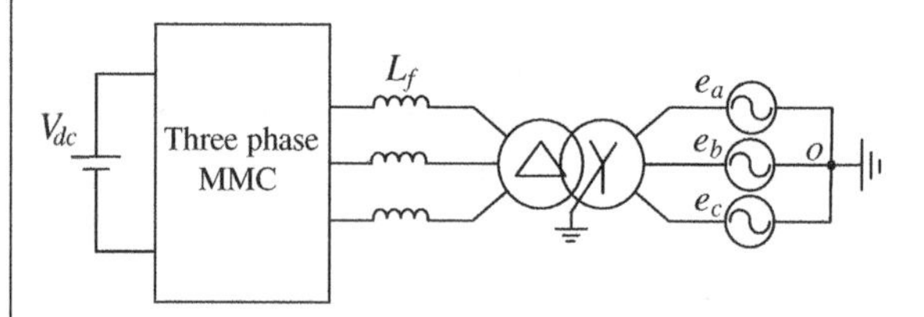

Figure 4.36 Schematic diagram of the simulation system.

Table 4.13 Main parameters of the simulation system.

Parameter	Value
DC-link voltage U_{dc}	6 kV
Load frequency	50 Hz
Carrier frequency	2 kHz
Number of SMs per arm	6
Nominal SM capacitance	12.5 mF
Arm inductance L_s	3 mH
Load inductance L_f	1 mH

the fundamental components are produced owing to the capacitor aging, which would affect the DC-side current. However, the optimal control can effectively eliminate the fundamental component in the circulating current and improve the DC-side current as well.

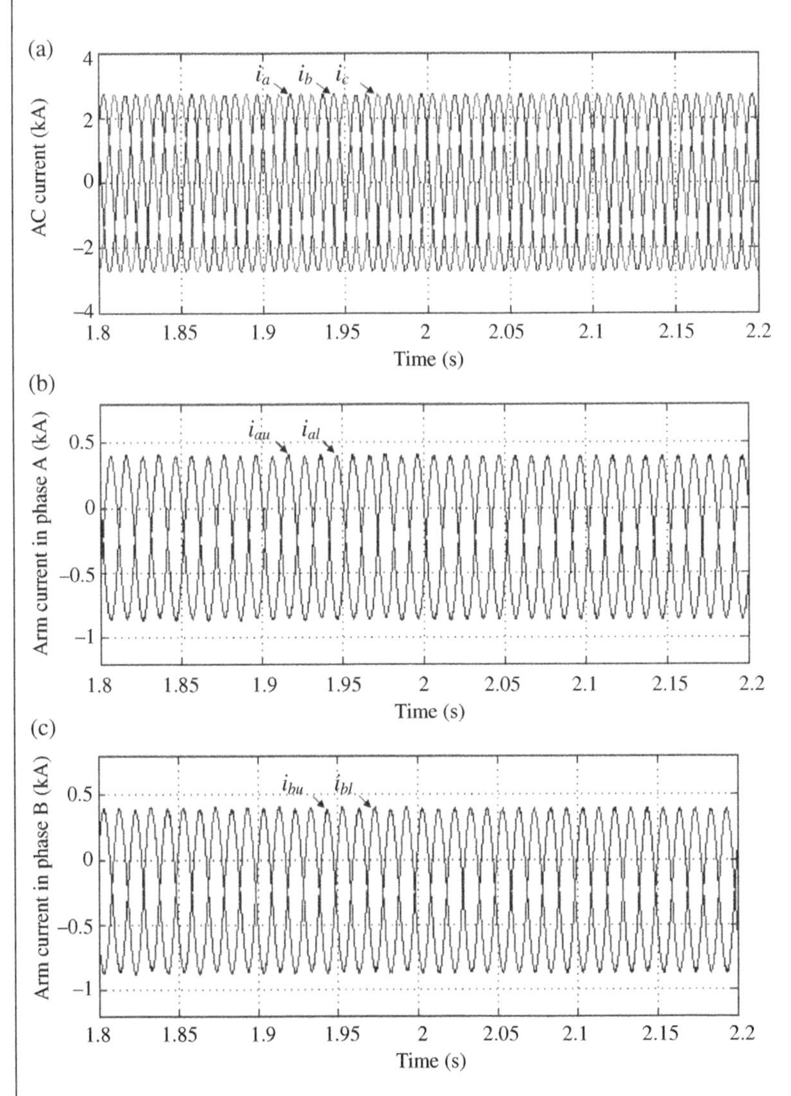

Figure 4.37 Simulated waveforms. (a) i_a, i_b, i_c. (b) i_{au}, i_{al}. (c) i_{bu}, i_{bl}. (d) i_{cu}, i_{cl}. (e) $i_{diff\_a}, i_{diff\_b}, i_{diff\_c}$. (f) i_{dc}.

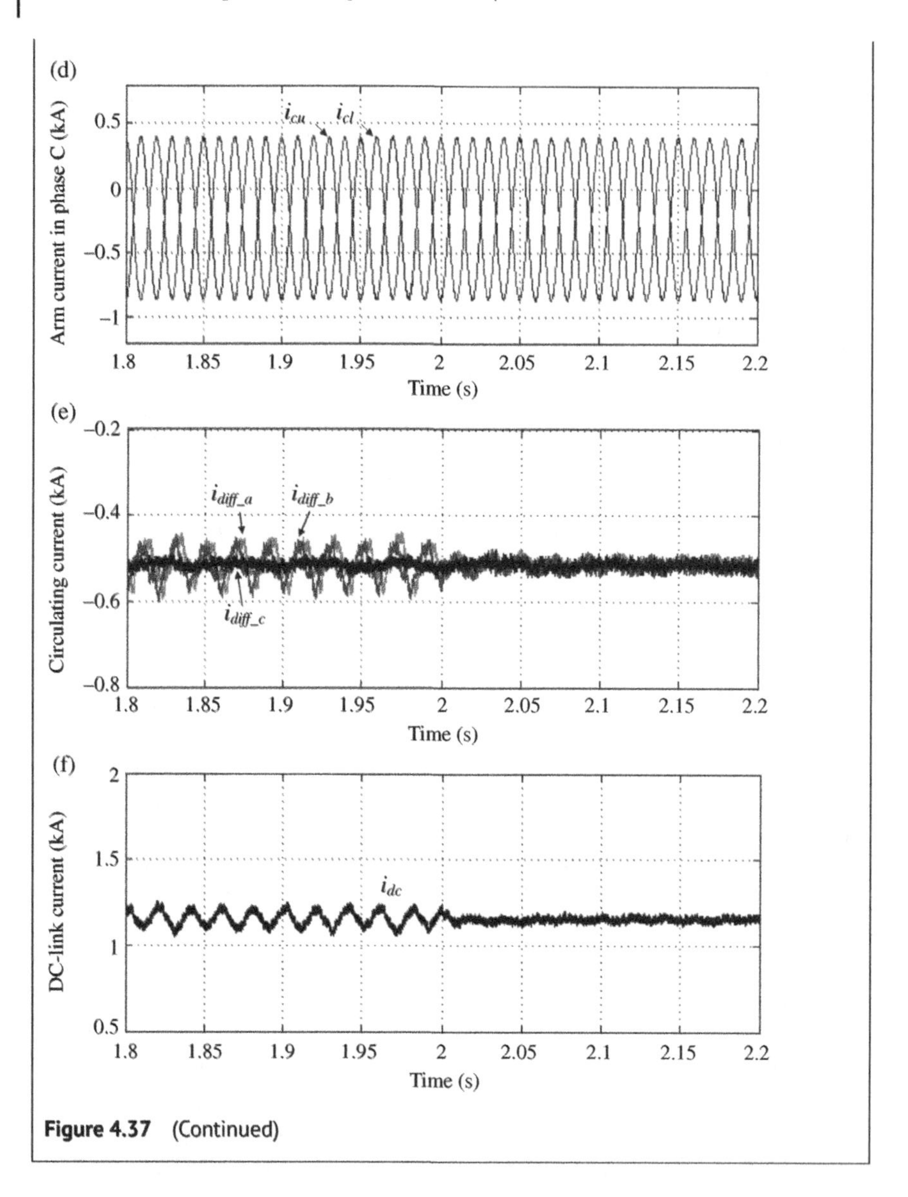

Figure 4.37 (Continued)

4.9 SM Power Losses Optimal Control Under Capacitor Aging

Owing to capacitor deterioration, the MMC would operate with different capacitances in different SMs. Apart from the rising circulating current discussed in the last section, capacitor aging would also affect the SM power losses in the arm of the

MMC. In this section, the power losses distribution of the MMC under capacitance deterioration is analyzed in detail, where the capacitor deterioration would cause different ERs for the SMs in the same arm. It would result in unbalanced SM power losses distribution in the same arm of the MMC and cause different aging speed of semiconductors, and therefore affect the reliability of the MMC. Then, an equivalent-reference control method is presented, which can effectively realize balanced SM power losses distribution in the same arm of the MMC through the VBC for the virtual capacitor voltages in the MMC under the capacitor deterioration.

4.9.1 Equivalent SM Reference

Based on equations (4.24) and (4.27), the ER y_{er_aui} for the i-th SM ($i = 1, 2, ..., n$) depends on the C_{aui}. Along with the drop of C_{aui}, the y_{er_aui} will be reduced, as shown in Table 4.14. On the other hand, according to equations (4.3) and (4.24), a DC component would be caused in y_{er_aui}, as shown in Table 4.15, as follows:

- *Situation I*: If $C_{aui} > C_{ave}$, the y_{er_aui} will be more than y_{au}. Here, a positive DC component would be caused in y_{er_aui}. If C_{aui} is far more than C_{ave}, the DC component in y_{er_aui} would be much bigger.
- *Situation II*: If $C_{aui} = C_{ave}$, the y_{er_aui} will be equal to y_{au} and there is no DC component in y_{er_aui}.
- *Situation III*: If $C_{aui} < C_{ave}$, the y_{er_aui} will be less than y_{au}. Here, a negative DC component would be caused in y_{er_aui}. If C_{aui} is far less than C_{ave}, the DC component in y_{er_aui} would be much smaller.

Figure 4.38 shows the performance of the MMC with various capacitances in the arm, which has 100 SMs per arm. Figure 4.38a shows the capacitances C_{au1}–C_{au6}. The other capacitances are all 15 mF. The average capacitance C_{ave} in the arm is obtained as

$$C_{ave} = \sum_{i=1}^{100} C_{aui} / 100 = 14.775\text{mF} \tag{4.53}$$

Among C_{au1}–C_{au6}, only C_{au1} is more than C_{ave} and the others are less than C_{ave}. Figure 4.38b shows the DC component in y_{er_au1}–y_{er_au6}, where only the DC component in y_{er_au1} is positive and the DC components in y_{er_au2}–y_{er_au6} are all negative, which verifies the analysis in Table 4.15. In addition, Figure 4.38c shows

Table 4.14 Relationship between C_{aui} and y_{er_aui}.

C_{aui}	ER y_{er_aui}
↓	↓

Table 4.15 DC component in ER.

Situation	C_{aui}	y_{er_aui}	DC component in y_{er_aui}
I	$>C_{ave}$	$>y_{au}$	>0
II	$=C_{ave}$	$=y_{au}$	=0
III	$<C_{ave}$	$<y_{au}$	<0

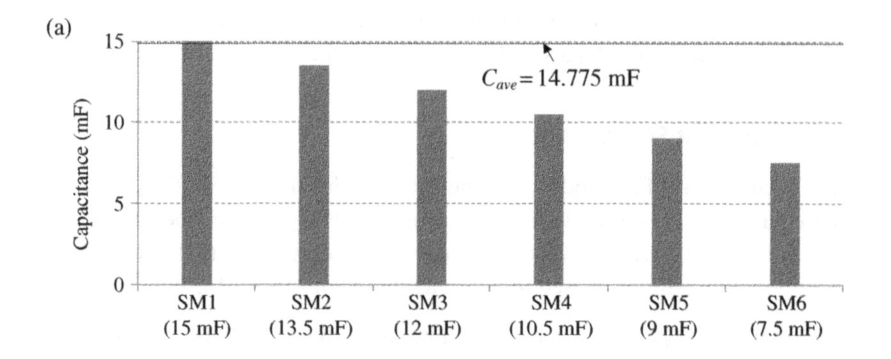

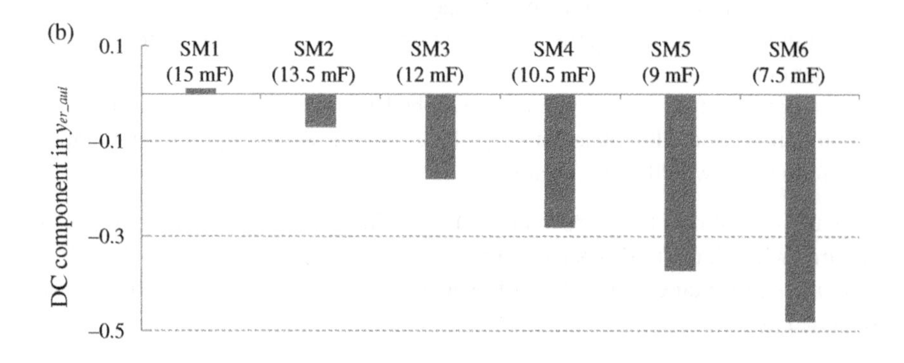

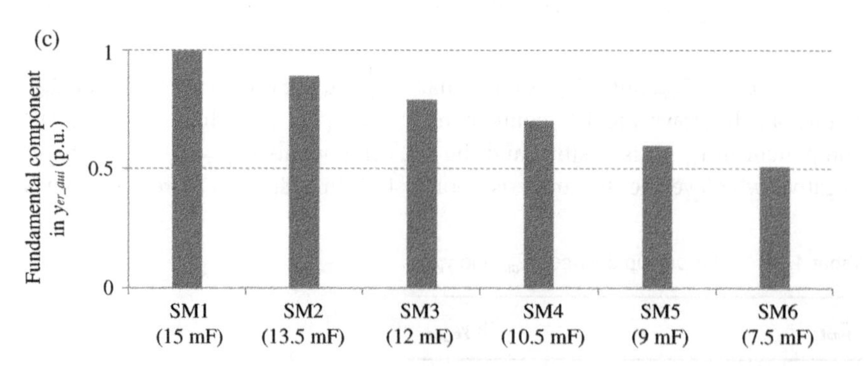

Figure 4.38 (a) SM capacitance. (b) DC component in y_{er_aui}. (c) Fundamental component in y_{er_aui}.

the relationship among the fundamental components in y_{er_au1}–y_{er_au6}, which meets with equation (4.24) and verifies the analysis in Table 4.14.

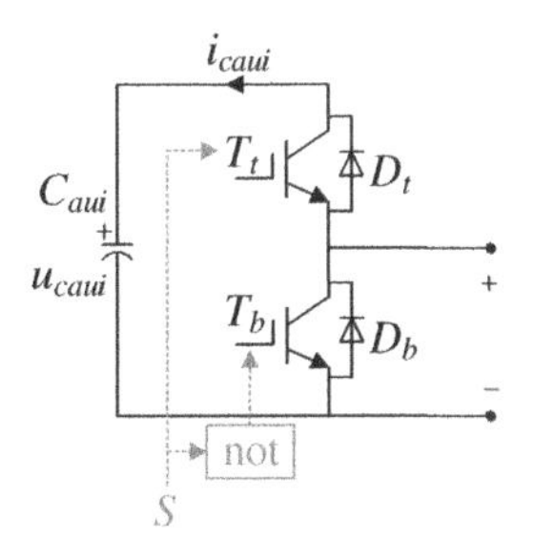

Figure 4.39 SM unit.

4.9.2 SM Conduction Losses

The conduction situation of the top and bottom switch/diode (T_t/D_t and T_b/D_b) in each SM, as shown in Figure 4.39, is listed in Table 4.16. The conduction losses of the switch/diode are analyzed in detail as follows:

- *Conduction losses P_{ctt} of T_t*: arm current i_{au} flows through T_t when $i_{au}<0$ and $S_{aui}=1$. The P_{ctt} can be expressed as [7]

$$P_{ctt\_i}=-i_{au}\cdot S_{aui}\cdot\left[u_{ceo}+r_c\cdot\left(-i_{au}\right)\cdot S_{aui}\right] \tag{4.54}$$

where u_{ceo} and r_c are switch on-state zero-current collector-emitter voltage and collector-emitter on-state resistance, respectively. Based on equations (4.21) and (4.54) can be rewritten as

$$P_{ctt\_i}=-i_{au}\cdot\frac{1+y_{er\_aui}}{2}\cdot\left[u_{ceo}+r_c\cdot\left(-i_{au}\right)\cdot\frac{1+y_{er\_aui}}{2}\right] \tag{4.55}$$

- *Conduction losses P_{cdt_i} of D_t*: i_{au} flows through D_t when $i_{au}\geq 0$ and $S_{aui}=1$. With the Fourier series expansion of S_{aui}, the P_{cdt_i} can be expressed as

$$P_{cdt\_i}=i_{au}\cdot\frac{1+y_{er\_aui}}{2}\cdot\left[u_{ceo}+r_c\cdot i_{au}\cdot\frac{1+y_{er\_aui}}{2}\right] \tag{4.56}$$

- *Conduction losses P_{ctb_i} of T_b*: i_{au} flows through T_b when $i_{au}\geq 0$ and $S_{aui}=0$. With the Fourier series expansion of S_{aui}, the P_{ctb_i} can be expressed as

$$P_{ctb\_i}=i_{au}\cdot\left(1-\frac{1+y_{er\_aui}}{2}\right)\cdot\left[u_{ceo}+r_c\cdot i_{au}\cdot\left(1-\frac{1+y_{er\_aui}}{2}\right)\right] \tag{4.57}$$

- *Conduction losses P_{cdb_i} of D_b*: i_{au} flows through D_b when $i_{au}<0$ and $S_{aui}=0$. With the Fourier series expansion of S_{aui}, the P_{cdb_i} can be expressed as

$$P_{cdb\_i}=i_{au}\cdot\left(1-\frac{1+y_{er\_aui}}{2}\right)\cdot\left[u_{ceo}+r_c\cdot\left(-i_{au}\right)\cdot\left(1-\frac{1+y_{er\_aui}}{2}\right)\right] \tag{4.58}$$

Table 4.16 Conduction situations in SM.

		Semiconductor current			
i_{au}	S	T_t	D_t	T_b	D_b
≥0	1	0	i_{au}	0	0
	0	0	0	i_{au}	0
<0	1	i_{au}	0	0	0
	0	0	0	0	i_{au}

Table 4.17 Conduction losses of SM.

		Top switch/diode		Bottom switch/diode	
C_{aui}	ER y_{er_aui}	T_t	D_t	T_b	D_b
		P_{ctt_i}	P_{cdt_i}	P_{ctb_i}	P_{cdb_i}
Drop	Reduced	Reduced	Reduced	Increased	Increased

The conduction loss of the T_t/D_t and T_b/D_b in the i-th SM is related to the ER y_{er_aui}, as shown in Table 4.17. Combining equations (4.24) and (4.55)–(4.58), the capacitor C_{aui} drop causes reduced y_{er_aui}, which reduces P_{ctt_i}, P_{cdt_i} and increases P_{ctb_i}, P_{cdb_i} in the i-th SM.

4.9.3 SM Switching Losses

Although all capacitor voltages in the arm are kept balanced with the VBC, the drop of the capacitance C_{aui} reduces ER y_{er_aui} for the SM, which would result in different switching frequencies and different switching losses for these SMs in the same arm. Under the VBC method in [14], it is possible that the SM with the dropped C_{aui} would reduce its switching times to reduce ER y_{er_aui}. Consequently, the capacitance drop would reduce SM switching frequency and decrease SM switching losses, as shown in Table 4.18.

Figure 4.40 shows the SM losses with various capacitances in the same arm, where the Infineon IGBT FZ1200R17HP4 is used, and the losses are calculated based on the simulated current waveforms and the semiconductor specifications from the manufacturer. The junction temperature is considered to be 125 °C. Along with the drop of the capacitance, as shown in Figure 4.40a, the conduction losses P_{ctt}, P_{cdt} in T_t, D_t are reduced and the conduction losses P_{ctb}, P_{cdb} in T_b, D_b are

Table 4.18 Trend of switching losses in SMs.

C_{aui}	ER y_{er_aui}	Trend of SM switching frequency	Switching losses
Drop	Reduced	Reduced	Reduced

increased, as shown in Figure 4.40a, which verifies the conduction losses analysis in Table 4.18. In addition, the drop of the capacitance causes the reduction of the SM switching frequency and the decrease of the switching losses P_{stt}, P_{sdt}, P_{stb}, and P_{sdb} in T_t, D_t, T_b, and D_b, respectively, as shown in Figure 4.40b,c, which verifies the switching losses analysis in Table 4.18. Figure 4.40d shows the power losses of the top switch/diode T_t/D_t and bottom switch/diode T_d/D_d. Along with the capacitance drop, the power losses of the top switch/diode T_t/D_t are reduced and the power losses of the bottom switch/diode T_d/D_d are increased, which results in that the error between the power losses of the top switch/diode T_t/D_t and the power losses of the bottom switch/diode T_d/D_d is increased along with the capacitance drop.

The different capacitance in the different SMs would cause different ERs for the SMs, which would cause different T_t loss, D_t loss, T_b loss, and D_b loss in the different SMs, respectively. It would cause different aging speeds of the T_t, D_t, T_b, and D_b in the different SMs, respectively, which results in different lifetimes of the T_t, D_t, T_b, and D_b in the different SMs, respectively, and therefore affects the reliability of the MMC.

4.9.4 SM Power Losses Optimal Control

Although the capacitor voltages are kept balanced, the capacitance drop causes different ERs y_{er_au1}–y_{er_aun} for the SMs in the same arm. The different ERs of the SMs would result in different SM losses in the same arm and affect the reliability of the MMC.

To improve the performance of the MMC, an ER control method is presented for the MMC, as shown in Figure 4.41, which can ensure that the ERs are close to each other in the same arm. In Figure 4.41, the capacitor voltage u_{caui} in the upper arm of phase A is monitored, which can be expressed as

$$u_{caui} = u_{dsm} + \Delta u_{caui} \tag{4.59}$$

where u_{dsm} is the DC component and Δu_{caui} is the ripple component. In the steady-state situation, the arm current does not affect the DC component u_{dsm} but the ripple component Δu_{caui} [27]. In the presented ER control, the virtual capacitor voltage (VCV) u'_{caui} is defined as

$$u'_{caui} = u_{dsm} + k_i \cdot \Delta u_{caui} \tag{4.60}$$

where k_i is the coefficient.

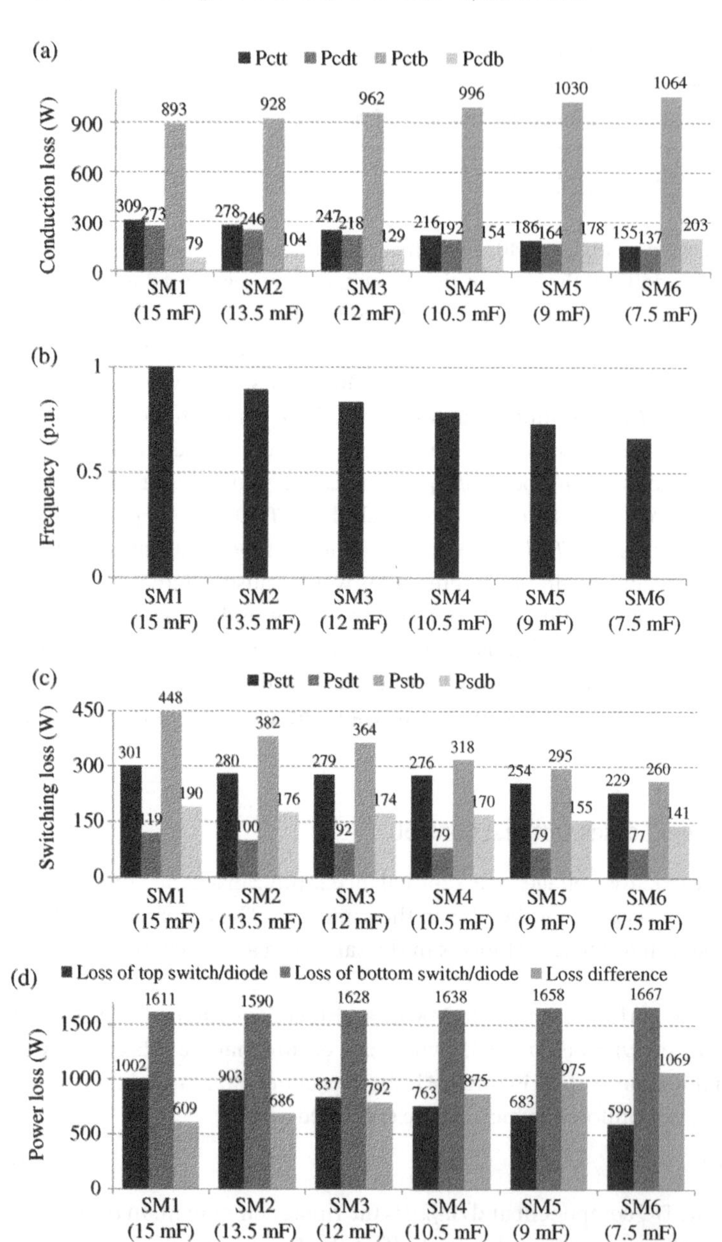

Figure 4.40 Power losses of the SM1–SM6. (a) Conduction losses. (b) Switching frequency. (c) Switching losses. (d) Top and bottom switch/diode losses.

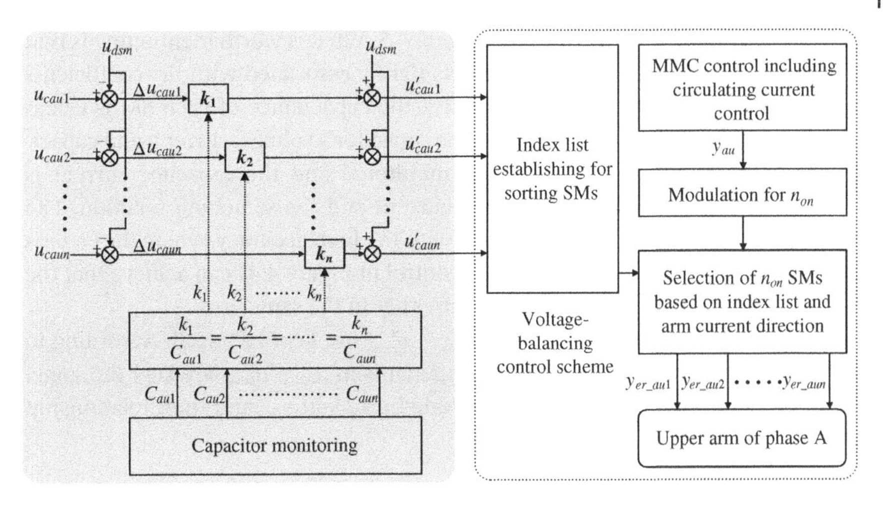

Figure 4.41 ER control for the MMC under capacitor deterioration.

Based on the VCV $u'_{cau1} - u'_{caun}$, the VBC [14] is implemented. The index list for the SMs in the arm is established through sorting VCV $u'_{cau1} - u'_{caun}$ in ascending order. The required on-state SM number n_{on} is obtained by the arm reference y_{au}, which is calculated based on not only the MMC control methods such as active power control, reactive power control, and DC-link voltage control but also the circulating current suppress control. According to the index list, required on-state SM number n_{on} and the arm current, appropriate SMs will be switched to the "On" state, and the ER y_{er_au1}–y_{er_aun} for the SMs in the same arm will be generated, which can ensure the VCV balancing as

$$u'_{cau1} = u'_{cau2} = \cdots\cdots = u'_{caun} \tag{4.61}$$

Combining equations (4.2), (4.3), (4.19), (4.59), and (4.61), (4.62) can be obtained as

$$k_1 \cdot \frac{1 + y_{er\_au1}}{C_{au1}} = k_2 \cdot \frac{1 + y_{er\_au2}}{C_{au2}} = \cdots\cdots = k_n \cdot \frac{1 + y_{er\_aun}}{C_{aun}} \tag{4.62}$$

In order to keep the ERs y_{er_au1}–y_{er_aun} close to each other even if the capacitors C_{au1}–C_{aun} are not the same in Figure 4.41, equation (4.63) should be satisfied.

$$\frac{k_1}{C_{au1}} = \frac{k_2}{C_{au2}} = \cdots\cdots = \frac{k_n}{C_{aun}} \tag{4.63}$$

The coefficients k_1–k_n in Figure 4.41 can be decided based on the SM capacitances in the arm. The capacitance in the MMC can be calculated based on the capacitance monitoring methods in Section 4.5. What is worth mentioning is that the accuracy of capacitance estimation is tightly associated with the coefficients k_1–k_n based on equation (4.63). Generally, the capacitance of each SM is calculated based on the relationship among the capacitor's voltage, current, and capacitance, where the capacitor voltage is monitored and the capacitor current is obtained based on the monitored arm current and the switching function. The capacitance estimation can be achieved with the high accuracy, where the error is less than 1%. As a result, the presented control in Figure 4.41 can achieve that the ERs y_{er_au1}–y_{er_aun} are nearly close to each other in the same arm.

With the control strategy, the VCV $u'_{cau1} - u'_{caun}$ are kept balanced. According to equations (4.59) and (4.60), the DC components in u_{cau1}–u_{caun} are kept the same. The only difference is the ripple amplitude in u_{cau1}–u_{caun}, and their relationship can be expressed as

$$k_1 \cdot \Delta u_{cau1} = k_2 \cdot \Delta u_{cau2} = \cdots\cdots = k_n \cdot \Delta u_{caun} \tag{4.64}$$

Case Study 4.6 Analysis of SM Power Losses Optimal Control under Capacitor Parameters Faults

Objective: In this case study, a three-phase MMC system is built with the time-domain simulation tool PSCAD/EMTDC to verify the power losses optimal control, as shown in Figure 4.42. The MMC works in inverter mode, and the power flow is from the DC side to the AC side.

Parameters: The main system parameters are listed in Table 4.19. The active power P and the reactive power Q are 100 MW and 0 MVar, respectively. In addition, to show the influence caused by capacitance drop clearly, the situation that only the capacitors within phase A are deteriorated is considered in this section. Here, the capacitances C_{au1}–C_{au6} drop, where C_{au1} = 15 mF, C_{au2} = 13.5 mF, C_{au3} = 12 mF, C_{au4} = 10.5 mF, C_{au5} = 9 mF, and C_{au6} = 7.5 mF, respectively, as shown in Figure 4.38a.

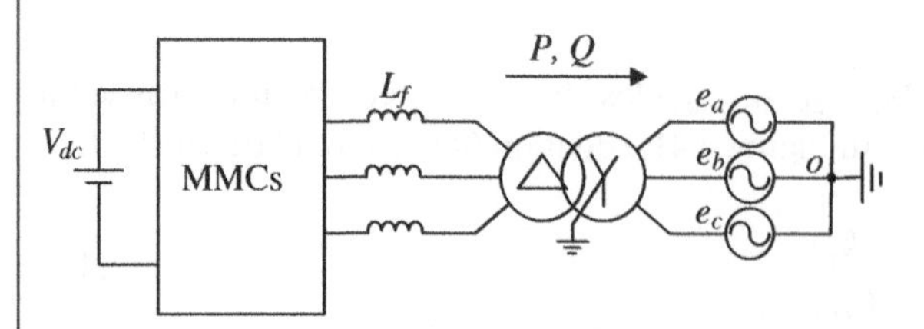

Figure 4.42 Block diagram of the simulation system.

Table 4.19 Simulation system parameters.

Parameters	Value
DC-link voltage U_{dc} (kV)	100
Grid line-to-line voltage (kV)	220
Grid frequency (Hz)	50
Transformer voltage rating	50 kV/220 kV
Number of SMs per arm n	100
Nominal SM capacitance C (mF)	15
Inductance L_s (mH)	10
Load inductance L_f (mH)	2

Simulation results and analysis:

Figures 4.43–4.46 show the performance of the three-phase MMC. In Figure 4.43, the presented control is enabled since 2 seconds. Figure 4.43a shows the arm current i_{au}. Figure 4.43b,c shows the capacitor voltage u_{cau1}–u_{cau10}, where the peak values of the u_{cau2}–u_{cau6} are increased to 1.08, 1.09, 1.1, 1.12, and 1.15 p.u., respectively, under the presented control so as to improve the SM power losses distribution, which meets equations (4.63) and (4.64) in the presented control. Suppose that the capacitor needs to be replaced when its capacitance is less than 80% of the rated value, the voltage peak value of the replaced capacitor is only 1.09 p.u., which is only increased by 1.8%. The ripple amplitudes in u_{cau1}–u_{cau6} under the presented control are shown in Figure 4.43d, which are far less than the capacitor voltages, as shown in Figure 4.43b. In addition, the THD of the arm current i_{au} without and with the presented control is shown in Figure 4.43e, respectively. The THD of the MMC's output voltage u_{ab} without and with the presented control is shown in Figure 4.43f, respectively. The THD of the MMC's output current i_a without and with the presented control is shown in Figure 4.43g, respectively. The THD of the i_{au}, u_{ab}, and i_a with the presented control is almost the same as that without the presented control, respectively, which shows that the presented control has little impact on the performance of the MMC because the capacitor voltage ripple is quite small in comparison with the capacitor voltage.

Figures 4.44 and 4.45 show the performance of the MMC under the optimal control. Figure 4.44a shows the DC component in y_{er_au1}–y_{er_au6}, which is quite small and can be negligible compared with the fundamental component. Figure 4.44b shows the amplitude of the fundamental component in

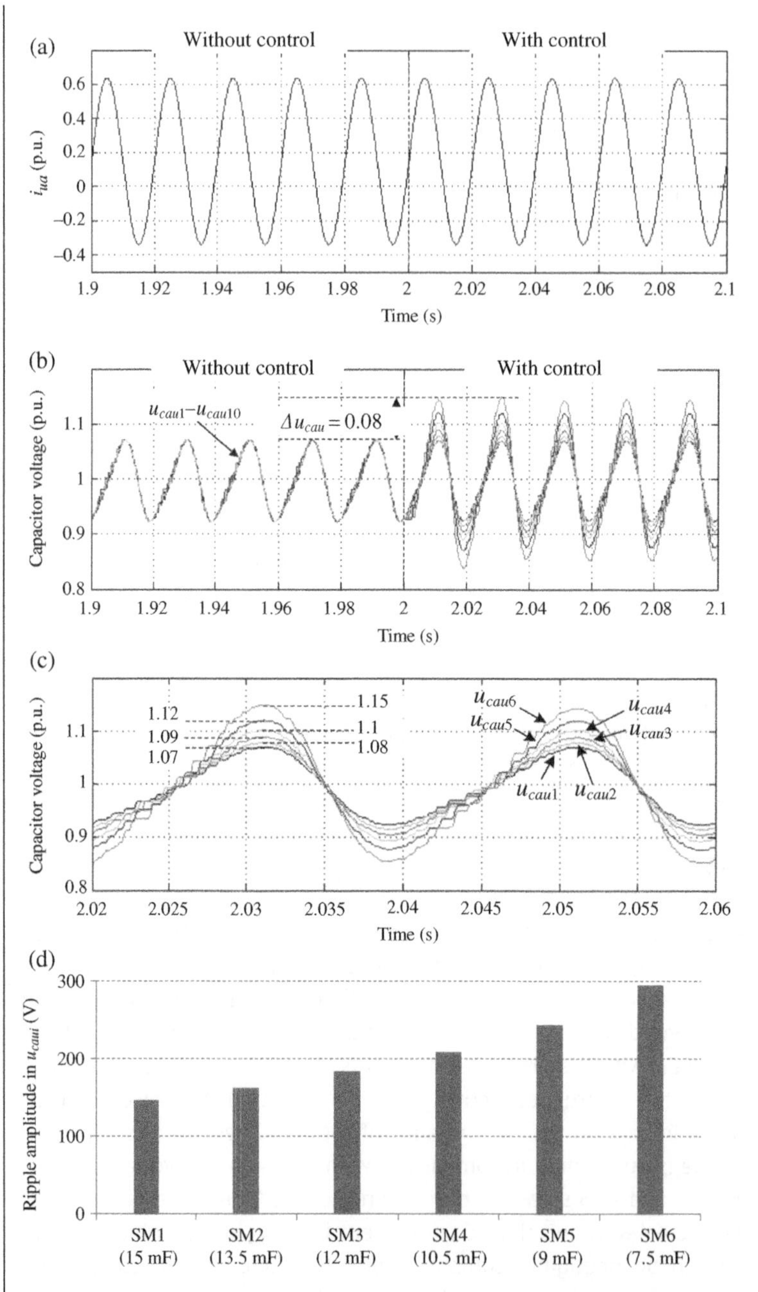

Figure 4.43 (a) i_{au}. (b) $u_{cau1}-u_{cau10}$. (c) $u_{cau1}-u_{cau6}$. (d) Ripple amplitudes in $u_{cau1}-u_{cau6}$. (e) THD analysis of i_{au}. (f) THD analysis of u_{ab}. (g) THD analysis of i_a.

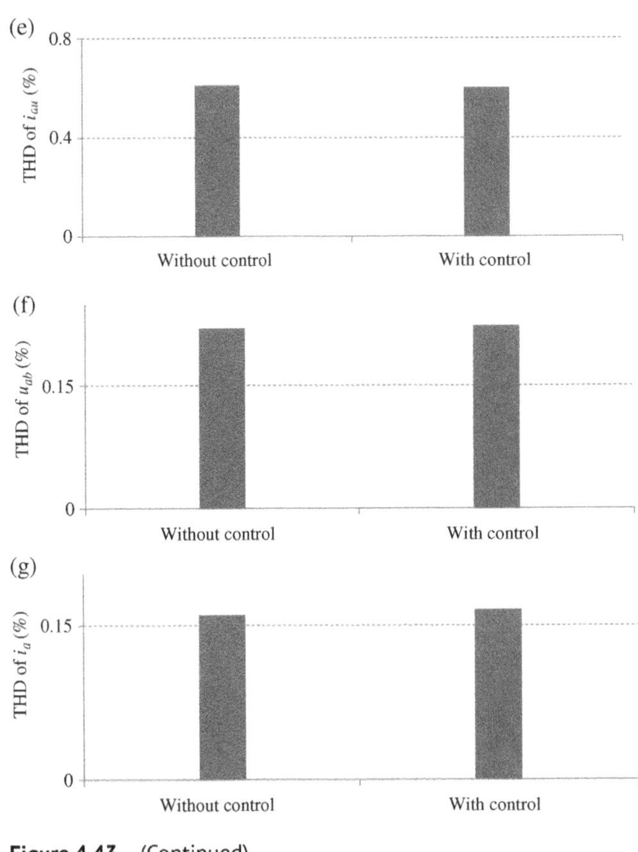

Figure 4.43 (Continued)

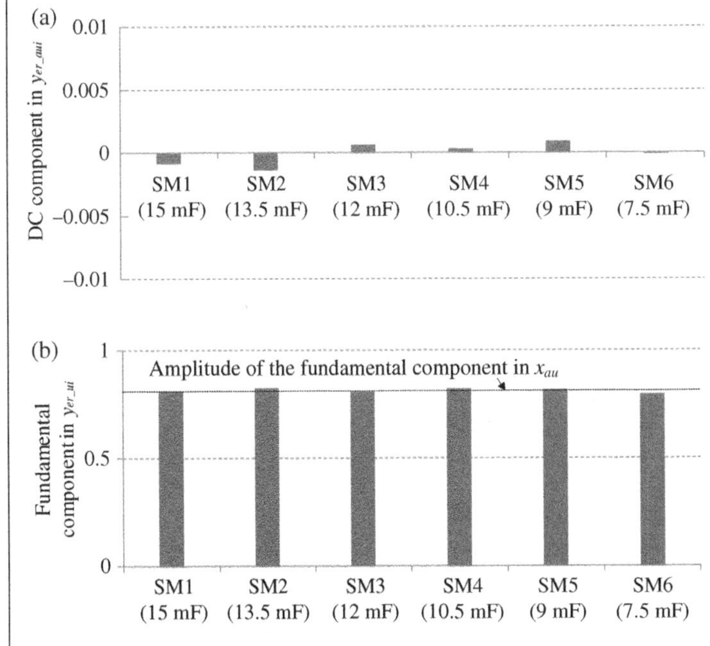

Figure 4.44 (a) DC components in y_{er_au1}–y_{er_au6}. (b) Fundamental components in y_{er_au1}–y_{er_au6}.

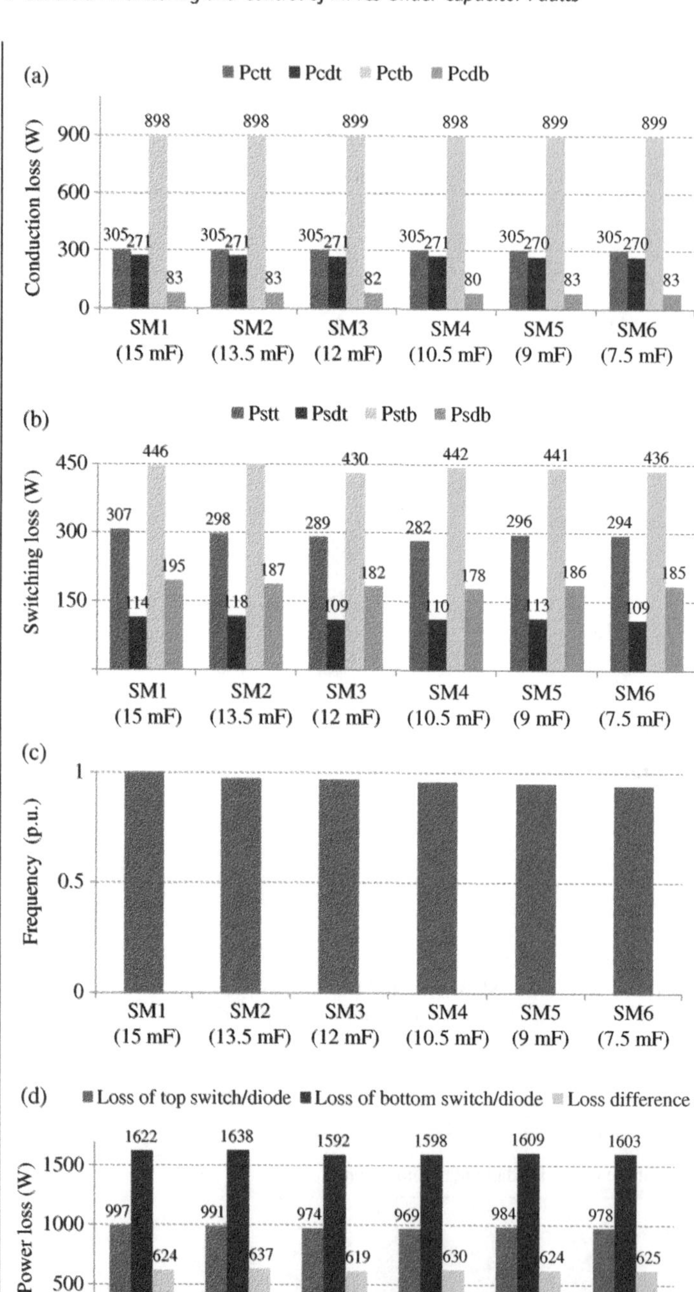

Figure 4.45 Power losses in SM1–SM6. (a) Conduction losses. (b) Switching losses. (c) Switching frequency. (d) Top and bottom switch/diode losses.

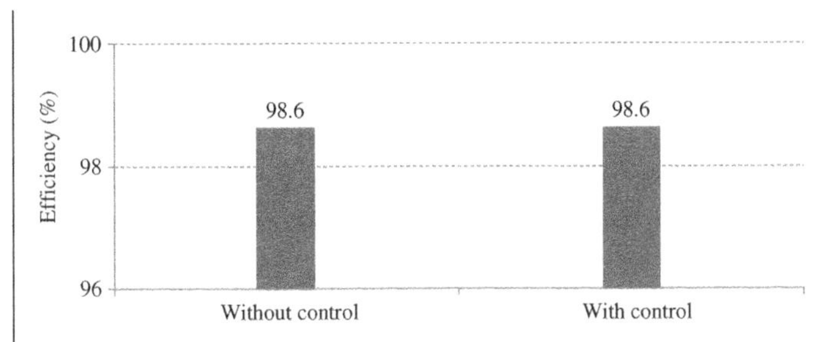

Figure 4.46 Efficiency of MMCs without and with the presented control.

$y_{er\_au1}-y_{er\_au6}$, which are almost the same with each other and nearly equal to that in y_{au}. As a result, the presented control effectively improves the ERs in the MMC. With the presented control, the conduction losses, switching losses, and switching frequency in SM1–SM6 are nearly close to each other, respectively, as shown in Figure 4.45a–c. The losses of the top switch/diode and the losses of the bottom switch/diode in SM1–SM6 are nearly close to each other, respectively, as shown in Figure 4.45d. Therefore, the losses difference between the top switch/diode and the bottom switch/diode in SM1–SM6 are nearly close to each other, as shown in Figure 4.45d, which improves the power losses distribution in the arm in comparison with that without the presented control shown in Figure 4.40. Figure 4.46 shows the efficiencies of the MMC without and with the presented control and they are quite close to each other.

4.10 Summary

Capacitor is one of the key components in the MMC, and aluminum electrolytic capacitors are the preferred choice due to desirable characteristics such as high volume and low cost. However, the aging effect of electrolytic capacitors, which is usually expressed as capacitance drop and ESR rise, would affect the normal operation of the MMC. Condition monitoring method is known as an effective method to estimate the health state of a capacitor and ensure the reliable operation of the MMC.

This chapter illustrates up-to-date developments in capacitor monitoring strategies and optimized operation schemes based on capacitor aging. Capacitance and ESR are two prominent indicators to represent capacitor's health state. Based on capacitance drop during operation, direct capacitance estimation scheme, RSM-based capacitance estimation scheme, and sorting-based capacitance estimation scheme are presented, respectively. Based on ESR rise during operation, direct ESR monitoring scheme and sorting-based ESR monitoring scheme are presented. Furthermore, temperature effect of electrolytic capacitors is analyzed. The circulating current would be introduced and SM switch loss would become unbalanced due to parameter change of capacitors. Accordingly, fault-tolerant optimized control can ensure the normal operation of the MMC via suppressing circulating current and keeping each SM's switch loss balanced caused by the capacitor's parameter change.

References

1 H. Liu, P. C. Loh, and F. Blaabjerg, "Review of fault diagnosis and fault-tolerant control for modular multilevel converter of HVDC," in *Proceedings of IECON 2013*, pp. 1242–1247, 2013.

2 P. Venet, F. Perisse, M.H. El-Husseini, and G. Rojat, "Realization of a smart electrolytic capacitor circuit," *IEEE Ind. Appl. Mag.*, vol. 8, no. 1, pp. 16–20, Jan–Feb. 2002.

3 M. Vogelsberger, T. Wiesinger, and H. Ertl, "Life-cycle monitoring and voltage-managing unit for DC-link electrolytic capacitors in PWM converters," *IEEE Trans. Power Electron.*, vol. 26, no. 2, pp. 493–503, Feb. 2011.

4 A. Amaral and A. Cardoso, "An experimental technique for estimating the ESR and reactance intrinsic values of aluminum electrolytic capacitors," in *2006 IEEE Instrumentation and Measurement Technology Conference Proceedings*, Sorrento, Italy, pp. 1820–1825, Apr. 2006.

5 H. Soliman, H. Wang and F. Blaabjerg, "A Review of the Condition Monitoring of Capacitors in Power Electronic Converters," *IEEE Trans. Ind. Appl.*, vol. 52, no. 6, pp. 4976–4989, Nov–Dec. 2016.

6 Q. Heng, F. Deng, C. Liu, Q. Wang, and J. Chen, "Circulating current control scheme under capacitor aging in modular multilevel converter," in *8th Renewable Power Generation Conference*, Shanghai, China, pp. 1–7, 2019.

7 F. Deng, Q. Heng, C. Liu, X. Cai, R. Zhu, Z. Chen, and W. Chen, "Power losses control for modular multilevel converters under capacitor deterioration," *IEEE Trans. Emerging Sel. Top. Power Electron.*, vol. 8, no. 4, pp. 4318–4332, Dec. 2020.

8 K. Lee, M. Kim, J. Yoon, S. B. Lee and J. Yoo, "Condition monitoring of DC-link electrolytic capacitors in adjustable-speed drives," *IEEE Trans. Ind. Appl.*, vol. 44, no. 5, pp. 1606–1613, Sept.-Oct. 2008.

9 K. Laadjal, M. Sahraoui, A. J. M. Cardoso, and A. M. R. Amaral, "Online estimation of aluminum electrolytic-capacitors parameters using a modified Prony's method." *IEEE Trans. Ind. Appl.*, vol. 54, no. 5, pp. 4764–4774, Sept.–Oct. 2018.

10 Aluminum Electrolytic Capacitor Application Guide, 2018. [Online]. Available: http://www.cde.com/resources/catalogs/AEappGUIDE.pdf.

11 J. Lyu, X. Cai, and M. Molinas, "Optimal design of controller parameters for improving the stability of MMC-HVDC for wind farm integration," *IEEE Trans. Emerging Sel. Top. Power Electron.*, vol. 6, no. 1, pp. 40–53, Mar. 2018.

12 F. Deng, Q. Yu, Q. Wang, R. Zhu, X. Cai, and Z. Chen, "Suppression of DC-link current ripple for modular multilevel converters under phase-disposition PWM," *IEEE Trans. Power Electron.*, vol. 35, no. 3, pp. 3310–3324, Mar. 2020.

13 M. Vasiladiotis, N. Cherix, and A. Rufer, "Impact of grid asymmetries on the operation and capacitive energy storage design of modular multilevel converters," *IEEE Trans. Ind. Electron.*, vol. 62, no. 11, pp. 6697–6707, Nov. 2015.

14 S. Debnath, J. Qin, B. Bahrani, M. Saeedifard, and P. Barbosa, "Operation, control, and applications of the modular multilevel converter: a review," *IEEE Trans. Power Electron.*, vol. 30, no. 1, pp. 37–53, Mar. 2014.

15 O. Abushafa, S. Gadoue, M. Dahidah, and D. Atkinson, "A new scheme for monitoring submodule capacitance in modular multilevel converter," in *Proceedings of PEMD 2016*, pp. 1–6, 2016.

16 F. Deng *et al.*, "Capacitor ESR and C monitoring in modular multilevel converters," *IEEE Trans. Power Electron.*, vol. 35, no. 4, pp. 4063–4075, Apr. 2020.

17 Z. Wang, Y. Zhang, H. Wang, and F. Blaabjerg, "A reference submodule based capacitor condition monitoring method for modular multilevel converters," *IEEE Trans. Power Electron.*, vol. 35, no. 7, pp. 6691–6696, Jul. 2020.

18 F. Deng, Q. Wang, D. Liu, Y. Wang, M. Cheng, and Z. Chen, "Reference submodule-based capacitor monitoring strategy for modular multilevel converters," *IEEE Trans. Power Electron.*, vol. 34, no. 5, pp. 4711–4721, May 2019.

19 J. Yun-jae, N. Thanh Hai, and L. Dong-Choon, "Condition monitoring of submodule capacitors in modular multilevel converters," in *Proceedings of ECCE 2014*, pp. 2121–2126, 2014.

20 Q. Tu, Z. Xu, and L. Xu, "Reduced switching-frequency modulation and circulating current suppression for modular multilevel converters," *IEEE Trans. Power Delivery*, vol. 26, no. 3, pp. 2009–2017, Jul. 2011.

21 A. M. R. Amaral and A. J. M. Cardoso, "A simple offline technique for evaluating the condition of aluminum–electrolytic–capacitors," *IEEE Trans. Ind. Electron.*, vol. 56, no. 8, pp. 3230–3237, Aug. 2009.

22 F. Argall and A. Jonscher, "Dielectric properties of thin films of aluminium oxide and silicon oxide," *Thin Solid Films*, vol. 2, no. 3, pp. 185–210, 1968.

23 X. S. Pu, T. H. Nguyen, D. C. Lee, K. B. Lee, and J. M. Kim, "Fault diagnosis of dc-link capacitors in three-phase ac/dc PWM converters by online estimation of equivalent series resistance," *IEEE Trans. Ind. Electron.*, vol. 60, no. 9, pp. 4118–4127, Sep. 2013.

24 P. Inttpat, P. Paisuwanna, and S. Khomfoi, "Capacitor lifetime monitoring for multilevel modular capacitor clamped DC to DC converter," in *The 8th Electrical Engineering/ Electronics, Computer, Telecommunications and Information Technology (ECTI) Association of Thailand - Conference*, pp. 719–722, 2011.

25 J. Salaet, S. Alepuz, A. Gilabert, and J. Bordonau, "Comparison between two methods of DQ transformation for single phase converters control. Application to a 3-level boost rectifier," in *2004 IEEE 35th Annual Power Electronics Specialists Conference (IEEE Cat. No.04CH37551)*, Aachen, Germany, vol. 1, pp. 214–220, 2004.

26 M. Saeedifard and R. Iravani, "Dynamic performance of a modular multilevel back-to-back HVDC system," in *IEEE Power Energy Society General Meeting*, pp. 1–1, 2011.

27 Q. Song, W. Liu, X. Li, H. Rao, S. Xu, and L. Li, "A steady-state analysis method for a modular multilevel converter," *IEEE Trans. Power Electron.*, vol. 28, no. 8, pp. 3702–3713, Aug. 2013.

5

Fault-Tolerant Control of MMCs Under SM Faults

5.1 Introduction

The modular multilevel converter (MMC) consists of a large number of SMs, and each SM may be considered as a potential failure point. The SM faults may distort the voltage and current and affect the normal operation of the MMC [1, 2]. Fault-tolerant operation is one of the most important challenges for the MMC, where it is desired that the MMC can continue operating without any interruption, despite some of the submodule (SMs') malfunction [3–5]. Normally, the faulty SMs are bypassed from the arm of the MMC [3]. However, the performance of the MMC would be affected if no extra control is used under SM faults. Therefore, an effective fault-tolerant approach is essential for the MMC under SM faults.

This chapter deals with the state-of-the-art and recent advancements in the SM redundant scheme and corresponding fault-tolerant control in the MMC. The SM protection circuit and redundant SMs for the MMC are presented in Sections 5.2 and 5.3, respectively. Based on the redundant SMs in the MMC, a brief description of various fault-tolerant schemes for the MMC is presented in Section 5.4. Section 5.5 discusses fault-tolerant approach for the MMC under the SM faults. At last, the summary of this chapter is discussed in Section 5.6.

5.2 SM Protection Circuit

An SM with protection circuit is shown in Figure 5.1. In the event of SM fault during operation, the faulty SM can be bypassed by a high-speed bypass switch, as shown in Figure 5.1. The bypass switch provides fail-safe functionality, as the arm

Modular Multilevel Converters: Control, Fault Detection, and Protection, First Edition.
Fujin Deng, Chengkai Liu, and Zhe Chen.

Published 2023 by John Wiley & Sons, Inc.

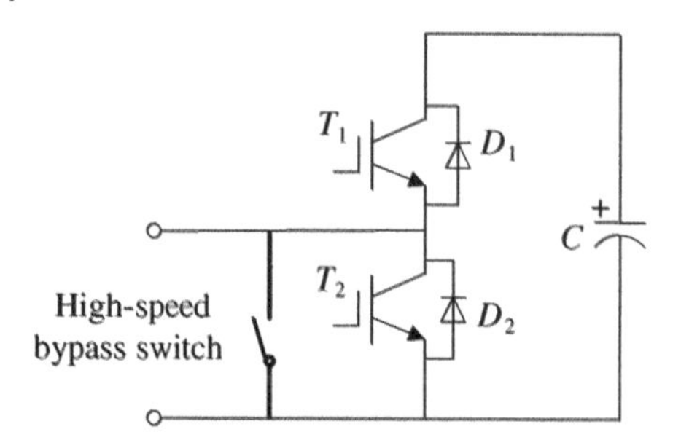

Figure 5.1 Structure of an MMC submodule with a bypass switch.

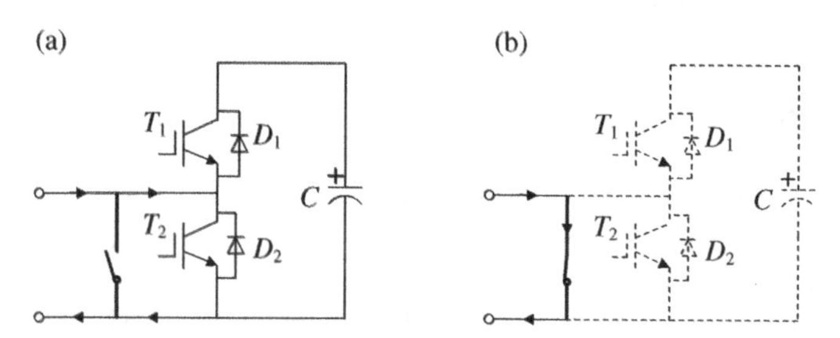

Figure 5.2 Current paths of an MMC submodule with a bypass switch. (a) Bypass switch is opened. (b) Bypass switch is closed.

current can flow through the bypass switch of the faulty SM and the converter continues to operate without being interrupted [6].

Figure 5.2a,b shows the current paths of the SM under normal operation and under faulty operation, respectively.

- *Normal operation*: when the SM is healthy and works normally, the bypass switch is opened, as shown in Figure. 5.2a. The arm current flows into the SM and does not flow through the bypass switch.
- *Faulty operation*: when the SM fault occurs, the bypass switch is closed and bypasses the faulty SM, as shown in Figure 5.2b. The arm current flows through the bypass switch.

5.3 Redundant Submodules

During the operation of the MMC, a part of SMs may be faulty. The faulty SMs will be bypassed and replaced by the redundant SMs in the MMC, so that the MMC can continue to operate without being interrupted. Obviously, the more redundant SMs are, the more reliable the system would be, but the higher the cost is. As a result, the redundant SMs cannot be added unlimitedly. Appropriate redundancy

Table 5.1 Redundant configuration of each system.

Project		Number of normal SMs per arm	Number of redundant SMs per arm	Redundant ratio
Trans Bay Cable Project		200	16	8%
Nan-ao Island MMC-HVDC Project	Sucheng Station	147	14	10%
	Jinniu Station	220	20	10%
	Qingao Station	220	20	10%
Luxi Back-back HVDC interconnector Project	Yunnan Station	310	25	8%
	Guangxi Station	438	30	6.8%
Xiamen DC Transmission Project	Pengcuo Station	200	16	8%
	Hubian Station	200	16	8%

is an important parameter in the MMC system. Normally, 8–10% redundant SMs are equipped in the MMC in practice.

Table 5.1 lists the redundant configurations of some existing MMC based projects. In the Trans Bay Cable Project in the United States, the MMC has 216 SMs in each arm with 16 redundant SMs [7, 8]. Three converter stations have been constructed in the Nan-ao Island MMC-HVDC Project [9]. In Sucheng Station, the MMC has 161 SMs in each arm with 14 redundant SMs. In Jinniu Station and Qingao Station, the MMC has 240 SMs in each arm with 20 redundant SMs. Two converter stations have been constructed in Luxi Back-back HVDC interconnector Project [10]. In Yunnan Station, the MMC has 335 SMs in each arm with 25 redundant SMs. In Guangxi Station, the MMC has 468 SMs in each arm with 30 redundant SMs. Two converter stations have been constructed in Xiamen DC Transmission Project [11]. In Pengcuo Station and Hubian Station, the MMC has 216 SMs in each arm with 16 redundant SMs.

5.4 Fault-Tolerant Scheme

For the MMC, it is normally demanded that the MMC can continue working under SM fault conditions [12], and therefore the fault-tolerant scheme is normally required for the MMC. For SM level fault-tolerant control, usually redundant SMs are integrated into arms [13]. A high-speed bypass switch is usually connected in parallel with the output port to bypass the SM if the SM fault occurs. And the redundant SMs are activated to maintain normal operation of the MMC. This section describes different fault-tolerant schemes under SM level faults.

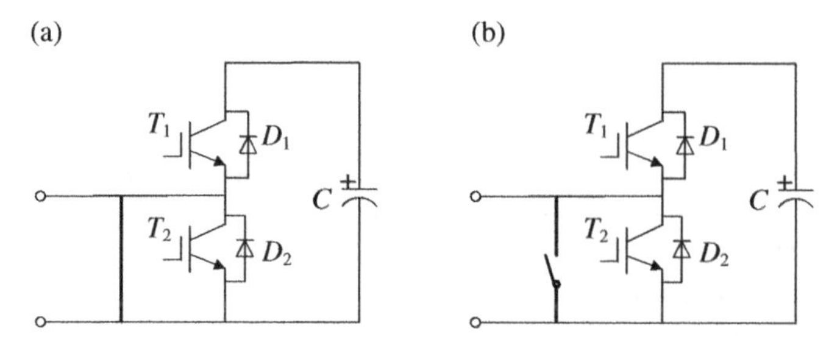

Figure 5.3 Reserved modes of redundant SMs. (a) Cold reserve mode. (b) Spinning reserve mode.

The fault-tolerant scheme can be divided into two modes according to different switching modes of SMs, including cold reserve mode and spinning reserve mode.

- *Cold reserve mode, as shown in Figure 5.3a*: The bypass switch keeps closed during the normal operation of the MMC, which means the redundant SMs are bypassed from the arm. The arm current will not flow through the SMs but flows through the bypass switch. The capacitor voltages in the cold reserve SMs are zero.
- *Spinning reserve mode, as shown in Figure 5.3b*: The bypass switch is opened during the normal operation of MMCs, which means that the redundant SMs are incorporated into the circuit and work as normal SMs. Spinning reserve mode can be divided into three situations, including spinning reserve modes-I, -II, and -III.

5.4.1 Cold Reserve Mode

Figure 5.3a shows a cold reserve mode of redundant SMs. They are inserted into the MMC only under fault conditions. The redundant SMs are bypassed by switches under normal operation.

Figure 5.4 shows the operation of cold reserve SMs. Figure 5.4a shows the operation during normal conditions, and Figure 5.4b shows the operation under fault conditions. When the SM faults occur, the faulty SMs are bypassed. At the same time, the same number of redundant SMs will be put into operation, in which the SM capacitors are charged to the same voltage and operated as normal SMs [14, 15].

Since the cold reserve SMs don't participate in the normal operation of MMCs, the control system is easy to apply and the power losses of the reserve SM are low. However, before the redundant SMs are put into operation, they should be charged and then inserted into the arms, which prolongs the fault-tolerant operation time

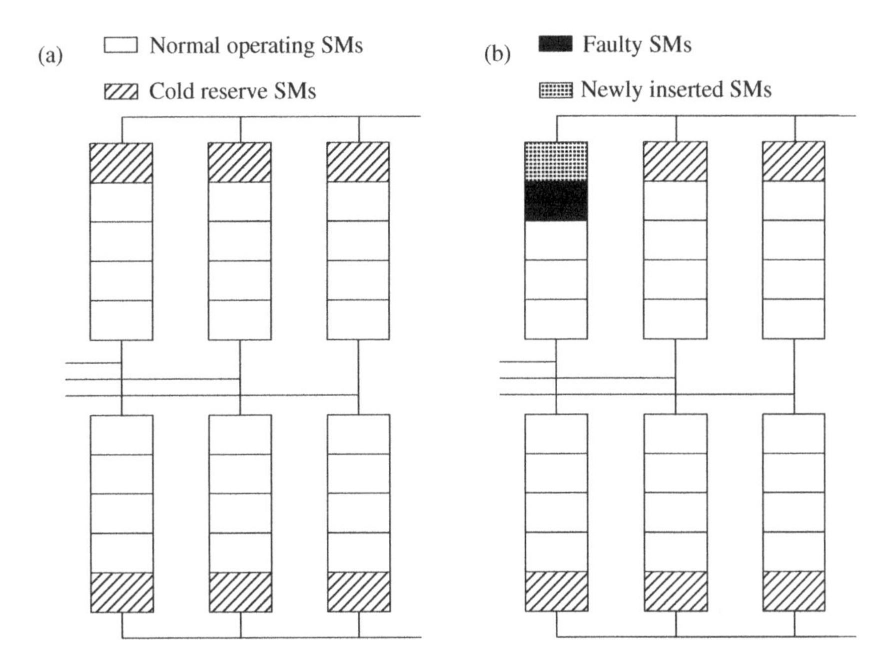

Figure 5.4 Operation of cold reserve SMs. (a) Normal conditions. (b) Fault conditions.

of the redundant SM. In addition, since the redundant SMs are always bypassed during normal operation, the SM's utilization rate is low.

5.4.2 Spinning Reserve Mode-I

There is no difference between the spinning reserve SMs and the normal SMs at normal operation of the MMC. Figure 5.5a shows a diagram of the MMC at normal operation, and Figure 5.5b shows the operation of spinning reserve mode-I under fault conditions. When the SM faults occur, the faulty SM is bypassed and the same number of healthy SMs are also bypassed in the other five arms.

In spinning reserve mode-I, the number of operated SMs in six arms is the same. Thus, this mode keeps the balance of the MMC system [14]. The symmetrical operation of the system avoids the additional circulating current. In addition, compared with the cold reserve SMs, the spinning reserve SMs avoid the process of charging, which reduces the action time. However, since the same number of healthy SMs should be bypassed in the other arms while bypassing the faulty SMs in the faulty arm, the SM utilization rate in the MMC is low in fault condition and the SM reduction may affect the number of output levels of the MMC.

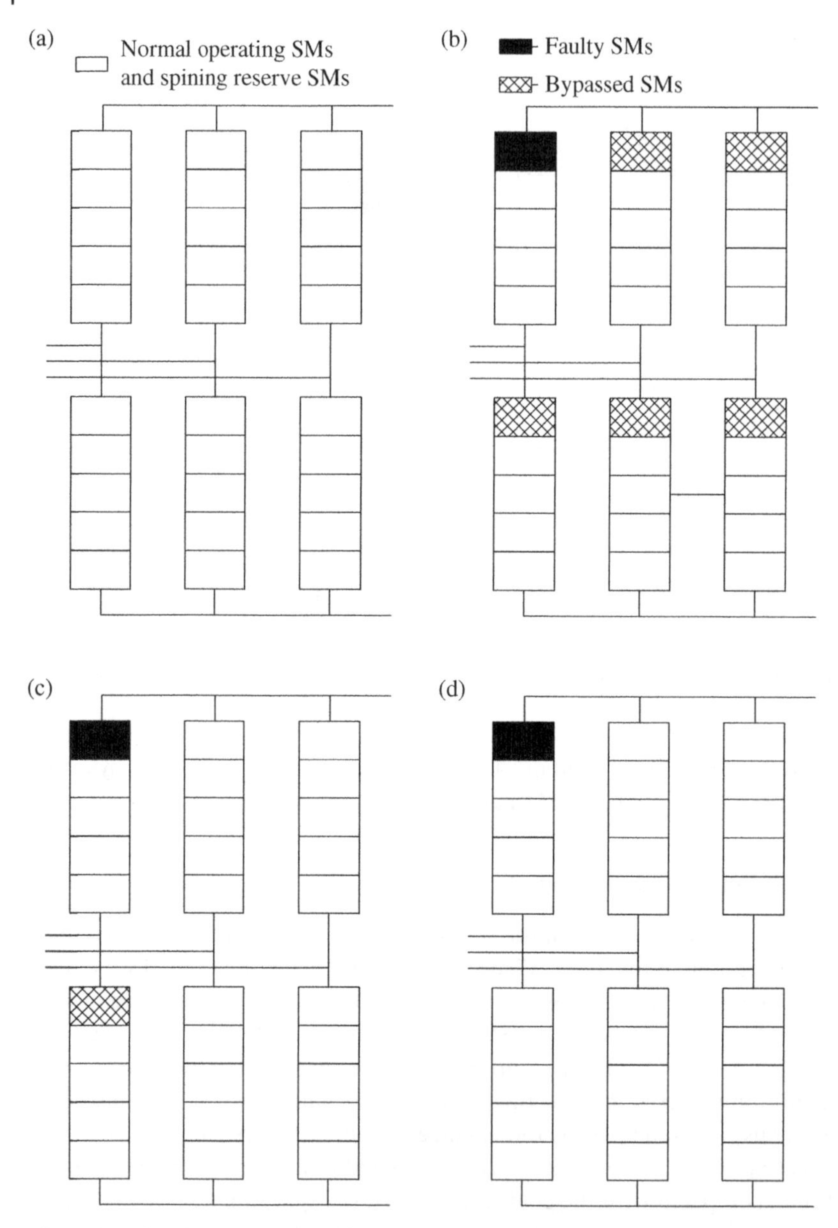

Figure 5.5 Spinning reserve mode. (a) Normal condition. (b) Spinning reserve mode-I. (c) Spinning reserve mode-II. (d) Spinning reserve mode-III.

5.4.3 Spinning Reserve Mode-II

Figure 5.5c shows the operation of spinning reserve mode-II under fault condition. When the SM faults occur, the faulty SM is bypassed and the same number of healthy SMs are only bypassed in another arm of the same phase.

In spinning reserve mode-II, since the number of remained SMs in the upper arm and the lower arm is the same, this mode keeps energy balancing in the phase unit with faulty SMs [16]. Compared with mode-I, mode-II has higher SM utilization rate. However, the asymmetry between the faulty phase and other phases breaks the balance of the whole system and then generates additional circulating current in the MMC.

5.4.4 Spinning Reserve Mode-III

Figure 5.5d shows the operation of spinning reserve mode-III under fault condition. When the SM faults occur, only the faulty SM is bypassed.

This mode achieves the highest SM utilization rate and highest economics in the MMC, but at the cost of unbalanced energy, which is distributed not only among the three phases of the MMC but also between the upper arm and the lower arm of the faulty phase unit. The spinning reserve mode-III also causes circulating current in the MMC.

5.4.5 Comparison of Fault-Tolerant Schemes

Table 5.2 shows the comparison of different reserve modes in terms of economy, submodule utilization, circulating current, and energy balance. More ★ means better performance.

Table 5.2 Advantages and disadvantages of different methods.

Scheme	Submodule utilization	Circulating current	Energy balance
Cold reserve mode	★	★★★★	★★★★
Spinning reserve mode-I	★★	★★★	★★★
Spinning reserve mode-II	★★★	★★	★★
Spinning reserve mode-III	★★★★	★	★

5.5 Fundamental Circulating Current Elimination Based Tolerant Control

The fault-tolerant operation is one of the important issues for the MMC. This section presents a fault-tolerant approach for the MMC under SM faults. The characteristic of the MMC with arms containing different number of healthy SMs is analyzed. Based on the characteristic, the approach can effectively keep the MMC operation as normal under SM faults. It can effectively improve the MMC performance under SM faults without the knowledge of the number of faulty SMs in the arm, without extra demand on communication systems.

5.5.1 Equivalent Circuit of MMCs

Figure 5.6a shows the upper and lower arms of phase A. The reference for phase A is y_a, and the references for upper and lower arms of phase A, respectively, are $-y_a$ and y_a. With the voltage balancing control [17, 18], the capacitor voltages of all healthy SMs in the upper and lower arms of phase A can be kept the same as u_{cau} and u_{cal}, respectively. The capacitor voltages u_{cau} and u_{cal} in the upper and lower arms of phase A can be expressed as

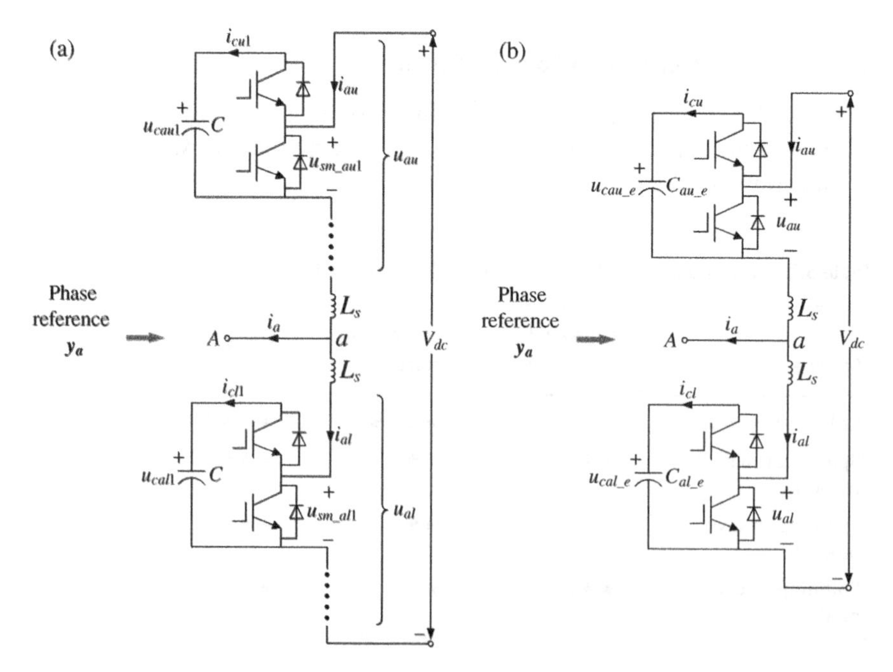

Figure 5.6 Phase A of the MMC. (a) Circuit of phase A. (b) Equivalent circuit of phase A.

$$\begin{cases} u_{cau} = \frac{1}{C}\int i_{au}\frac{1-y_a}{2}dt \\ u_{cal} = \frac{1}{C}\int i_{al}\frac{1+y_a}{2}dt \end{cases} \tag{5.1}$$

with

$$y_a = m \cdot \sin(\omega t + \phi) \tag{5.2}$$

Except for the bypassed faulty SMs, suppose that the number of healthy SMs in the upper and lower arms of phase A is n_{au} and n_{al}, respectively, and according to [19], the output voltage u_{au} and u_{al} of the series-connected SMs in the upper and lower arms of phase A, as shown in Figure 5.6a, can be described as

$$\begin{cases} u_{au} = n_{au} \cdot u_{cau}\frac{1-y_a}{2} \\ u_{al} = n_{al} \cdot u_{cal}\frac{1+y_a}{2} \end{cases} \tag{5.3}$$

Owing to that the SM capacitor voltage is kept the same in each arm with the voltage balancing control, the SMs in each arm operate similarly and the series-connected SMs in each arm can be simplified to one equivalent SM [14]. Based on equations (5.1)–(5.3), the equivalent circuit for phase A of the MMC can be obtained, as shown in Figure 5.6b, with the relationship of

$$\begin{cases} u_{cau\_e} = n_{au} \cdot u_{cau} \\ u_{cal\_e} = n_{al} \cdot u_{cal} \end{cases} \tag{5.4}$$

$$\begin{cases} C_{au\_e} = \frac{C}{n_{au}} \\ C_{al\_e} = \frac{C}{n_{al}} \end{cases} \tag{5.5}$$

According to Kirchhoff's voltage law and Figure 5.6b, the voltage relationship of phase A can be expressed as

$$V_{dc} = u_{au} + u_{al} + 2L_s\frac{di_{diff\_a}}{dt} \tag{5.6}$$

Referring to equations (5.1) and (5.3), the output voltage u_{au} and u_{al} of the equivalent upper and lower arm SMs in Figure 5.6b can be expressed as

$$\begin{cases} u_{au} = \frac{1}{C_{au\_e}}\int i_{au} \cdot \frac{1-y_a}{2}dt \cdot \frac{1-y_a}{2} \\ u_{al} = \frac{1}{C_{al\_e}}\int i_{al} \cdot \frac{1+y_a}{2}dt \cdot \frac{1+y_a}{2} \end{cases} \tag{5.7}$$

5.5.2 Fundamental Circulating Current

The upper arm current i_{au} is composed of the DC component $\bar{i}_{au}$ and the AC component $\tilde{i}_{au}$, the lower arm current i_{al} is composed of the DC component $\bar{i}_{al}$ and the AC component $\tilde{i}_{al}$, as

$$\begin{cases} i_{au} = \bar{i}_{au} + \tilde{i}_{au} \\ i_{al} = \bar{i}_{al} + \tilde{i}_{al} \end{cases} \tag{5.8}$$

Suppose that the second-order harmonic component in the arm current is suppressed with the control [20] and only considering the fundamental component, the arm current i_{au} and i_{al} in the equation can be rewritten as

$$\begin{cases} i_{au} = \bar{i}_{au} + I_{aum}\sin\left(\omega t + \alpha_r\right) \\ i_{al} = \bar{i}_{al} + I_{alm}\sin\left(\omega t + \beta_r\right) \end{cases} \tag{5.9}$$

where I_{aum}, I_{alm} and α_r, β_r are the amplitude and the phase angle of $\tilde{i}_{au}$ and $\tilde{i}_{al}$, respectively. Since the components higher than 3rd harmonics is small, only the component up to 3rd harmonics are considered. Substituting equations (5.8) and (5.9) into equation (5.7), the AC component in the summation of u_{au} and u_{al} can be expressed as

$$\begin{aligned} \tilde{u}_{au} + \tilde{u}_{al} = {} & \frac{1}{2C_{au\_e}}\Bigg[\underbrace{-\frac{m^2+8}{16} I_{aum}\cos(\omega t + \alpha_r) + \frac{m\bar{i}_{au}}{2}\cos(\omega t + \phi)}_{\text{Fundamental components}} \\ & \underbrace{+ \frac{3m}{8} I_{aum}\sin(2\omega t + \phi + \alpha_r) - \frac{m^2\bar{i}_{au}}{4}\sin(2\omega t + 2\phi)}_{\text{2nd harmonic components}} \\ & \underbrace{+ \frac{m^2}{16} I_{aum}\cos(3\omega t + 2\phi + \alpha)}_{\text{3rd harmonic component}}\Bigg] \\ & + \frac{1}{2C_{al\_e}}\Bigg[\underbrace{\frac{m^2+8}{16} I_{alm}\cos(\omega t + \beta_r) - \frac{m\bar{i}_{al}}{2}\cos(\omega t + \phi)}_{\text{Fundamental components}} \\ & \underbrace{+ \frac{3mI_{alm}}{8}\sin(2\omega t + \phi + \beta_r) - \frac{m^2\bar{i}_{al}}{4}\sin(2\omega t + 2\phi)}_{\text{2nd harmonic components}} \\ & \underbrace{- \frac{m^2}{16} I_{alm}\cos(3\omega t + 2\phi + \beta_r)}_{\text{3rd harmonic component}}\Bigg] \end{aligned} \tag{5.10}$$

Substituting equation (5.10) into equation (5.6), a fundamental component would be caused in the inner differential current i_{diff_a} if C_{au_e} and C_{al_e} are not the same as each other, which can be caused by the SM fault. According to equation (5.5), if the upper and lower arms in phase A are operated with the different numbers of healthy SMs, the C_{au_e} would be not equal to C_{al_e} and a fundamental component would be caused in the inner difference current i_{diff_a} of phase A. With the same analysis, the fundamental component may also be caused in the inner difference current of the phases B and C, respectively, which are not repeated here. The caused fundamental component in the inner difference current would affect the performance of the MMC.

5.5.3 Fundamental Circulating Current Elimination Control

The caused fundamental component i_{1f_a} in the inner difference current of phase A derived from the SM faults would cause the corresponding voltage $u_{diff\_1f\_a}$ in phase A. According to equation (5.6), the relationship for i_{1f_a} and $u_{diff\_1f\_a}$ in phase A of Figure 5.6 can be expressed as

$$u_{diff\_1f\_a} = 2L_s \frac{di_{1f\_a}}{dt} \tag{5.11}$$

Applying the transformation matrix given in [21], which transfers variants from static single-phase axis to synchronous rotating axis, to equation (5.11), the behavior of the phase A system in the synchronous rotating reference frame can be expressed as

$$\begin{bmatrix} u_{diff\_1f\_a\_d} \\ u_{diff\_1f\_a\_q} \end{bmatrix} = 2L_s \frac{d}{dt}\begin{bmatrix} i_{1f\_a\_d} \\ i_{1f\_a\_q} \end{bmatrix} + 2L_s \begin{bmatrix} 0 & \omega \\ -\omega & 0 \end{bmatrix} \cdot \begin{bmatrix} i_{1f\_a\_d} \\ i_{1f\_a\_q} \end{bmatrix} \tag{5.12}$$

where $u_{diff\_1f\_a\_d}$, $u_{diff\_1f\_a\_q}$ and $i_{1f\_a\_d}$, $i_{1f\_a\_q}$ are the dq components of the voltage $u_{diff\_1f\_a}$ and the current i_{1f_a} in the rotating reference frame, respectively.

Figure 5.7 shows a fault-tolerant approach for the MMC under SM faults, which is used to eliminate the caused fundamental component i_{1f_a} in the inner difference current of phase A under SM faults. The fundamental component i_{fa} in the circuit can be obtained based on the upper arm current, lower arm current, and the band-pass filter (BPF) tuned at the fundamental frequency. i_{1f_a} is transformed into dq domain by a single-phase Park transform [21], where one product by 2sin(ωt) and the other one by 2cos(ωt). Then, a notch filter (NF) tuned at twice the fundamental frequency. Finally, the dq component $i_{1f\_a\_d}$ and $i_{1f\_a\_q}$ of the i_{1f_a} can be obtained. In Figure 5.7a, the BPF aims to let the fundamental frequency pass and block other frequencies, while

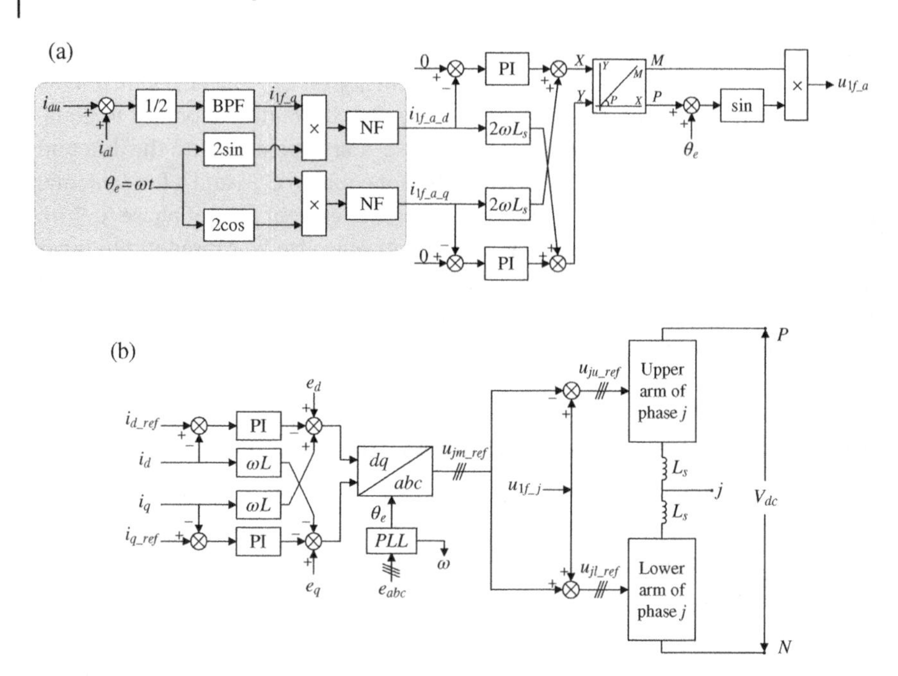

Figure 5.7 Block diagram of the fault-tolerant approach for MMCs. (a) Control in phase A. (b) MMCs control.

the NF is used to block the double fundamental frequency. The BPF and NF can be expressed as

$$\mathrm{BPF}(s)=\frac{h\omega s}{s^2+h\omega s+\omega^2} \tag{5.13}$$

$$\mathrm{NF}(s)=\frac{s^2+(2\omega)^2}{s^2+2h\omega s+(2\omega)^2} \tag{5.14}$$

where h is a real positive constant that determines the sharpness of the filter. Here, h is designed as 1/0.606. In Figure 5.7, the fundamental phase angle θ_e and angular frequency ω are obtained by the phase locked loop.

With the PI controller, the fundamental component i_{1f_a} can be eliminated when the $i_{1f\_a\_d}$ and $i_{1f\_a\_q}$ are both regulated to zero by the voltage reference u_{1f_a} in phase A. The control for phases B and C is the same to that for phase A, which is not repeated here. Figure 5.7b shows the control structure of the MMC with the presented fault-tolerant approach. According to the control objective such as active power, reactive power, and DC-link voltage control, the current reference

i_{d_ref} and i_{q_ref} can be obtained [20, 22]. The vector control method is used here for grid current control [20, 22] and to produce the three-phase reference voltage u_{jm_ref} $(j = a, b, c)$. To realize fault-tolerant control of the MMC, the references for the upper and lower arms of phase j can be obtained as $u_{ju\_ref} = -u_{jm\_ref} + u_{1f\_j}$ and $u_{jl\_ref} = u_{jm\_ref} + u_{1f\_j}$, respectively. The fault-tolerant approach can effectively keep the MMC operation as normal and improve the MMC performance under SM faults without the knowledge of the number of faulty SMs in the arm, without extra demand on communication systems, which potentially increases the reliability of the MMC system.

5.5.4 Control Analysis

Figure 5.8 shows the average capacitor voltage and average energy in the upper and lower arms of phase A with the fault-tolerant control in Section 5.5

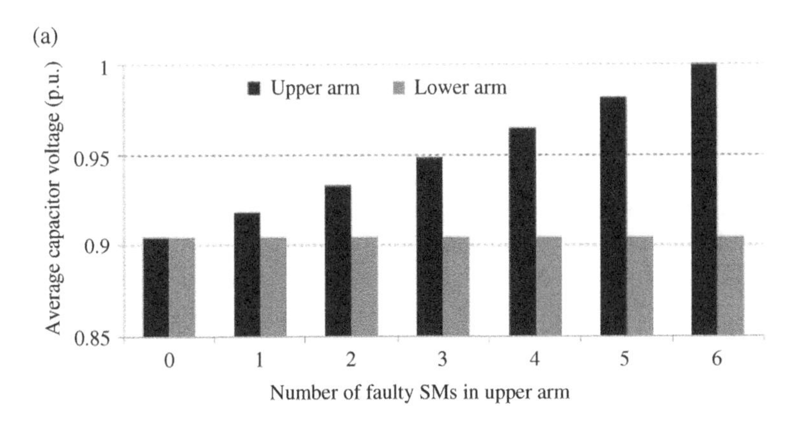

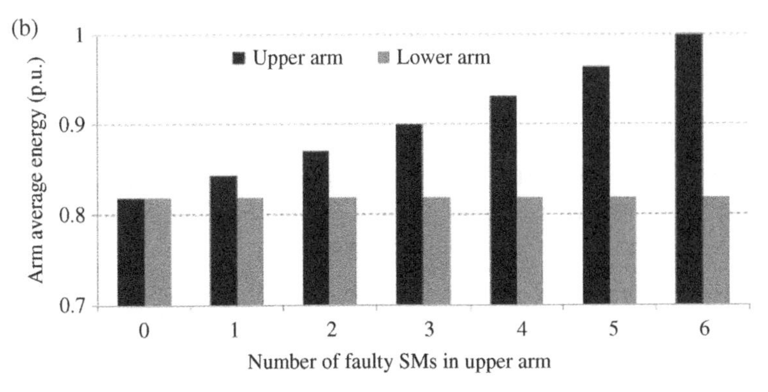

Figure 5.8 Performance with fault-tolerant control under different number of faulty SMs in phase A. (a) Average capacitor voltage. (b) Arm average energy.

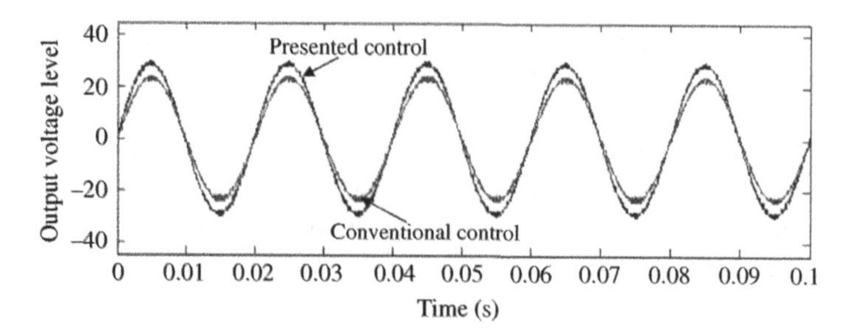

Figure 5.9 Output phase voltage level of the MMC under normal situation.

under different number of faulty SMs, which is derived from the simulation in Case Study 5.1. The fault-tolerant control in Section 5.5 can produce lower capacitor voltage in comparison with the conventional control under the normal situation, which reduces the voltage stress on the power devices inside the SMs and increases the reliability. Furthermore, the lower capacitor voltage of the fault-tolerant control in Section 5.5 could increase the utilization ratio of the SMs in comparison with the conventional control under the normal situation, which could produce more voltage levels for a lower total harmonic distortion and improve power quality. It can be verified by Figure 5.9, derived from the simulation in Case Study 5.1, with the conventional control and the presented fault-tolerant control in Section 5.5, respectively, under the same situation, where the presented fault-tolerant control in Section 5.5 produces more output phase voltage level for the MMC in comparison with the conventional control.

Case Study 5.1 Analysis of the Fundamental Circulating Current Elimination Based Tolerant Control

Objective: To verify the fundamental circulating current elimination, a three-phase MMC system, as shown in Figure 5.10, is simulated with the time-domain simulation tool power systems computer aided design/electromagnetic transients including DC (PSCAD/EMTDC). The three-phase MMC system is linked to a three-phase AC system.

Parameters: In the simulation, the active power P and the reactive power Q is 22 W and 0 MVar, respectively. Six SMs are simultaneously bypassed from the upper arm of phase A. The other system parameters are shown in the Table 5.3.

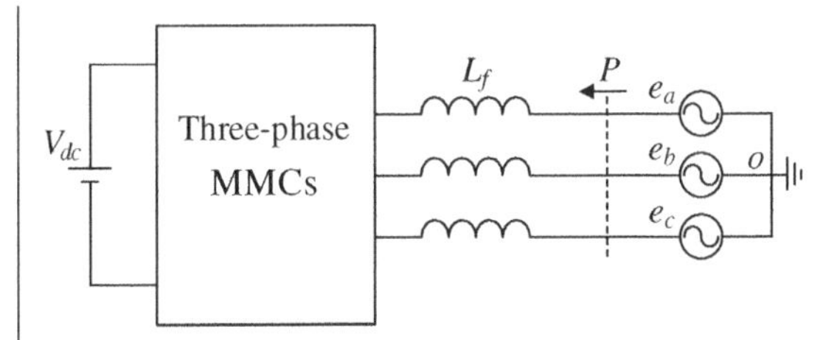

Figure 5.10 Block diagram of the simulation system.

Table 5.3 Simulation system parameters.

Parameters	Value
Rated power (MW)	22
DC-link voltage V_{dc} (kV)	30
Number of regular SMs per arm n	26
Number of redundant SMs per arm m	6
Nominal SM capacitance C (mF)	9
Filter inductance L_f (mH)	1
Carrier frequency f_s (kHz)	1

Simulation results and analysis:

The performance of the MMC is shown in Figure 5.11, where six SMs are simultaneously bypassed from the upper arm of phase A at 1.6 seconds and the fault-tolerant control is enabled since 2.1 seconds. Figure 5.11a shows the three-phase current i_a, i_b, i_c of the MMC. Figure 5.11b,c show the arm current and the inner difference current in phase A of the MMC. A fundamental component is caused in the inner difference current of phase A when the faulty SMs are bypassed, whose peak-to-peak value is approximately 72 A and nearly 30% of the DC component in the inner difference current. The arm currents and inner difference currents in phases B and C are slightly affected, as shown in Figure 5.11d–g. The caused fundamental component in the inner difference current would flow into the DC link of the MMC and cause DC-link current ripple about 9.7%, as shown in Figure 5.11h. In addition, the bypass of the faulty SMs results in the increase of the upper arm capacitor voltage by 20%, as shown in Figure 5.11i. After the fault-tolerant control is enabled since 2.1 seconds, the fundamental component in the inner difference current of

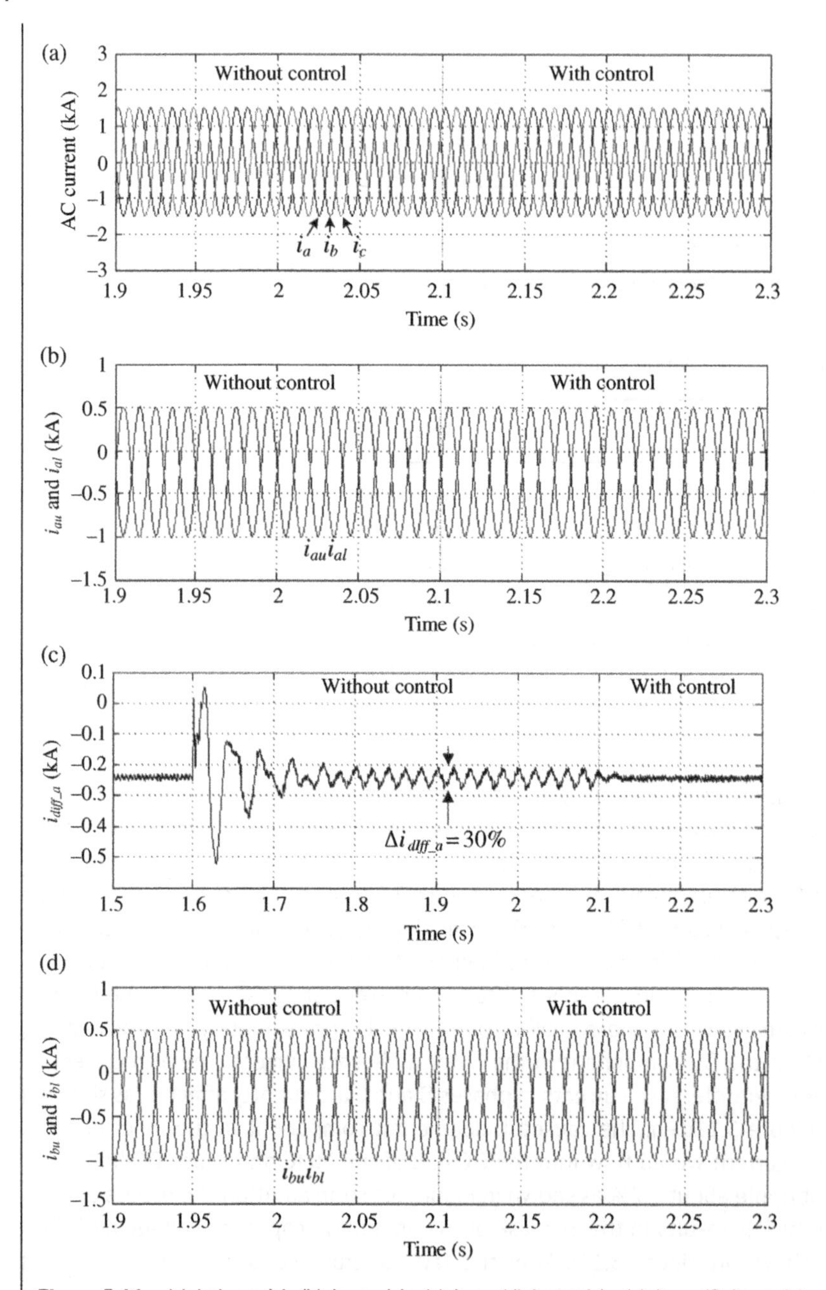

Figure 5.11 (a) i_a, i_b, and i_c. (b) i_{au} and i_{al}. (c) i_{diff_a}. (d) i_{bu} and i_{bl}. (e) i_{diff_b}. (f) i_{cu} and i_{cl}. (g) i_{diff_c}. (h) i_{dc}. (i) Capacitor voltage in phase A. (j) Upper and lower arm capacitor energy. (k) Reference y_{au} and y_{fa}.

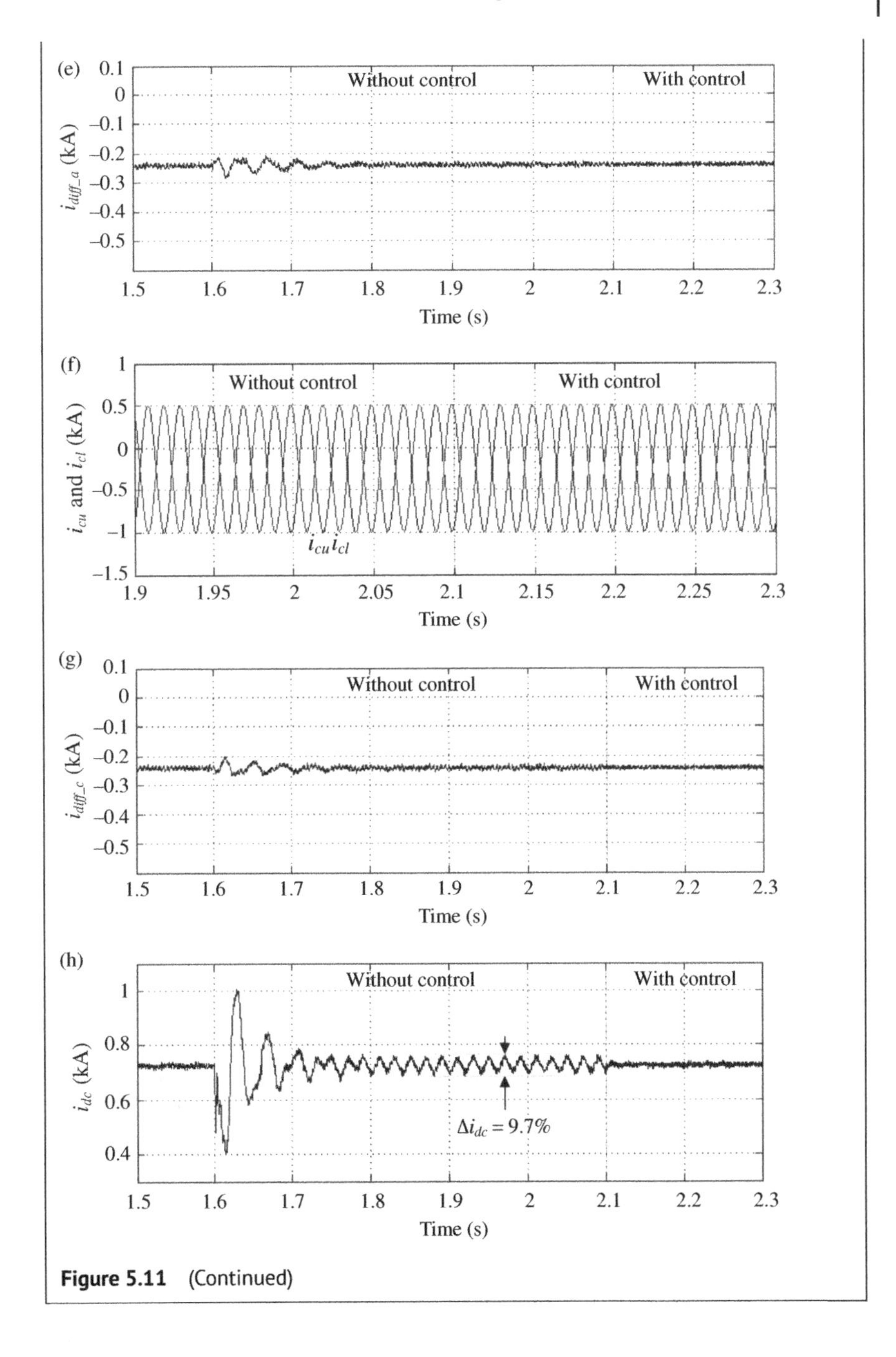

Figure 5.11 (Continued)

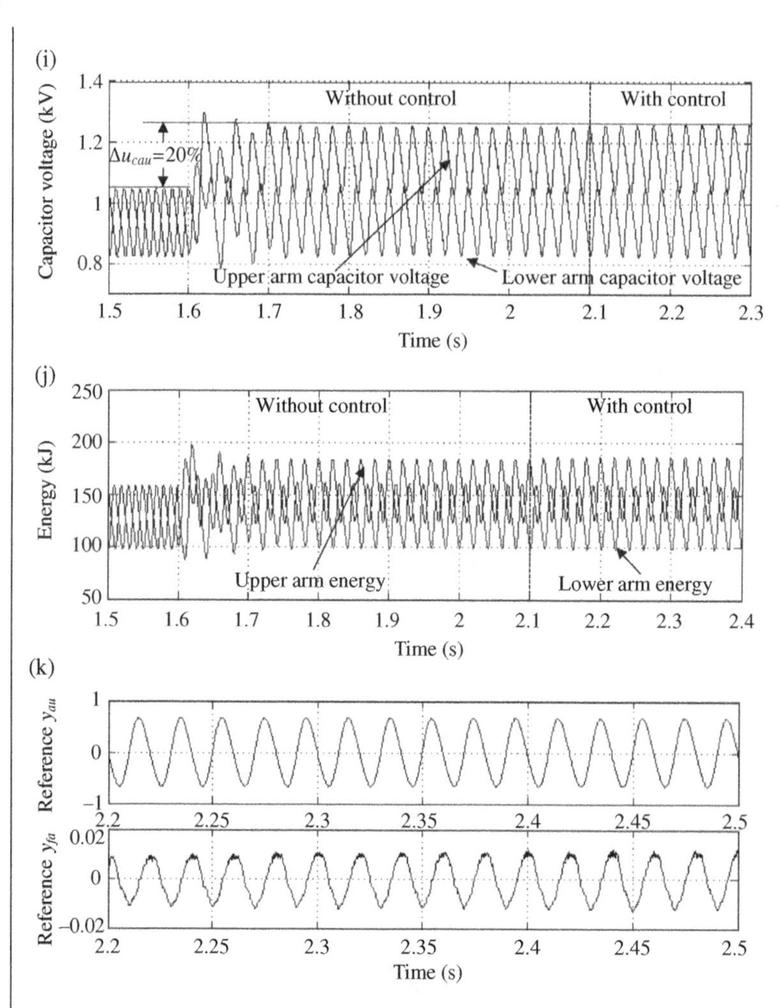

Figure 5.11 (Continued)

phase A disappears and the DC-link current ripple is eliminated, which effectively improves the performance of the MMC, as shown in Figure 5.11c,h. The capacitor voltage in phase A is kept the same as shown in Figure 5.11i. Although the number of healthy SMs in the upper arm is smaller than that in the lower arm, the energy stored in the upper arm capacitor is higher than that stored in the lower arm capacitor in phase A under faults, as shown in Figure 5.11j. The reference $y_{fa} = 2u_{1f\_a}/V_{dc}$ produced by the presented control is not in phase with $y_{au} = -2u_{au\_ref}/V_{dc}$, as shown in Figure 5.11k, which shows that both the amplitude and the phase angle of the references y_{au} and y_{al} for the upper and lower arms of the faulty phase are regulated, respectively.

5.6 Summary

The fault-tolerant control under SM faults is one of the important issues for the MMC. This chapter illustrates state-of-the-art and recent advancements in SM redundant scheme and corresponding fault-tolerant control in the MMC. Based on the SM protection circuit and redundant SMs in the MMC, the fault-tolerant control is presented. Based on fault-tolerant schemes, the fundamental circulating current elimination based fault-tolerant control for the MMC under SM faults is introduced to keep the MMC operating normally under SM fault conditions.

References

1 F. Richardeau and T. Pham, "Reliability calculation of multilevel converters: theory and applications," *IEEE Trans. Ind. Electron.*, vol. 60, no. 10, pp. 4225–4233, Oct. 2013.
2 H. Liu, P. C. Loh, and F. Blaabjerg, "Review of fault diagnosis and fault-tolerant control for modular multilevel converter of HVDC," in *Proceedings of 39th Annual Conference of the IEEE Industrial Electronics Society (IECON'13)*, 2013, pp. 1242–1247.
3 B. Gemmell, J. Dorn, D. Retzmann, and D. Soerangr, "Prospects of multilevel VSC technologies for power transmission," in *Proceedings* of *IEEE/PES Transmission and Distribution Conference & Exposition*, Apr. 21–24, 2008, pp. 1–16.
4 B. Li, S. Shi, B. Wang, G. Wang, W. Wang, and D. Xu, "Fault diagnosis and tolerant control of single IGBT open-circuit failure in modular multilevel converters," *IEEE Trans. Power Electron.*, vol. 31, no. 4, pp. 3165–3176, Apr. 2015.
5 Y. Feng, J. Zhou, Y. Qiu, and K. Feng, "Fault tolerance for wind turbine power converter," in *Proceedings* of *2nd IET Renewable Power Generation Conference*, Beijing, China, Sep. 9–11, 2013, pp. 1–4.
6 B. Li, Y. Zhang, R. Yang, R. Xu, D. Xu, and W. Wang, "Seamless transition control for modular multilevel converters when inserting a cold-reserve redundant submodule," *IEEE Trans. Power Electron.*, vol. 30, no. 8, pp. 4052–4057, Aug. 2015.
7 G. Liu, Z. Xu, Y. Xue, and G. Tang, "Optimized control strategy based on dynamic redundancy for the modular multilevel converter," *IEEE Trans. Power Electron.*, vol. 30, no. 1, pp. 339–348, Jan. 2015.
8 S. P. Teeuwsen, "Modeling the trans bay cable project as voltage-sourced converter with modular multilevel converter design," in *Proceedings of IEEE Power and Energy Society General Meeting*, pp. 1–8, 2011.
9 L. Yang, X. Li, and S. Xu, "The integrated system design scheme of Nan'ao VSC-MTDC demonstration project," *South. Power Syst. Technol.*, vol. 09, no. 01, pp. 63–67, 2015.

10 W. Huang *et al.*, "Technology of Back-to-Back DC transmission system and its application in Luxi asynchronous interconnection project," *South. Power Syst. Technol.*, vol. 12, no. 04, pp. 1–6, 2018.

11 X. Guo, M. Deng, and K. Wang, "Characteristics and performance of Xiamen VSC-HVDC transmission demonstration project," in *Proceedings of IEEE International Conference on High Voltage Engineering and Application*, Chengdu, China, 2016, pp. 1–4.

12 L. Yang, Y. Li, Z. Li, P. Wang, and F. Xu, "A low-loss energy rebalancing control scheme of MMC under submodule fault conditions," in *Proceedings of IECON 2017*, Beijing, China, 2017, pp. 4472–4476.

13 S. Sedghi, A. Dastfan, and A. Ahmadyfard, "Fault detection of a seven level modular multilevel inverter via voltage histogram and neutral network," in *International Conference on Power Electronics*, pp. 1005–1012, Shilla Jeju, Korea, Asia, 30 May–3 June 2011.

14 P. Hu, D. Jiang, Y. Zhou, Y. Liang, J. Guo, and Z. Lin, "Energy-balancing control strategy for modular multilevel converters under submodule fault conditions," *IEEE Trans. Power Electron.*, vol. 29, no. 9, pp. 5021–5030, 2014.

15 T. S. Gum *et al.*, "Design and control of a modular multilevel HVDC converter with redundant power modules for non-interruptible energy transfer," *IEEE Trans. Power Delivery*, vol. 27, no. 3, pp. 1611–1619, Jul. 2012.

16 W. Wu, X. Wu, L. Jing, and J. Li, "A fault-tolerated control strategy for sub-module faults of modular multilevel converters," *Power Syst. Technol.*, vol. 40, no. 1, pp. 11–18, 2016.

17 P. Meshram and V. Borghate, "A simplified nearest level control (NLC) voltage balancing method for modular multilevel converter (MMC)," *IEEE Trans. Power Electron.*, vol. 30, no. 1, pp. 450–462, Jan. 2015.

18 J. Pou, S. Ceballos, G. Konstantinou, V. G. Agelidis, R. Picas, and J. Zaragoza, "Circulating current injection methods based on instantaneous information for the modular multilevel converter," *IEEE Trans. Ind. Electron.*, vol. 62, no. 2, pp. 777–788, Feb. 2015.

19 Q. Song, W. Liu, X. Li, H. Rao, S. Xu, and L. Li, "A steady-state analysis method for a modular multilevel converter," *IEEE Trans. Power Electron.*, vol. 28, no. 8, pp. 3702–3713, Aug. 2013.

20 Q. Tu, Z. Xu, and L. Xu, "Reduced switching-frequency modulation and circulating current suppression for modular multilevel converters," *IEEE Trans. Power Delivery*, vol. 26, no. 3, pp. 2009–2017, Jul. 2011.

21 J. Salaet, S. Alepuz, A. Gilabert, and J. Bordonau, "Comparison between two methods of DQ transformation for single phase converters control. Application to a 3-level boost rectifier," in *2004 IEEE 35th Annual Power Electronics Specialists Conference (IEEE Cat. No.04CH37551)*, Aachen, Germany, 2004, pp. 214–220.

22 M. Saeedifard and R. Iravani, "Dynamic performance of a modular multilevel back-to-back HVDC system," *IEEE Trans. Power Delivery*, vol. 25, no. 4, pp. 2903–2912, Oct. 2010.

6

Control of MMCs Under AC Grid Faults

6.1 Introduction

The high-voltage direct current (HVDC) transmission system based on modular multilevel converters (MMCs) is a promising solution for efficient bulk power transmission over long distances. In grid-connected converters, the occurrence of grid-side faults is rather common, resulting in imbalances and distortions of the AC-side voltages [1–6]. The operation and control of the MMC-HVDC system in the presence of grid imbalances therefore become very important in order to satisfy grid codes and reliability requirements.

Under balanced voltage conditions, the AC-side current of the MMC contains only positive-sequence component. However, under unbalanced voltage conditions, the AC-side current will consist of negative-sequence component [7–17]. Generally, due to the presence of the converter transformer connected in Y/Δ configuration, the zero-sequence current does not exist. However, in transformerless applications of the MMC, the zero-sequence current may exist in the AC-side current [13]. The unbalanced AC-side current causes overcurrent of the MMC system and the oscillation of active and reactive power, which is harmful to the system.

Except for the AC-side current, the circulating current is also influenced by the asymmetric AC-side voltage. Under balanced grid-side voltage conditions, only negative-sequence circulating current exists in the MMC [14]. However, under unbalanced voltage conditions, there are positive-, negative-, and zero-sequence circulating current in the arm current [15]. The circulating currents not only increase the power losses of the whole system but also exacerbate the fluctuations of the submodule capacitor voltage.

The purpose of this chapter is to provide a comprehensive review of the operation and control method applied to the MMC based HVDC systems under

Modular Multilevel Converters: Control, Fault Detection, and Protection, First Edition.
Fujin Deng, Chengkai Liu, and Zhe Chen.

Published 2023 by John Wiley & Sons, Inc.

unbalanced AC-grid conditions. Section 6.2 analyzes the mathematical model of MMC under AC grid faults, including the AC-side mathematical model in Section 6.2.1 and the instantaneous power mathematical model in Section 6.2.2. Section 6.3 introduces the AC current control strategies under AC grid faults. Section 6.3.1 gives the positive- and negative-sequence current control, while Section 6.3.2 gives the zero-sequence current control. Section 6.3.3 introduces a proportional resonant (PR) controller based control, which is used to track the reference current in the stationary-$\alpha\beta$ frame. Section 6.4 offers an analysis on circulating current of MMCs under AC grid faults in Section 6.4.1 and then introduces three methods to suppress positive-, negative-, and zero-sequence components of circulating currents under unbalanced conditions, including single-phase vector based control in Section 6.4.2, $\alpha\beta 0$ stationary frame based control in Section 6.4.3, and three-phase stationary frame based control in Section 6.4.4.

6.2 Mathematical Model of MMCs under AC Grid Faults

The mathematical model of the MMC under balanced grid conditions only contain positive-sequence components. However, the AC-side components contain not only positive-sequence components but also negative-sequence components and even zero-sequence components under asymmetrical grid faults. This section presents the AC-side mathematical model and the instantaneous power mathematical model of MMCs under asymmetrical grid faults.

6.2.1 AC-Side Mathematical Model

6.2.1.1 MMC with AC-Side Transformer

Figure 6.1 depicts the circuit configuration of a three-phase MMC connected with the grid through a converter transformer. When an asymmetrical fault occurs on the grid side, the asymmetrical grid voltage can be decomposed into positive-, negative-, and zero-sequence voltage components according to the symmetrical component method. However, as the converter transformer in the MMC system is connected in the Y/Δ configuration, with the Δ-connection on the converter side, the zero-sequence current does not exist on the converter side. Generally, the transformer in the MMC based HVDC system is connected in the Y/Δ configuration to exclude the zero-sequence components from the converter.

Balanced Grid Conditions

Under balanced grid condition, the AC-side voltages and currents at fundamental frequency only contain positive-sequence components. According to Section 1.5, the AC-side mathematical model in the stationary-*abc* frame can be presented as

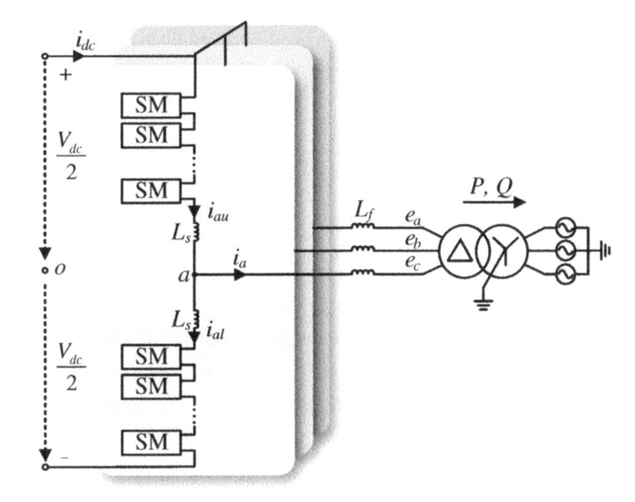

Figure 6.1 Circuit configuration of three-phase MMC with AC-side transformer.

$$L\begin{bmatrix}\frac{di_a(t)}{dt}\\ \frac{di_b(t)}{dt}\\ \frac{di_c(t)}{dt}\end{bmatrix}=\begin{bmatrix}u_{am}(t)+u_o(t)\\ u_{bm}(t)+u_o(t)\\ u_{cm}(t)+u_o(t)\end{bmatrix}-\begin{bmatrix}e_a(t)\\ e_b(t)\\ e_c(t)\end{bmatrix} \tag{6.1}$$

where u_o is the potential to ground of the DC-side neutral point. In the balanced MMC system, the u_o is regarded as zero.

Unbalanced Grid Conditions

Under unbalanced grid conditions (e.g. asymmetrical faults), the unbalanced AC-side voltages and currents at fundamental frequency can be represented as the orthogonal sum of positive- and negative-sequence components. The method used in [13] to separate the positive- and negative-sequence components is adopted here. Then, the system described by equation (6.1) can be decomposed into decoupled positive subsystem and negative subsystem as

$$L\begin{bmatrix}\frac{di_a^p(t)}{dt}\\ \frac{di_b^p(t)}{dt}\\ \frac{di_c^p(t)}{dt}\end{bmatrix}=\begin{bmatrix}u_{am}^p(t)\\ u_{bm}^p(t)\\ u_{cm}^p(t)\end{bmatrix}-\begin{bmatrix}e_a^p(t)\\ e_b^p(t)\\ e_c^p(t)\end{bmatrix} \tag{6.2}$$

$$L\begin{bmatrix} \frac{di_a^n(t)}{dt} \\ \frac{di_b^n(t)}{dt} \\ \frac{di_c^n(t)}{dt} \end{bmatrix} = \begin{bmatrix} u_{am}^n(t) \\ u_{bm}^n(t) \\ u_{cm}^n(t) \end{bmatrix} - \begin{bmatrix} e_a^n(t) \\ e_b^n(t) \\ e_c^n(t) \end{bmatrix} \tag{6.3}$$

where i_j^p, u_{jm}^p, and e_j^p ($j = a, b, c$) are the positive-sequence components of the AC-side currents, the output voltages of the MMC, and the AC-grid voltages, respectively; i_j^n, u_{jm}^n, and e_j^n ($j = a, b, c$) are the negative-sequence components of the AC-side currents, the output voltages of MMC, and the grid-side voltages, respectively.

The three-phase components in the stationary-*abc* frame can be transformed to the synchronous-*dq* frame by using the Park's transformation as

$$L\frac{d}{dt}\begin{bmatrix} i_d^p \\ i_q^p \end{bmatrix} = \begin{bmatrix} u_{dm}^p \\ u_{qm}^p \end{bmatrix} - \begin{bmatrix} e_d^p \\ e_q^p \end{bmatrix} + \begin{bmatrix} 0 & \omega L \\ -\omega L & 0 \end{bmatrix} \cdot \begin{bmatrix} i_d^p \\ i_q^p \end{bmatrix} \tag{6.4}$$

$$L\frac{d}{dt}\begin{bmatrix} i_d^n \\ i_q^n \end{bmatrix} = \begin{bmatrix} u_{dm}^n \\ u_{qm}^n \end{bmatrix} - \begin{bmatrix} e_d^n \\ e_q^n \end{bmatrix} + \begin{bmatrix} 0 & -\omega L \\ \omega L & 0 \end{bmatrix} \cdot \begin{bmatrix} i_d^n \\ i_q^n \end{bmatrix} \tag{6.5}$$

where i_d^p, i_q^p, u_{dm}^p, u_{qm}^p, e_d^p, e_q^p are the positive-sequence *dq*-axis components of the AC-side currents, the AC-side voltages of MMC, and the grid-side voltages, respectively; i_d^n, i_q^n, u_{dm}^n, u_{qm}^n, e_d^n, e_q^n are the negative-sequence *dq*-axis components of the AC-side currents, the AC-side voltages of MMC, and the grid-side voltages, respectively.

6.2.1.2 MMCs without AC-Side Transformer

Figure 6.2 depicts a circuit configuration of a three-phase MMC connected with the grid in a transformerless manner. In this case, the asymmetrical faults would not only cause undesired negative-sequence components but also cause zero-sequence components in the converter. Here, the AC-side mathematical model in the stationary-*abc* frame can refer to equation (6.1), and the positive and negative components can refer to equations (6.2)–(6.5), which are not repeated here. The zero-sequence component can be extracted from three-phase quantities by

$$f^z = \frac{f_a + f_b + f_c}{3} \tag{6.6}$$

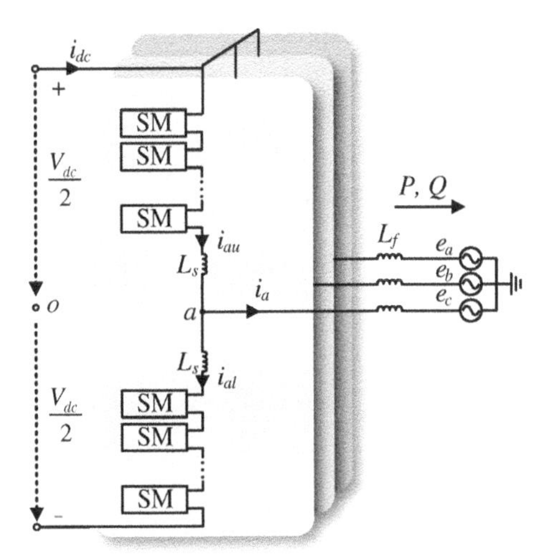

Figure 6.2 Circuit configuration of three-phase MMC without AC-side transformer.

where f^z represents the zero-sequence AC component, and f_a, f_b, and f_c represent the three-phase AC components.

The zero-sequence subsystem is derived based on equations (6.1) and (6.6) as

$$L\frac{di^z(t)}{dt} = \left[u_m^z(t) + u_o(t)\right] - e^z(t) \tag{6.7}$$

where i^z, u_m^z, and e^z are the zero-sequence components of the AC-side current, the AC-side voltage of MMC, the grid-side voltage, respectively.

6.2.2 Instantaneous Power Mathematical Model

Under unbalanced grid conditions, the three-phase grid voltage and grid current without zero-sequence component can be represented as the orthogonal sum of positive- and negative-sequence components [16], respectively, which can be expressed as

$$\begin{cases} \boldsymbol{E}_{dq} = e^{j\omega t}\boldsymbol{E}_{dq}^p + e^{-j\omega t}\boldsymbol{E}_{dq}^n \\ \boldsymbol{I}_{dq} = e^{j\omega t}\boldsymbol{I}_{dq}^p + e^{-j\omega t}\boldsymbol{I}_{dq}^n \end{cases} \tag{6.8}$$

where $\boldsymbol{E}_{dq}^p = e_d^p + je_q^p, \boldsymbol{E}_{dq}^n = e_d^n + je_q^n, \boldsymbol{I}_{dq}^p = i_d^p + ji_q^p$, and $\boldsymbol{I}_{dq}^n = i_d^n + ji_q^n$.

The instantaneous apparent power $\boldsymbol{S}_{ac}$ [16] can be expressed as

$$\boldsymbol{S}_{ac} = P_{ac} + jQ_{ac} = \boldsymbol{E}_{dq}\boldsymbol{I}_{dq}^{*} \tag{6.9}$$

where P_{ac} and Q_{ac} are the instantaneous active and reactive power, respectively. Accordingly, the active and reactive power are obtained as

$$\begin{cases} P_{ac} = P + P_c\cos(2\omega t) + P_s\sin(2\omega t) \\ Q_{ac} = Q + Q_c\cos(2\omega t) + Q_s\sin(2\omega t) \end{cases} \tag{6.10}$$

with

$$\begin{bmatrix} P \\ Q \\ P_c \\ P_s \\ Q_c \\ Q_s \end{bmatrix} = \frac{3}{2}\begin{bmatrix} e_d^p & e_q^p & e_d^n & e_q^n \\ e_q^p & -e_d^p & e_q^n & -e_d^n \\ e_d^n & e_q^n & e_d^p & e_q^p \\ e_q^n & -e_d^n & -e_q^p & e_d^p \\ e_q^n & -e_d^n & e_q^p & -e_d^p \\ -e_d^n & -e_q^n & e_d^p & e_q^p \end{bmatrix}\begin{bmatrix} i_d^p \\ i_q^p \\ i_d^n \\ i_q^n \end{bmatrix} \tag{6.11}$$

Equations (6.10) and (6.11) show that the asymmetric AC-grid voltages cause double-line frequency oscillations in active power and reactive power. Generally, the d-axis is aligned with the positive-sequence grid voltage component, thus the q-axis component of positive-sequence grid voltage e_q^p is zero in steady state. The negative-sequence current is not preferred from the perspective of power quality, which is usually controlled to be zero to ensure symmetrical AC-side current [13, 17]. Thus, the active power and reactive power can be simplified as

$$\begin{cases} P = 1.5e_d^p i_d^p \\ Q = -1.5e_d^p i_q^p \end{cases} \tag{6.12}$$

6.3 AC-Side Current Control of MMCs under AC Grid Faults

Under unbalanced voltage conditions, the AC-side current can be decomposed into positive-, negative-, and zero-sequence currents. The unbalanced AC-side current would cause overcurrent of the MMC system, as well as the active and reactive power oscillation. This section presents three AC-side current control methods under unbalanced conditions, including the positive- and negative-sequence current control, zero-sequence current control, and PR based current control.

6.3.1 Positive- and Negative-Sequence Current Control

Figure 6.3 illustrates the positive- and negative-sequence current control of the MMC based on equations (6.4) and (6.5). The control scheme is divided into two separate loops including an inner current loop and an outer power loop [13, 17].

6.3.1.1 Inner Loop Current Control

The inner current control is developed to regulate the positive- and negative-sequence currents i_d^p, i_q^p, i_d^n, and i_q^n. The control inputs are generated with proportional-integral (PI) controllers as

$$u_{dm}^p = e_d^p - \omega L i_q^p + \left[K_p\left(i_{d\_ref}^p - i_d^p\right) + K_i \int \left(i_{d\_ref}^p - i_d^p\right) dt\right] \tag{6.13}$$

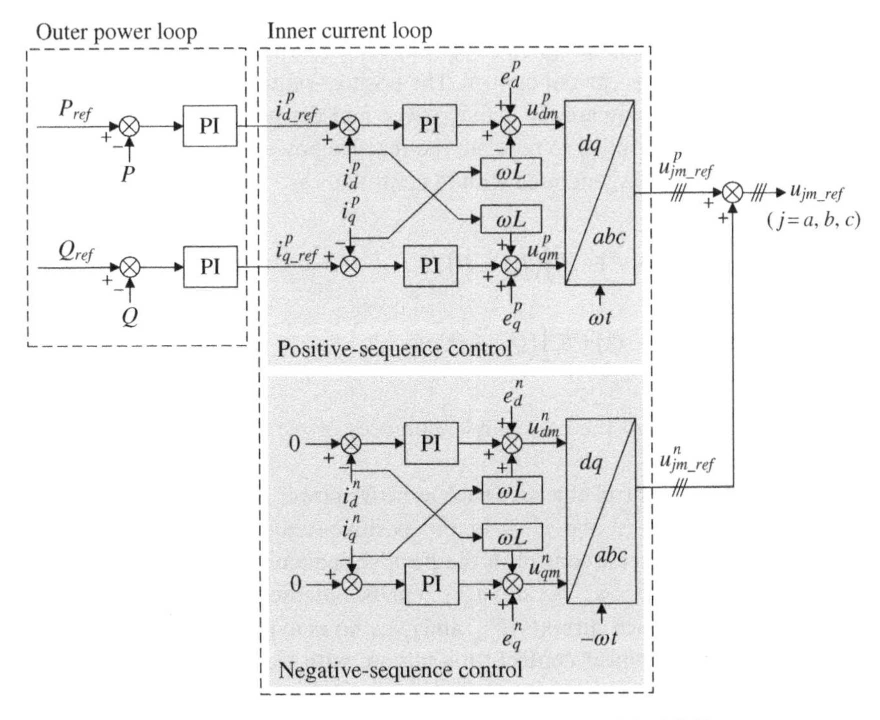

Figure 6.3 Positive- and negative-sequence current control of the MMC.

$$u_{qm}^{p} = e_{q}^{p} + \omega L i_{d}^{p} + \left[K_{p}\left(i_{q\_ref}^{p} - i_{q}^{p}\right) + K_{i}\int\left(i_{q\_ref}^{p} - i_{q}^{p}\right)dt \right] \tag{6.14}$$

$$u_{dm}^{n} = e_{d}^{n} + \omega L i_{q}^{n} + \left[K_{p}\left(i_{d\_ref}^{n} - i_{d}^{n}\right) + K_{i}\int\left(i_{d\_ref}^{n} - i_{d}^{n}\right)dt \right] \tag{6.15}$$

$$u_{qm}^{n} = e_{q}^{n} - \omega L i_{d}^{n} + \left[K_{p}\left(i_{q\_ref}^{n} - i_{q}^{n}\right) + K_{i}\int\left(i_{q\_ref}^{n} - i_{q}^{n}\right)dt \right] \tag{6.16}$$

where $i_{d\_ref}^{p}$ and $i_{q\_ref}^{p}$ are command references of dq-axis positive-sequence current components, and $i_{d\_ref}^{n}$ and $i_{q\_ref}^{n}$ are command references of dq-axis negative-sequence current components.

To eliminate negative-sequence current components, the command references for negative-sequence current are set as

$$\begin{cases} i_{d\_ref}^{n} = 0 \\ i_{q\_ref}^{n} = 0 \end{cases} \tag{6.17}$$

6.3.1.2 Outer Power Control

The outer loop control is designed to provide the command references $i_{d\_ref}^{p}$ and $i_{q\_ref}^{p}$ for the inner loop current control. The positive-sequence d-axis current i_{d}^{p} can be controlled to regulate the active power, and the positive-sequence q-axis current i_{q}^{p} can be controlled to regulate the reactive power. The command references $i_{d\_ref}^{p}$ and $i_{q\_ref}^{p}$ are generated with PI controllers as

$$i_{d\_ref}^{p} = K_{p}\left(P_{ref} - P\right) + K_{i}\int\left(P_{ref} - P\right)dt \tag{6.18}$$

$$i_{q\_ref}^{p} = K_{p}\left(Q_{ref} - Q\right) + K_{i}\int\left(Q_{ref} - Q\right)dt \tag{6.19}$$

where P_{ref} and Q_{ref} are the command references of active power and reactive power, respectively.

According to the control objective such as active power and reactive power, the reference currents $i_{d\_ref}^{p}$ and $i_{q\_ref}^{p}$ in the positive-sequence reference can be obtained. With the vector control in the positive-sequence reference frame, the reference voltages $u_{am\_ref}^{p}$, $u_{bm\_ref}^{p}$, and $u_{cm\_ref}^{p}$ can be obtained to control the i_{d}^{p} and i_{q}^{p} to follow the reference currents $i_{d\_ref}^{p}$ and $i_{q\_ref}^{p}$, so as to realize the active power control and reactive power control. In addition, with the vector control in the negative-sequence reference frame, the reference voltages $u_{am\_ref}^{n}$, $u_{bm\_ref}^{n}$, and $u_{cm\_ref}^{n}$ are obtained to control the i_{d}^{n} and i_{q}^{n} to zero so as to eliminate the negative-sequence components of the AC-side current. Afterwards, the reference for the three phases of the MMC can be obtained as

$$\begin{cases} u_{am\_ref} = u_{am\_ref}^{p} + u_{am\_ref}^{n} \\ u_{bm\_ref} = u_{bm\_ref}^{p} + u_{bm\_ref}^{n} \\ u_{cm\_ref} = u_{cm\_ref}^{p} + u_{cm\_ref}^{n} \end{cases} \tag{6.20}$$

6.3.2 Zero-Sequence Current Control

The controller described in Section 6.3.1 takes no consideration of the zero-sequence component, since the converter transformer connected in Y/Δ configuration can exclude the zero-sequence current in the converter side. However, in the transformerless application of MMCs, the zero-sequence components are unavoidable. To eliminate the zero-sequence current component, a PI feedback controller with a zero-sequence grid-voltage feedforward compensator is applied [13], as illustrated in Figure 6.4. According to the dynamics of zero-sequence components in equation (6.7), the control input is generated with PI controller as

$$u_m^z(t) = e^z(t) + \left[K_p\left(i_{ref}^z - i^z\right) + K_i \int \left(i_{ref}^z - i^z\right) dt \right] \tag{6.21}$$

where i_{ref}^z is the command reference of zero-sequence current. To eliminate the zero-sequence component, the command reference i_{ref}^z is set as zero.

Figure 6.5 shows an overview of the unbalanced control scheme, where the zero-sequence current controller coordinates well with the positive- and negative-sequence controller, as follows.

- Reference voltages $u_{am\_ref}^p$, $u_{bm\_ref}^p$, and $u_{cm\_ref}^p$ are obtained to control the active power and reactive power with the vector control in the positive-sequence reference frame.

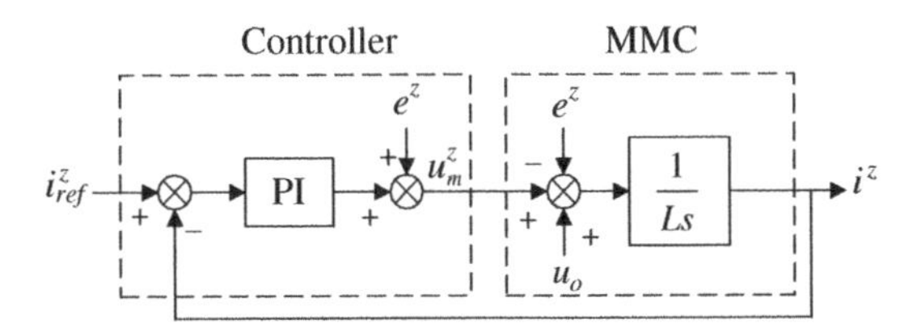

Figure 6.4 Block diagram of the zero-sequence current controller.

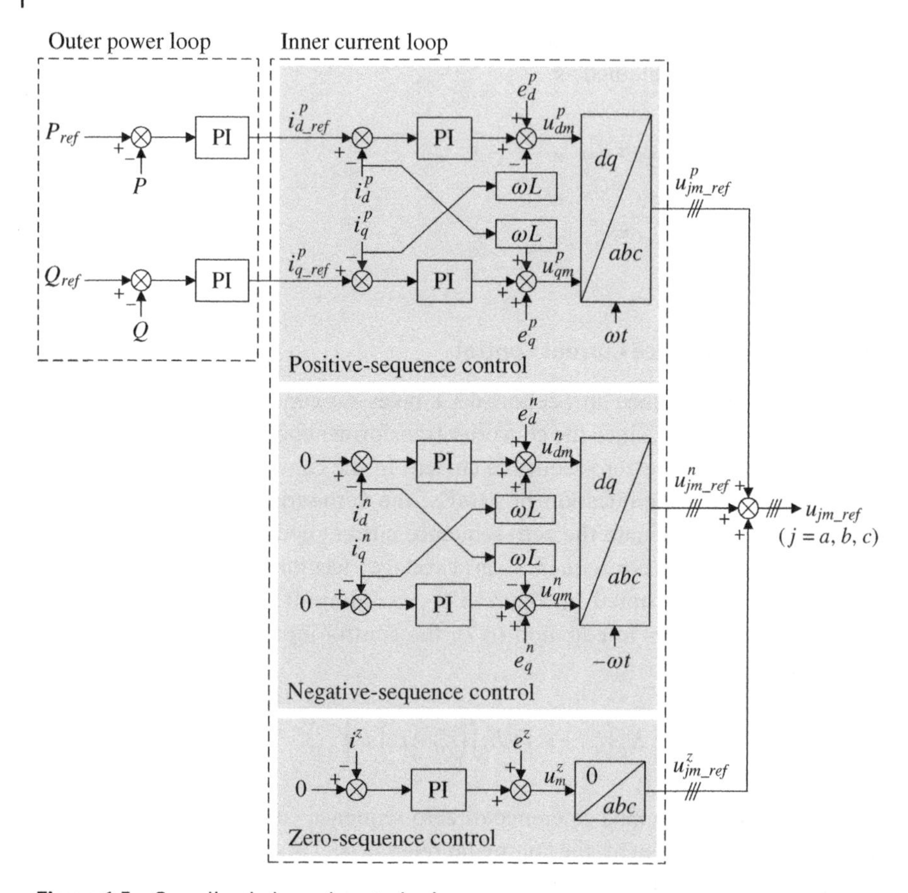

Figure 6.5 Overall unbalanced control scheme.

- Reference voltages $u_{am\_ref}^{n}$, $u_{bm\_ref}^{n}$, and $u_{cm\_ref}^{n}$ are obtained to eliminate the negative-sequence components of the AC-side current with the vector control in the negative-sequence reference frame.
- Reference voltages $u_{am\_ref}^{z}$, $u_{bm\_ref}^{z}$, and $u_{cm\_ref}^{z}$ are obtained to eliminate the zero-sequence components of the AC-side current with the zero-sequence reference frame.

The reference voltage for the three phases of the MMC can be expressed as

$$\begin{cases} u_{am\_ref} = u_{am\_ref}^{p} + u_{am\_ref}^{n} + u_{am\_ref}^{z} \\ u_{bm\_ref} = u_{bm\_ref}^{p} + u_{bm\_ref}^{n} + u_{bm\_ref}^{z} \\ u_{cm\_ref} = u_{cm\_ref}^{p} + u_{cm\_ref}^{n} + u_{cm\_ref}^{z} \end{cases} \tag{6.22}$$

6.3.3 Proportional Resonant Based Current Control

Unlike the positive- and negative-sequence current control, the PR based current control is conducted in the stationary-$\alpha\beta$ frame [18, 19]. The AC components are transformed and controlled in the stationary-$\alpha\beta$ frame. Thus, the complex analysis of the positive-sequence and negative-sequence components in the synchronous-dq frame is omitted.

The three-phase variables in the stationary-abc frame can be transformed to the stationary-$\alpha\beta$ frame by using the Clark's transformation as

$$T_{abc/\alpha\beta} = \frac{2}{3}\begin{bmatrix} 1 & -\frac{1}{2} & -\frac{1}{2} \\ 0 & \frac{\sqrt{3}}{2} & -\frac{\sqrt{3}}{2} \end{bmatrix} \tag{6.23}$$

Based on the equations (6.1) and (6.23), the mathematic model of AC-loop in stationary $\alpha\beta$-frame can be obtained as

$$L\begin{bmatrix} \frac{di_\alpha(t)}{dt} \\ \frac{di_\beta(t)}{dt} \end{bmatrix} = \begin{bmatrix} u_{\alpha m}(t) \\ u_{\beta m}(t) \end{bmatrix} - \begin{bmatrix} e_\alpha(t) \\ e_\beta(t) \end{bmatrix} \tag{6.24}$$

where e_α, e_β, $u_{\alpha m}$, $u_{\beta m}$, i_α, i_β are the αβ-axis components of the AC-grid voltages, the AC-side voltages of MMC, and the AC-side currents, respectively.

Figure 6.6 illustrates the PR based current control of the MMC based on equation (6.24), where the PR controller is constituted by the proportional regulator and resonant controller [20–22]. The transfer function of the PR controller is given by

$$G_{PR}(s) = K_p + \frac{K_i s}{s^2 + \omega^2} \tag{6.25}$$

The resonant controller ensures the gain is infinite at the resonant frequency, and this is necessary to enforce the steady-state error to zero.

According to equation (6.24), the control inputs can be generated with PR controllers as

$$u_{\alpha m\_ref} = \left(K_p + \frac{K_i s}{s^2 + \omega^2}\right)\left(i_{\alpha\_ref} - i_\alpha\right) + e_\alpha \tag{6.26}$$

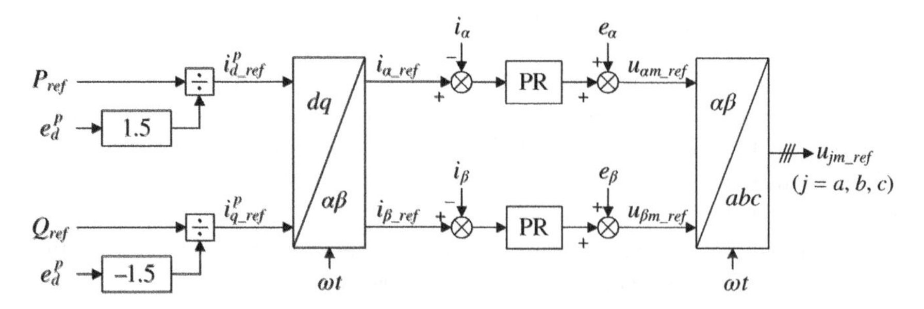

Figure 6.6 Proportional resonant based current control of the MMC.

$$u_{\beta m\_ref} = \left(K_p + \frac{K_i s}{s^2 + \omega^2}\right)\left(i_{\beta\_ref} - i_\beta\right) + e_\beta \tag{6.27}$$

where i_{α_ref} and i_{β_ref} are the command references of α- and β-axis current components.

In order to keep the AC-side current balance under asymmetrical faults, the negative-sequence current must be suppressed to zero. In this case, the current command references in synchronous-dq frame can be set based on equation (6.12) as

$$\begin{bmatrix} i_{d\_ref}^p \\ i_{q\_ref}^p \\ i_{d\_ref}^n \\ i_{q\_ref}^n \end{bmatrix} = \begin{bmatrix} \dfrac{P_{ref}}{1.5e_d^p} \\ -\dfrac{Q_{ref}}{1.5e_d^p} \\ 0 \\ 0 \end{bmatrix} \tag{6.28}$$

The reference currents i_{α_ref} and i_{β_ref} in equations (6.26) and (6.27) are obtained by transforming the reference currents in synchronous-dq frame into stationary αβ-frame with Park's reverse transformation shown as

$$\begin{bmatrix} i_{\alpha\_ref} \\ i_{\beta\_ref} \end{bmatrix} = \begin{bmatrix} \cos\omega t & -\sin\omega t \\ \sin\omega t & \cos\omega t \end{bmatrix}\begin{bmatrix} i_{d\_ref}^p \\ i_{q\_ref}^p \end{bmatrix} + \begin{bmatrix} \cos\omega t & \sin\omega t \\ -\sin\omega t & \cos\omega t \end{bmatrix}\begin{bmatrix} i_{d\_ref}^n \\ i_{q\_ref}^n \end{bmatrix} \tag{6.29}$$

The reference currents $i_{d\_ref}^p$, $i_{q\_ref}^p$, $i_{d\_ref}^n$, and $i_{q\_ref}^n$ are first calculated with equation (6.28) as depicted in Figure 6.6. And then, these reference currents are transformed into the stationary-α_β frame to obtain the reference currents i_{α_ref} and i_{β_ref}. With the PR current controllers, the reference voltages $u_{\alpha m\_ref}$ and $u_{\beta m\_ref}$ are obtained in stationary-α_β frame. Afterwards, the reference voltages in stationary-α_β frame are transformed into the stationary-abc frame by $T_{\alpha\beta/abc}$ to obtain the reference voltages u_{am_ref}, u_{bm_ref}, and u_{cm_ref}.

$$T_{\alpha\beta/abc} = \begin{bmatrix} 1 & 0 \\ -\frac{1}{2} & \frac{\sqrt{3}}{2} \\ -\frac{1}{2} & -\frac{\sqrt{3}}{2} \end{bmatrix} \tag{6.30}$$

6.4 Circulating Current Suppression Control of MMCs under AC Grid Faults

Under balanced AC-grid voltage conditions, only negative-sequence circulating currents exist in the MMC. However, under unbalanced AC-grid voltage conditions, there are positive-, negative-, and zero-sequence circulating currents in the MMC. The circulating current not only increases the power losses of the whole system but also aggravates the fluctuation of SM capacitor voltage. This section presents several methods to suppress the circulating currents under unbalanced conditions, including single-phase vector based control, $\alpha\beta 0$ stationary frame based control, and three-phase stationary frame based control.

6.4.1 Circulating Current of MMCs Under AC Grid Faults

In case of balanced AC-grid conditions, the second-order circulating currents in the three phases of the MMC are balanced and in the form of negative sequence. However, according to equations (6.10) and (6.11), the unbalanced AC-grid voltages cause double-line frequency oscillations in both the active power and the reactive power, which would cause unbalanced circulating currents in the MMC. Therefore, two kinds of second-order circulating currents may appear in each phase of the MMC [23–24].

- One is the second-order circulating current component produced by the MMC itself, which only contains negative-sequence component and would not affect the AC side and the DC side of the MMC.
- One is generated by the unbalanced AC-grid voltages, which contains positive-, negative-, and zero-sequence components and would affect both the AC side and the DC side of the MMC.

The circulating current in phase j ($j = a, b, c$) under the unbalanced AC-grid condition can be expressed as

$$i_{diff\_j} = \frac{i_{dc}}{3} + i_{2f\_j} \tag{6.31}$$

with

$$i_{2f\_j} = i^p_{2f\_j} + i^n_{2f\_j} + i^z_{2f\_j} \tag{6.32}$$

where $i^p_{2f\_j}$, $i^n_{2f\_j}$, and $i^z_{2f\_j}$ are the positive-, negative-, and zero-sequence circulating current components, respectively.

According to equation (6.32), the i_{2f_a}, i_{2f_b}, and i_{2f_c} in three phases can be expressed as

$$\begin{cases} i_{2f\_a} = I^p_{2f}\cos\left(2\omega t + \varphi^p\right) + I^n_{2f}\cos\left(2\omega t + \varphi^n\right) + I^z_{2f}\cos\left(2\omega t + \varphi^z\right) \\ i_{2f\_b} = I^p_{2f}\cos\left(2\omega t + \varphi^p - 120^\circ\right) + I^n_{2f}\cos\left(2\omega t + \varphi^n + 120^\circ\right) + I^z_{2f}\cos\left(2\omega t + \varphi^z\right) \\ i_{2f\_c} = I^p_{2f}\cos\left(2\omega t + \varphi^p + 120^\circ\right) + I^n_{2f}\cos\left(2\omega t + \varphi^n - 120^\circ\right) + I^z_{2f}\cos\left(2\omega t + \varphi^z\right) \end{cases} \tag{6.33}$$

where I^p_{2f}, I^n_{2f}, and I^z_{2f} are the amplitudes of positive-, negative- and zero-sequence circulating current components, respectively. φ^p, φ^n, and φ^z are the phase angles of positive-, negative-, and zero-sequence circulating current components, respectively.

The second-order circulating current i_{2f_j} in phase j would cause the corresponding unbalanced voltage u_{diff_j} imposed on the arm inductor [17], as shown in Figure 6.7. The relationship between the second-order circulating current i_{2f_j} and the unbalanced voltage u_{diff_j} in phase j can be expressed as

$$u_{diff\_j} = 2L_s \frac{di_{2f\_j}}{dt} \tag{6.34}$$

6.4.2 Single-Phase Vector Based Control

According to equation (6.34) and the method from single-phase system to dq transferring given in [25, 26], the behavior of the phase j in the synchronous rotating reference frame can be described as

$$\begin{bmatrix} u^d_{diff\_j} \\ u^q_{diff\_j} \end{bmatrix} = 2L_s \frac{d}{dt}\begin{bmatrix} i^d_{2f\_j} \\ i^q_{2f\_j} \end{bmatrix} + 2L_s \begin{bmatrix} 0 & -2\omega \\ 2\omega & 0 \end{bmatrix}\begin{bmatrix} i^d_{2f\_j} \\ i^q_{2f\_j} \end{bmatrix} \tag{6.35}$$

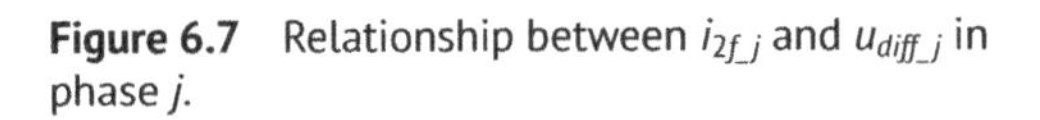

Figure 6.7 Relationship between i_{2f_j} and u_{diff_j} in phase j.

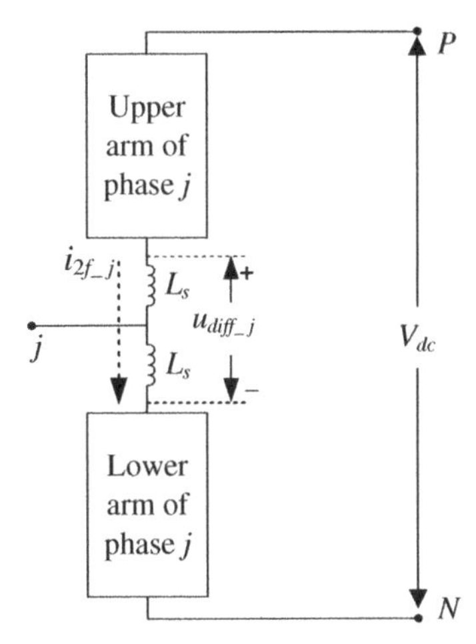

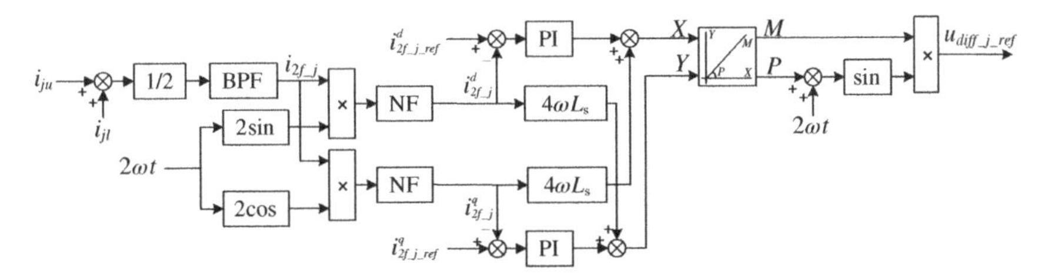

Figure 6.8 Single-phase vector based circulating current suppression control.

where $u^d_{diff\_j}$ and $u^q_{diff\_j}$ are the dq-axis components of the voltage u_{diff_a}, u_{diff_b}, u_{diff_c}, respectively. i_{2f_jd} and i_{2f_jq} are the dq-axis components of the current i_{2f_a}, i_{2f_b}, i_{2f_c}, respectively.

Figure 6.8 shows the single-phase vector based control to eliminate the second-order circulating current in phase j based on equation (6.35). According to equation (6.31), the second-order circulating current i_{2f_j} is the main component of the current i_{diff_j}, which can be obtained by the band-pass filter tuned at the double fundamental frequency, as shown in Figure 6.8. Then, the i_{2f_j} in phase j is transformed into dq domain through single-phase Park transformer [26], where one is multiplied by 2sin(2ωt) and the other one is multiplied by 2cos(2ωt). And then, a notch filter tuned at twice the second-order harmonic frequency. Accordingly, the dq-axis components $i^d_{2f\_j}$ and $i^q_{2f\_j}$ of the i_{2f_j} in phase j can be obtained. As a result, with the PI controller, the second-order circulating current i_{2f_j} in phase j would be eliminated by the voltage reference $u_{diff\_j\_ref}$, when the reference of the dq-axis components $i^d_{2f\_j}$ and $i^q_{2f\_j}$ are both regulated to zero.

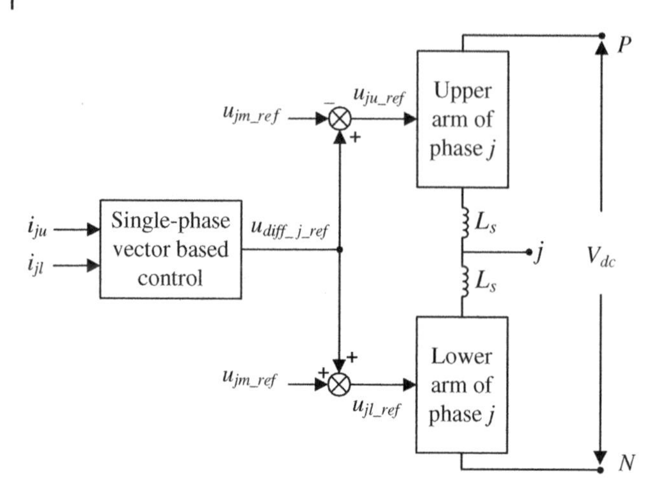

Figure 6.9 Overall control for phase j of the MMC.

The overall control for the MMC with the single-phase vector based circulating current suppression control in phase j is shown in Figure 6.9, where the voltage reference for the upper arm and the lower arm of phase j is produced as $u_{ju\_ref} = -u_{jm\_ref} + u_{diff\_j\_ref}$ and $u_{jl\_ref} = u_{jm\_ref} + u_{diff\_j\_ref}$, respectively. u_{jm_ref} is the reference voltage for phase j, as shown in Figures 6.3, 6.5, and 6.6. Consequently, the second-order circulating current can be effectively eliminated by the single-phase vector based control.

6.4.3 $\alpha\beta 0$ Stationary Frame Based Control

A $\alpha\beta 0$ stationary frame based control is presented for second-order circulating current suppression under AC-grid faults, where the positive-, negative-, and zero-sequence components of the second-order circulating current can be reduced.

The three-phase components in the stationary-abc frame can be transformed to the stationary $\alpha\beta 0$ frame by using the Clarke's transformation [15] as

$$T_{abc/\alpha\beta 0} = \frac{2}{3}\begin{bmatrix} 1 & -\frac{1}{2} & -\frac{1}{2} \\ 0 & \frac{\sqrt{3}}{2} & -\frac{\sqrt{3}}{2} \\ \frac{1}{2} & \frac{1}{2} & \frac{1}{2} \end{bmatrix} \tag{6.36}$$

According to equations (6.33), (6.34), and (6.36), the behavior of the phase j ($j = a, b, c$) system in the stationary $\alpha\beta 0$ frame can be described as

$$\begin{cases} u_{diff}^{\alpha} = 2L_s \dfrac{di_{2f}^{\alpha}}{dt} \\ u_{diff}^{\beta} = 2L_s \dfrac{di_{2f}^{\beta}}{dt} \\ u_{diff}^{0} = 2L_s \dfrac{di_{2f}^{0}}{dt} \end{cases} \tag{6.37}$$

with

$$\begin{cases} i_{2f}^{\alpha} = I_{2f}^{p} \cos\left(2\omega t + \varphi^{p}\right) + I_{2f}^{n} \cos\left(2\omega t + \varphi^{n}\right) \\ i_{2f}^{\beta} = I_{2f}^{p} \sin\left(2\omega t + \varphi^{p}\right) + I_{2f}^{n} \sin\left(2\omega t + \varphi^{n}\right) \\ i_{2f}^{0} = I_{2f}^{z} \cos\left(2\omega t + \varphi^{z}\right) \end{cases} \tag{6.38}$$

where u_{diff}^{α}, u_{diff}^{β}, and u_{diff}^{0} are the $\alpha\beta0$ components of the voltage u_{diff_j} in the $\alpha\beta0$ stationary frame, respectively. i_{2f}^{α}, i_{2f}^{β}, and i_{2f}^{0} are the $\alpha\beta0$ components of the second-order circulating current i_{2f_j} in the $\alpha\beta0$ stationary frame, respectively.

Figure 6.10 shows the two-phase stationary based control to eliminate the second-order circulating current based on equation (6.37). According to equation (6.31), the second-order circulating current i_{2f_j} in phase j is the main component of the current i_{diff_j}, which can be obtained by the band-pass filter tuned at the double fundamental frequency. Then, the circulating currents in three phases are transformed into the $\alpha\beta0$ stationary frame through Clarke transformer, and accordingly, the i_{2f}^{α}, i_{2f}^{β}, i_{2f}^{0} can be obtained. In order to suppress the positive-, negative-, and zero-sequence components of the second-order circulating current, the command references $i_{2f\_ref}^{\alpha}$, $i_{2f\_ref}^{\beta}$, and $i_{2f\_ref}^{0}$ are all set to zero [15]. The voltage references $u_{diff\_ref}^{\alpha}$, $u_{diff\_ref}^{\beta}$, and $u_{diff\_ref}^{0}$ in the $\alpha\beta0$ stationary frame can be generated with PR controller as

$$\begin{cases} u_{diff\_ref}^{\alpha} = \left(K_p + \dfrac{K_i s}{s^2 + (2\omega)^2}\right)\left(i_{2f\_ref}^{\alpha} - i_{2f}^{\alpha}\right) \\ u_{diff\_ref}^{\beta} = \left(K_p + \dfrac{K_i s}{s^2 + (2\omega)^2}\right)\left(i_{2f\_ref}^{\beta} - i_{2f}^{\beta}\right) \\ u_{diff\_ref}^{0} = \left(K_p + \dfrac{K_i s}{s^2 + (2\omega)^2}\right)\left(i_{2f\_ref}^{0} - i_{2f}^{0}\right) \end{cases} \tag{6.39}$$

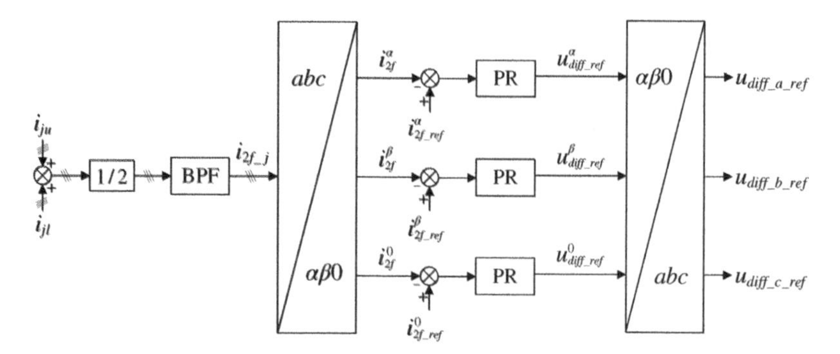

Figure 6.10 $\alpha\beta 0$ frame based circulating current suppression control.

Afterwards, the voltage references in the $\alpha\beta 0$ stationary frame are transformed into the stationary-*abc* frame by $T_{\alpha\beta 0/abc}$ to obtain the voltage references $u_{diff\_a\_ref}$, $u_{diff\_b\_ref}$, and $u_{diff\_c\_ref}$, as shown in Figure 6.10.

$$T_{\alpha\beta 0/abc} = \begin{bmatrix} 1 & 0 & 1 \\ -\frac{1}{2} & \frac{\sqrt{3}}{2} & 1 \\ -\frac{1}{2} & -\frac{\sqrt{3}}{2} & 1 \end{bmatrix} \tag{6.40}$$

The overall control for the MMC with $\alpha\beta 0$ stationary frame based circulating current suppression control is shown in Figure 6.11, where the voltage reference for the upper arm and the lower arm of phase *j* are produced as $u_{ju\_ref} = -u_{jm\_ref} + u_{diff\_j\_ref}$ and $u_{jl\_ref} = u_{jm\_ref} + u_{diff\_j\_ref}$, respectively. Consequently, the positive-, negative-, and zero-sequence components of the second-order circulating current in phase *j* can be effectively eliminated by the $\alpha\beta 0$ stationary frame based control.

6.4.4 Three-Phase Stationary Frame Based Control

A three-phase stationary frame based control is presented to suppress the second-order circulating current under AC-grid faults, which is implemented through a proportional-integral-resonant (PIR) controller, and can effectively eliminate the positive-, negative-, and zero-sequence circulating current components. Comparing with the control method based on $\alpha\beta 0$ stationary frame, the three-phase stationary frame based control is simpler because it is able to eliminate each circulating current component without decomposition into separate parts. The three-phase stationary

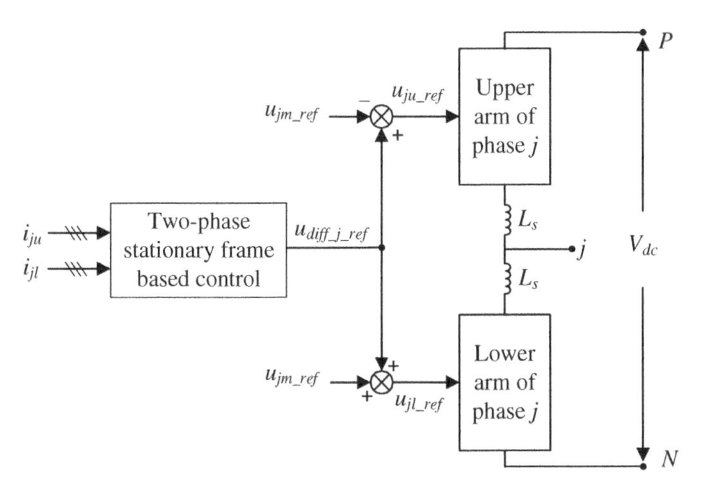

Figure 6.11 Overall control for phase *j* of the MMC.

frame based circulating current suppression control mainly contains positive- and negative-sequence controller as well as zero-sequence controller, as follows.

6.4.4.1 Positive- and Negative-Sequence Controller

The sum of the three-phase circulating currents can be expressed as

$$i_{diff\_sum} = i_{diff\_a} + i_{diff\_b} + i_{diff\_c} \tag{6.41}$$

Substituting equations (6.31) and (6.33) into equation (6.41), there is

$$i_{diff\_sum} = i_{dc} + 3I_{2f}^{z}\cos\left(\varphi^{z}\right) \tag{6.42}$$

Since the positive-sequence and negative-sequence components of the three-phase circulating currents are symmetrical, the i_{diff_sum} only contains the DC component and the zero-sequence component currents. To eliminate the positive- and negative-sequence circulating current components, the reference value of the i_{diff_a}, i_{diff_b}, and i_{diff_c} can all be set to i_{diff_ref} [23] as

$$i_{diff\_ref} = \frac{i_{diff\_sum}}{3} \tag{6.43}$$

The PIR controller is adopted to force the three-phase circulating currents to follow their reference values. The voltage references $u_{diff\_a\_ref}^{pn}$, $u_{diff\_b\_ref}^{pn}$, and $u_{diff\_c\_ref}^{pn}$ generated with PIR controllers can be expressed as

$$\begin{cases} u^{pn}_{diff\_a\_ref} = G_{PIR}(s)\cdot(i_{diff\_ref} - i_{diff\_a}) \\ u^{pn}_{diff\_b\_ref} = G_{PIR}(s)\cdot(i_{diff\_ref} - i_{diff\_b}) \\ u^{pn}_{diff\_c\_ref} = G_{PIR}(s)\cdot(i_{diff\_ref} - i_{diff\_c}) \end{cases} \tag{6.44}$$

with

$$G_{PIR}(s) = K_p + \frac{K_i}{s} + \frac{2K_r\omega_r s}{s^2 + 2\omega_r s + (2\omega)^2} \tag{6.45}$$

where $G_{PIR}(s)$ is the transfer function of the PIR controller, and the PIR controller's cut-off frequency is set to the double-line frequency 2ω. As a result, the positive-sequence and negative-sequence components of second-order circulating current would be eliminated by the voltage reference $u^{pn}_{diff\_j\_ref}$. However, according to equations (6.42) and (6.43), the reference values of the i_{diff_a}, i_{diff_b}, and i_{diff_c} still contain zero-sequence component current. Thus, a controller to reduce the zero-sequence circulating current component is necessary.

6.4.4.2 Zero-Sequence Controller

Neglecting the power loss in the MMC, the MMC's AC-side active power P would be equal to the DC-side active power, and the DC-side current i_{dc} is

$$i_{dc} = \frac{P}{V_{dc}} \tag{6.46}$$

Since the i_{dc} does not contain zero-sequence component, the i_{dc} can be applied to reduce the zero-sequence circulating current component. The PIR controller is adopted to force the i_{diff_sum} to follow its reference i_{dc}. The voltage references $u^{z}_{diff\_ref}$ generated with PIR controllers can be expressed as

$$u^{z}_{diff\_ref} = G_{PIR}(s)\cdot\left(i_{dc} - \sum_{j=a,b,c} i_{diff\_j}\right) \tag{6.47}$$

With the positive-sequence and negative-sequence controller as well as the zero-sequence controller, the total control output value to eliminate positive-, negative-, and zero-sequence circulating current components can be obtained as

$$u_{diff\_j\_ref} = u^{pn}_{diff\_j\_ref} + u^{z}_{diff\_ref} \tag{6.48}$$

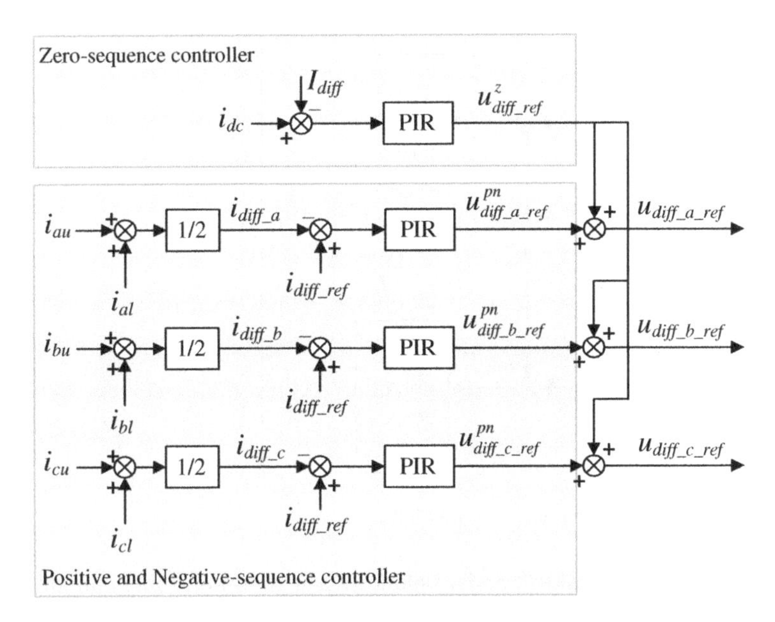

Figure 6.12 Three-phase stationary frame based control.

Figure 6.12 shows a block diagram of the three-phase stationary frame based circulating current suppression control, which contains three PIR controllers to suppress the positive-sequence and negative-sequence circulating currents and one PIR controller to suppress the zero-sequence circulating current.

The overall control for MMCs with the three-phase stationary frame based circulating current suppression control of phase j is shown in Figure 6.13. As a result, the positive-, negative-, and zero-sequence components of the circulating current in phase j can be effectively eliminated.

6.5 Summary

In grid-connected MMCs, the occurrence of grid-side faults is rather common, resulting in imbalances and distortions of the AC-side voltages. Therefore, the operation and control of MMC-HVDC system in the presence of grid imbalances become very important in order to satisfy grid codes and reliability requirements.

In this chapter, recent advancements in the operation and control methods applied to MMC-HVDC systems under unbalanced AC-grid conditions are introduced and analyzed.

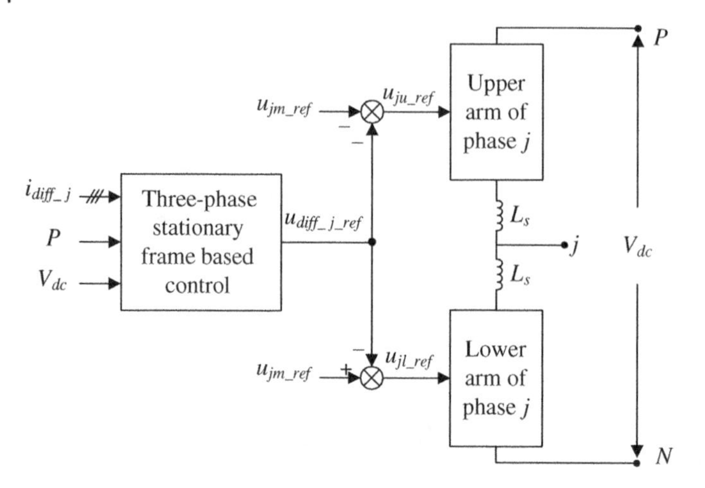

Figure 6.13 Overall control for phase j of the MMC.

The AC-side currents of MMCs under balanced grid conditions only contain positive-sequence components. However, the AC-side current of MMCs contain not only positive-sequence components but also negative-sequence components and even zero-sequence components under asymmetrical grid faults. According to the AC-side mathematical model of MMCs under asymmetrical grid faults, three AC-side current control methods under unbalanced conditions are presented, including the positive- and negative-sequence current control, zero-sequence current control, and PR based current control.

Except for the AC-side current, the circulating current is also influenced by the asymmetric AC-side voltage. Under balanced AC-grid voltage conditions, only negative-sequence circulating currents exist in the MMC. However, under unbalanced AC-grid voltage conditions, there are positive-, negative-, and zero-sequence circulating currents in the MMC. The circulating currents not only increase the power losses of the whole system but also aggravate the fluctuation of SM capacitor voltage. According to the instantaneous power mathematical model of MMCs under asymmetrical grid faults, several methods are presented to suppress the circulating currents under unbalanced conditions, including single-phase vector based control, $\alpha\beta 0$ stationary frame based control, and three-phase stationary frame based control.

References

1 J. Li, G. Konstantinou, H. R. Wickramasinghe, and J. Pou, "Operation and control methods of modular multilevel converters in unbalanced ac grids: a review," *IEEE J. Emerging Sel. Top. Power Electron.*, vol. 7, no. 2, pp. 1258–1271, Jun. 2019.

2 J. Wang, Y. Tang, and X. Liu, "Arm current balancing control for modular multilevel converters under unbalanced grid conditions," *IEEE Trans. Power Electron.*, vol. 35, no. 3, pp. 2467–2479, Mar. 2020.

3 J. Li, G. Konstantinou, H. R. Wickramasinghe, C. D. Townsend, and J. Pou, "Capacitor voltage reduction in modular multilevel converters under grid voltages unbalances," *IEEE Trans. Power Delivery*, vol. 35, no. 1, pp. 160–170, Feb. 2020.

4 Z. Gong, X. Wu, P. Dai, and R. Zhu, "Modulated model predictive control for mmc-based active front-end rectifiers under unbalanced grid conditions," *IEEE Trans. Ind. Electron.*, vol. 66, no. 3, pp. 2398–2409, Mar. 2019.

5 J. Xu, Y. Yu, and C. Zhao, "The predictive closed-loop averaging control of mmc phase-unit losses under unbalanced conditions," *IEEE Trans. Power Delivery*, vol. 34, no. 1, pp. 198–207, Feb. 2019.

6 Q. Hao, J. Man, F. Gao, and M. Guan, "Voltage limit control of modular multilevel converter based unified power flow controller under unbalanced grid conditions," *IEEE Trans. Power Delivery*, vol. 33, no. 3, pp. 1319–1327, Jun. 2018.

7 Y. Liang, J. Liu, T. Zhang, and Q. Yang, "Arm current control strategy for mmc-hvdc under unbalanced conditions," *IEEE Trans. Power Delivery*, vol. 32, no. 1, pp. 125–134, Feb. 2017.

8 A. E. Leon and S. J. Amodeo, "Energy balancing improvement of modular multilevel converters under unbalanced grid conditions," *IEEE Trans. Power Electron.*, vol. 32, no. 8, pp. 6628–6637, Aug. 2017.

9 S. Isik, M. Alharbi, and S. Bhattacharya, "An optimized circulating current control method based on pr and pi controller for mmc applications," *IEEE Trans. Ind. Appl.*, vol. 57, no. 5, pp. 5074–5085, Sept.–Oct. 2021.

10 J. Li, G. Konstantinou, H. R. Wickramasinghe, J. Pou, X. Wu, and X. Jin, "Impact of circulating current control in capacitor voltage ripples of modular multilevel converters under grid imbalances," *IEEE Trans. Power Delivery*, vol. 33, no. 3, pp. 1257–1267, Jun. 2018.

11 F. Wang, J. L. Duarte, and M. A. M. Hendrix, "Pliant active and reactive power control for grid-interactive converters under unbalanced voltage dips," *IEEE Trans. Power Electron.*, vol. 26, no. 5, pp. 1511–1521, May 2011.

12 A. Dekka, B. Wu, and N. R. Zargari, "Minimization of dc-bus current ripple in modular multilevel converter under unbalanced conditions," *IEEE Trans. Power Electron.*, vol. 32, no. 6, pp. 4125–4131, Jun. 2017.

13 M. Guan and Z. Xu, "Modeling and control of a modular multilevel converter-based hvdc system under unbalanced grid conditions," *IEEE Trans. Power Electron.*, vol. 27, no. 12, pp. 4858–4867, Dec. 2012.

14 Q. Yu, F. Deng, C. Liu, J. Zhao, F. Blaabjerg, and S. Abulanwar, "DC-link high-frequency current ripple elimination strategy for mmcs using phase-shifted double-group multicarrier-based phase-disposition pwm," *IEEE Trans. Power Electron.*, vol. 36, no. 8, pp. 8872–8886, Aug. 2021.

15 Y. Zhou, D. Jiang, J. Guo, P. Hu, and Y. Liang, "Analysis and control of modular multilevel converters under unbalanced conditions," *IEEE Trans. Power Delivery*, vol. 28, no. 4, pp. 1986–1995, Oct. 2013.

16 W. Tao, Z. Gu, L. Wang, and J. Li, "Research on control strategy of grid-connected inverter under unbalanced voltage conditions," in *2016 IEEE 8th International Power Electronics and Motion Control Conference (IPEMC-ECCE Asia)*, 2016, pp. 915–919.

17 F. Deng, D. Liu, Y. Wang, Q. Wang, and Z. Chen, "Modular multilevel converters based variable speed wind turbines for grid faults," in *IECON 2016 - 42nd Annual Conference of the IEEE Industrial Electronics Society*, 2016, pp. 2420–2425.

18 N. Quach, J. Ko, D. Kim, and E. Kim, "An application of proportional-resonant controller in mmc-hvdc system under unbalanced voltage conditions," *J. Electr. Eng. Technol.*, vol. 9, no. 5, pp. 1746–1752, Apr. 2014.

19 Q. Tu, Z. Xu, Y. Chang, and L. Guan, "Suppressing dc voltage ripples of mmc-hvdc under unbalanced grid conditions," *IEEE Trans. Power Delivery*, vol. 27, no. 3, pp. 1332–1338, Jul. 2012.

20 C. Xia, Z. Wang, T. Shi, and X. He, "An improved control strategy of triple line-voltage cascaded voltage source converter based on proportional–resonant controller," *IEEE Trans. Ind. Electron.*, vol. 60, no. 7, pp. 2894–2908, Jul. 2013.

21 G. Shen, X. Zhu, J. Zhang, and D. Xu, "A new feedback method for pr current control of lcl-filter-based grid-connected inverter," *IEEE Trans. Ind. Electron.*, vol. 57, no. 6, pp. 2033–2041, Jun. 2010.

22 S. Li, X. Wang, Z. Yao, T. Li, and Z. Peng, "Circulating current suppressing strategy for mmc-hvdc based on nonideal proportional resonant controllers under unbalanced grid conditions," *IEEE Trans. Power Electron.*, vol. 30, no. 1, pp. 387–397, Jan. 2015.

23 J. W. Moon, C. S. Kim, J. W. Park, D. W. Kang, and J. M. Kim, "Circulating current control in mmc under the unbalanced voltage," *IEEE Trans. Power Delivery*, vol. 28, no. 3, pp. 1952–1959, Jul. 2013.

24 J. Moon, J. Park, D. Kang, and J. Kim, "A control method of hvdc-modular multilevel converter based on arm current under the unbalanced voltage condition," *IEEE Trans. Power Delivery*, vol. 30, no. 2, pp. 529–536, Apr. 2015.

25 C. H. Ng, L. Ran, and J. Bumby, "Unbalanced-grid-fault ride-through control for a wind turbine inverter," *IEEE Trans. Ind. Appl.*, vol. 44, no. 3, pp. 845–856, May–Jun. 2008.

26 J. Salaet, S. Alepuz, A. Gilabert, and J. Bordonau, "Comparison between two methods of dq transformation for single phase converters control. Application to a 3-level boost rectifier," in *2004 IEEE 35th Annual Power Electronics Specialists Conference (IEEE Cat. No.04CH37551)*, 2004, pp. 214–220.

7

Protection Under DC Short-Circuit Fault in HVDC System

7.1 Introduction

The modular multilevel converter (MMC) based high-voltage direct current (HVDC) transmission system has been a preferred choice for bulky and flexible power transferring since the MMC offers remarkable advantages, such as easy scalability and high-quality output voltage.

The MMC-HVDC system with overhead line is restricted due to the occasional occurrence of the DC short-circuit faults, which can be caused by lightning strikes or mechanical faults caused by broken branches or bushes. As a low inertia system, the HVDC system is prone to having a high fault current rising rate after the DC fault occurs. The fault current is required to be interrupted in several milliseconds to protect the semiconductor devices and other components of the MMC. The half-bridge submodule (HB SM) based MMC is lack of fault-current handling capability because the MMC forms an uncontrollable rectifier circuit after all switches are turned off, which continuously feeds DC current to the short-circuit point.

This chapter focuses on the DC fault protection methods for the MMC based HVDC system. Section 7.2 presents the protection for HB SM based MMC under DC fault, where the AC circuit breaker (CB) plays an important role in interrupting the DC fault current. Section 7.3 introduces several protection methods based on the DC CBs. Section 7.4 gives a brief description about the MMC topology with fault blocking ability. Sections 7.5 and 7.6 present the DC fault protection based on bypass thyristor (BT) MMC and the crossing-thyristor branch based hybrid MMC, respectively.

Modular Multilevel Converters: Control, Fault Detection, and Protection, First Edition.
Fujin Deng, Chengkai Liu, and Zhe Chen.

Published 2023 by John Wiley & Sons, Inc.

7.2 MMC Under DC Short-Circuit Fault

The DC short-circuit fault is an important issue the MMC based HVDC must deal with. The MMC constituted with HB SM cannot clear the fault current. Employing a DC CB is an approach to quickly extinguish DC-line fault current. However, in real applications, the availability of DC CB is limited due to its immaturity and high cost [1, 2]. The AC CB may be adopted to achieve the required DC-line protection for the HB SM based MMC. However, the freewheeling diodes in HB SM should withstand the short-circuit fault current until the AC CB trips. Hence, there is a risk in depending on AC CB protection only since the semiconductor devices may be damaged due to overcurrent. To enhance the reliability of the MMC-HVDC system, a single-thyristor switch can be used in conjunction with the AC CB.

7.2.1 System Configuration

The HVDC transmission is an effective means to achieve energy balance between regions [3]. Nowadays, the HB SM based MMC (HB-MMC) is widely adopted in the HVDC system. The configuration of the HB-MMC HVDC is depicted as Figure 7.1.

To enhance the reliability of the MMC-HVDC system, the DC short-circuit fault current should be limited. However, the HB-MMC HVDC is defenseless against DC short-circuit fault since its freewheeling diodes function as an uncontrolled rectifier bridge and feed the DC fault [4], even if the semiconductor devices are turned off. During the DC faults, the AC-side current is contributed into the DC fault point through the freewheeling diodes in each SM, as depicted in Figure 7.2. As a result, the semiconductor devices may be damaged due to high fault current.

7.2.2 AC Circuit Breaker

The AC CB is equipped at the AC side of the MMC, which can be used to interrupt the AC-side current of the MMC. The AC CB is the most economical way to

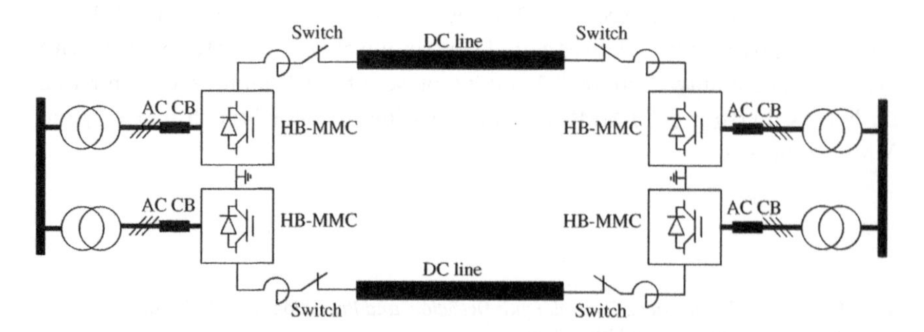

Figure 7.1 HB-MMC HVDC configuration.

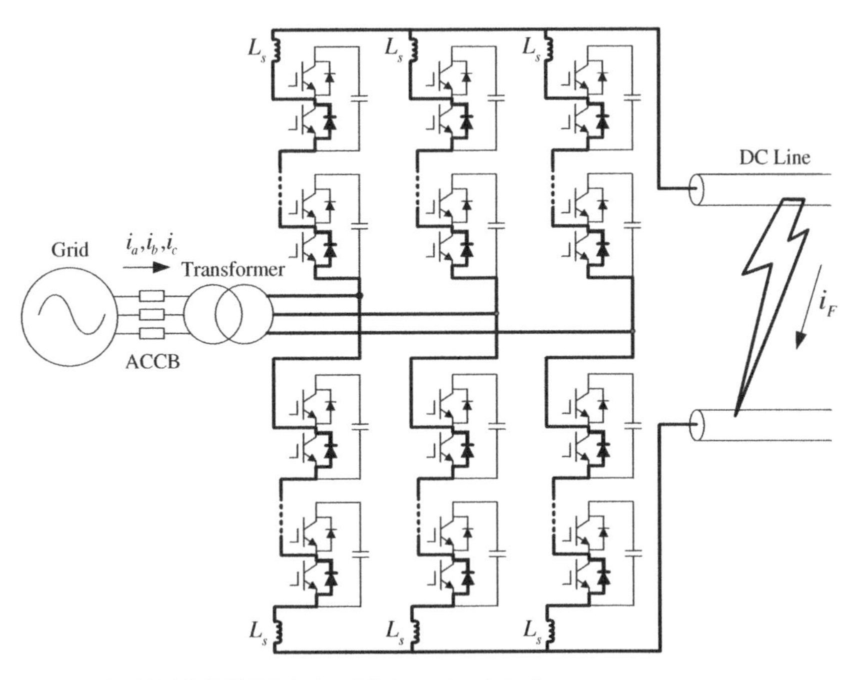

Figure 7.2 HB-MMC HVDC during DC short-circuit fault.

prevent the AC grid current contribution and protect the system under the DC short-circuit fault [5], as shown in Figure 7.1.

When using an AC CB to protect the system against the DC short-circuit fault, the voltage and the current of the DC line are monitored, which will be fed back to a standard relay to monitor overcurrent. If the DC short-circuit fault occurs, the fault current will be continuously fed from AC grid to the fault point through freewheeling diodes in the HB SMs of the MMC, as shown in Figure 7.2. Accordingly, the DC-line current of the MMC will rise over the rated value. Once the relay senses the overcurrent condition, it will trip the AC CB for protection.

Due to its mechanical restrictions, the tripping of the AC CB requires about 60–100 ms [5], which would cause a long interruption time of the fault current at the AC side of the MMC.

7.2.3 Protection Thyristor

During the DC-line short-circuit fault, the AC-side currents contributed into the DC fault flow through the freewheeling diodes, as shown in Figure 7.3a. Although the AC CB is tripped after the DC-line short-circuit fault occurrence, the tripping of AC CB requires about 60–100 ms [5], which causes a long interruption time of the fault current at the AC side of the MMC. And the diodes have to withstand a

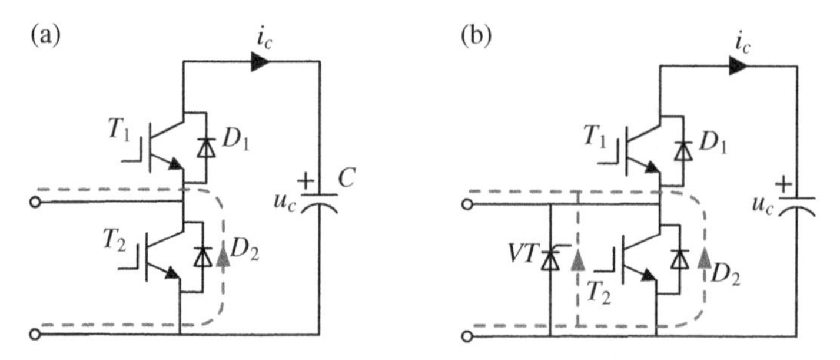

Figure 7.3 The path of fault current flowing through the SM. (a) HB SM. (b) HB SM with protection thyristor.

surge fault current until the AC CB trips, as shown in Figure 7.1. The freewheeling diodes used in the MMC have a low capacity for withstanding the surge current events related to their silicon surface [6]. Therefore, the freewheeling diodes in the MMC may be damaged due to high fault current and thermal overstress.

To protect the diodes in the MMC, a single-thyristor switch VT is parallel connected with the freewheeling diodes in each SM, as depicted in Figure 7.3b, which will share the arm current and protect the SM.

During normal operation, the thyristor VT is kept in an off-state condition. When the DC-side short-circuit fault is detected, the thyristor VT is switched on to force a part of the surge fault current to flow through it, as shown in Figure 7.3b. Since the thyristor has a higher capability for withstanding the surge current compared to the diode, the diode in the SM is protected from being damaged and the reliability of HB-MMC HVDC system under DC faults is enhanced.

Figure 7.4 demonstrates the fault current path in the MMC equipped with single-thyristor SM during the DC fault. The single-thyristor SM does not change the rectifier path. A single thyristor can be adopted to prevent the damage of diodes, but it is not able to prevent the AC grid current contribution.

7.2.4 Protection Operation

The protection operation for DC-line short-circuit fault in HB-MMC HVDC system is illustrated in Figure 7.5, which can be divided into five stages. The protection operations are explained in detail as follows.

1) *Stage 1: MMC Blocking*
   When the DC-line short-circuit fault is detected, all insulated gate bipolar transistors (IGBTs) at HB-MMC of the faulty pole will be blocked. In addition, the protection thyristor parallel connected with freewheeling diodes should be fired to protect the semiconductor devices in each SM.

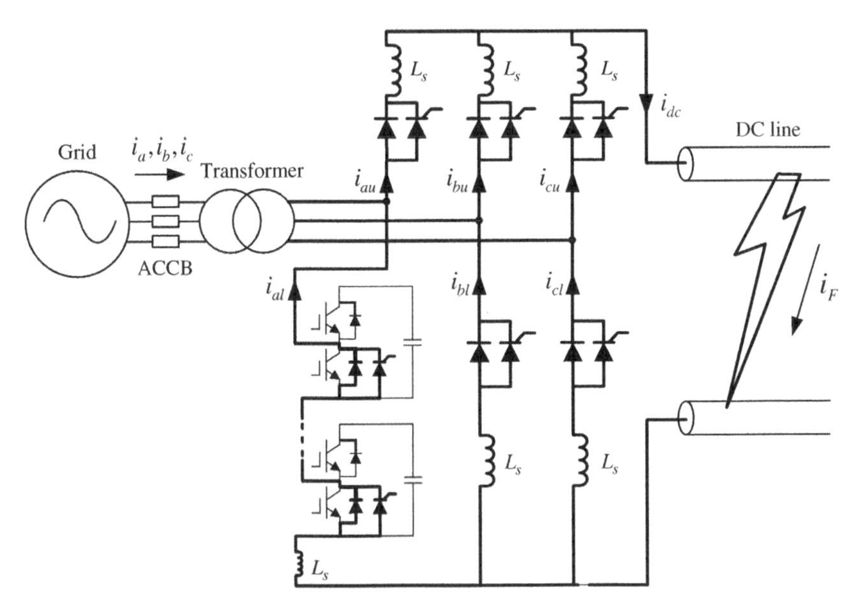

Figure 7.4 MMC with single-thyristor submodules under DC-line fault.

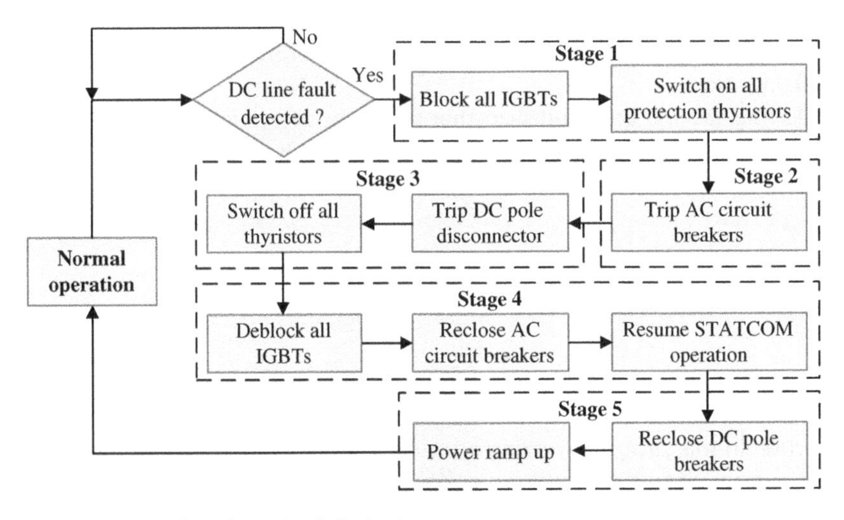

Figure 7.5 DC-line short-circuit fault clearance process.

2) *Stage 2: AC CB Tripping*
Due to their mechanical restrictions, the response time t_{ac} of AC CBs requires about 60 to 100 ms. During this period, the AC-side current feeds into the DC fault point through the freewheeling diodes and the protection thyristors.

Here, the converter can be regarded as a nonlinear inductive load that consumes massive amount of reactive power and lowers the AC voltage [7].

3) *Stage 3: Fault Arc Extinguishment*
   After the AC CBs are tripped, the AC side stops injecting fault current to the DC side through the blocked MMC. The residual DC fault currents will gradually decay through the damping loop consisted of the freewheeling diodes, the protection thyristors, the overhead lines, and the fault arc. Because the DC fault currents and the time constants in the damping loop are large, it may take an extinguishing time t_{ex} for the residual DC fault currents to freely decay to the value less than the maximum DC breaking current I_{off} of the DC-side mechanical switch. Then the DC-side mechanical switches (MSs) can be tripped so as to utterly cut off the DC fault currents. Similarly, the DC-side MSs require t_{dc} (60–100 ms) to be completely tripped. In the meantime, the driving signals for thyristors are removed to switch off all thyristors.
4) *Stage 4: DC line deionization and STATCOM operation*
   The overhead line requires deionization that lasts about t_{io} (150–500 ms) to restore its insulation. At the meantime, the faulty pole HB-MMCs can resume static synchronous compensator (STATCOM) operation after switching off all protection thyristors, deblocking all IGBTs and reclosing the AC CBs.
5) *Stage 5: DC power recovery*

   The DC-side MSs are reclosed and the faulty pole of the system resumes normal operation.

Neglecting the DC line fault detecting time, the time interval for HB-MMC HVDC configuration to clear a nonpermanent DC-line short-circuit fault T_{DCF} can be obtained as

$$T_{DCF} = t_{ac} + t_{ex} + t_{dc} + t_{io} \tag{7.1}$$

which is about 570–1500 ms in practical projects.

Case Study 7.1 Analysis of DC-Line Short-Circuit Faults

Objective: In this case, the working principle and the performance of the DC-line short-circuit fault protection is studied through professional time-domain simulation software power systems computer aided design/electromagnetic transients including DC (PSCAD/EMTDC). A 300 MVA/250 kV HVDC system shown in Figure 7.6 is simulated. The MMC is controlled to transmit an active power of 200 MW under steady state. An overhead line with the length of 150 km is adopted. At t = 1 second, a DC-line short-circuit fault occurs on the overhead line 10 km away from the MMC.

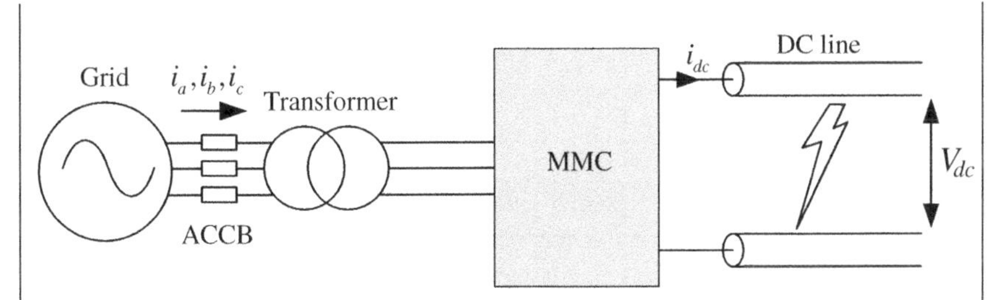

Figure 7.6 Block diagram of the simulation system.

Table 7.1 Simulation system parameters.

Parameters	Value
DC-link voltage V_{dc} (kV)	250
AC-link voltage (kV, RMS)	150
Number of SMs per arm n	8
Nominal SM capacitance C (mF)	2
Inductance L_s (mH)	56
DC-line short-circuit resistance (Ω)	1
Resistance of overhead line (Ω/km)	0.0917
Inductance of overhead line (mH/km)	1.66

Parameters: The simulation system parameters are shown in Table 7.1.

Simulation results and analysis:
Figure 7.7 shows the simulation results of the MMC during a DC-line short-circuit fault. The tripping time of the AC circuit breakers is set to be 100 ms after the DC-line short-circuit fault is detected. The DC-line fault results in severe DC voltage drop as illustrated in Figure 7.7a. Before the AC CB is tripped, all IGBTs in MMC are turned off and the MMC forms an uncontrolled rectifier. There will be a large short-circuit current in the DC line, AC line, and three-phase arms of the MMC, as depicted in Figure 7.7b–d, respectively. Hence, the freewheeling diode in each SM without protection thyristor has to withstand huge overcurrent. Figure 7.7e shows that the protection thyristor takes a great part of current from the freewheeling diode and relieves overcurrent stress as well as overheat stress. At t = 1.1 second, the AC CB is tripped and the AC grid currents are cut off, as shown in Figure 7.7c. Then, the DC-line current gradually decays to zero as shown in Figure 7.7b.

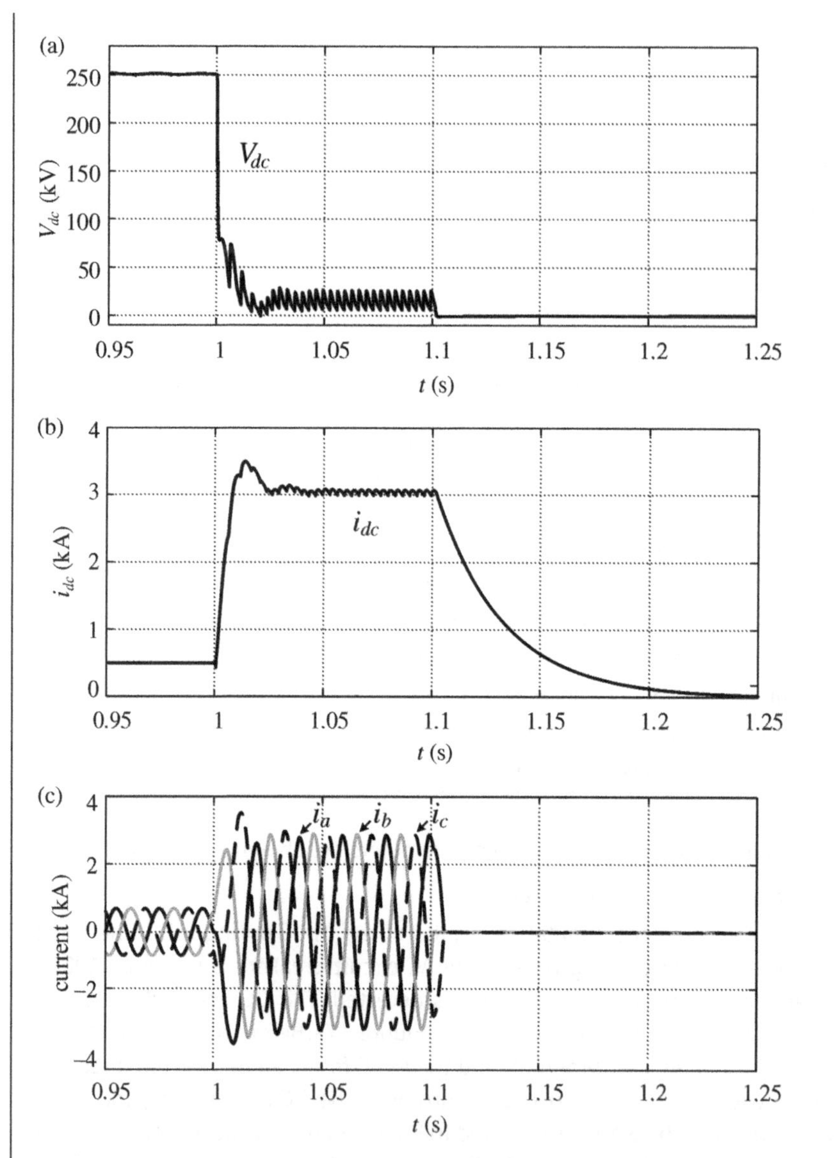

Figure 7.7 Waveforms of the MMC during a DC-line short-circuit fault. (a) DC-line voltage V_{dc}. (b) DC-line current i_{dc}. (c) AC grid currents. (d) Arm currents in phase A. (e) Lower diode current and protection thyristor current.

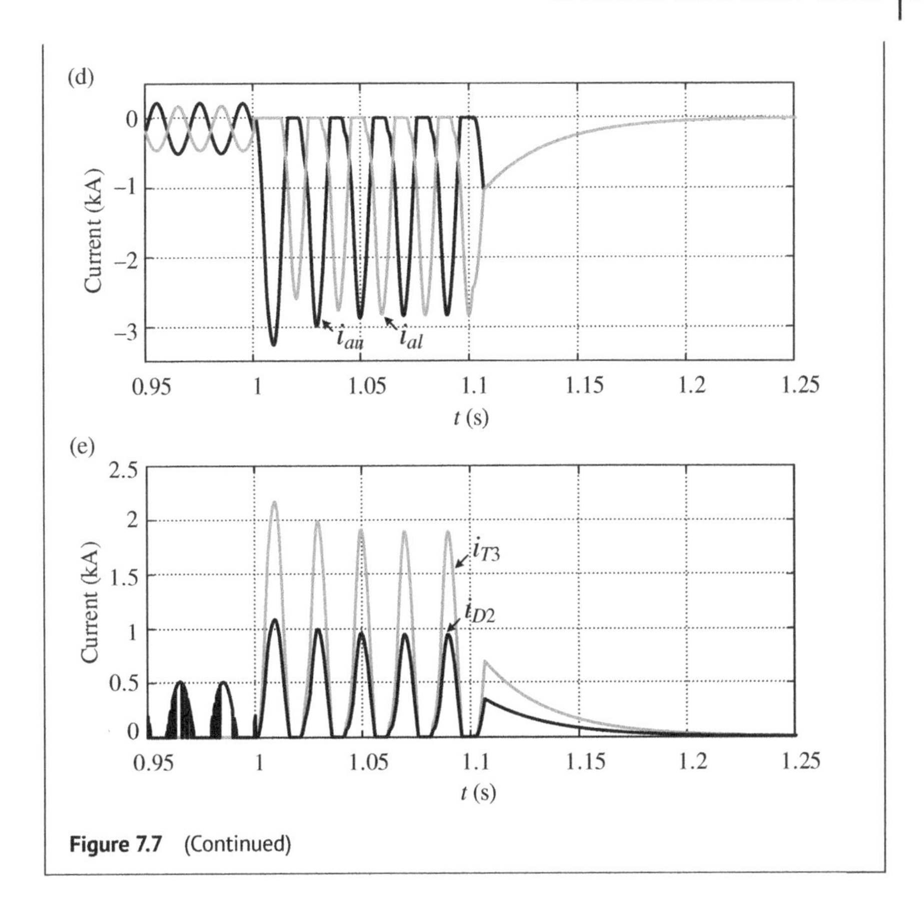

Figure 7.7 (Continued)

7.3 DC Circuit Breaker Based Protection

With the development of HVDC transmission technology, the isolation of DC faults is becoming increasingly important. Even the switches in the MMC are all turned off, the AC system still continues to supply a steady fault current to the short-circuit point. Besides, the fault current has a high rising rate, which needs to be interrupted within several milliseconds, or it will seriously affect the normal operation of the whole grid. Thus, the clearance of DC faults is a key issue in DC power grids. As a key component of the HVDC grid, the DC CB plays an important role in closing, loading, and interrupting normal current and breaking the fault current. According to the differences of breaking principles and methods, this section presents several types of DC CBs, i.e. mechanical circuit breaker (MCB), semiconductor circuit breaker (SCB), hybrid circuit breaker (HCB), multiterminal CB, and superconducting fault current limiter-based current breaker.

7.3.1 Mechanical Circuit Breaker

Figure 7.8 shows the basic topology of the mechanical circuit breaker (MCB), which consists of three parallel branches, including main branch, oscillation branch, and energy absorption branch. The main branch is mainly composed of series-connected MSs, which conducts the DC current in normal operation. The oscillation branch is mainly composed of a precharged capacitor C, an inductor L, and a trigger switch K. The energy-absorbing branch is mainly composed of series–parallel metal oxide arresters (MOA).

In normal operation, the MS is closed, and the system current flows through the main branch. In case of DC fault detection, MS is quickly released by the electromagnetic repulsion mechanism. When the contacts of MS separate, the arc across MS is generated. At this time, the DC fault current still flows through the main branch. Afterwards, the K in the oscillation branch is closed to generate a high-frequency oscillating current by the resonance of capacitor C and inductor L. The direction of the oscillating current is opposite to the faulty current, and it is superimposed on the current of the main branch. The current of the main branch drops to zero rapidly, and the arc is extinguished. When the MS's current is forced to zero, the DC fault current is completely transferred to the oscillation branch. In the meantime, the capacitor C is reversely charged by the fault current. As the C is charged to the knee voltage of MOA, the fault current primarily flows through the energy absorption branch and the voltage on DC CB is clamped by the MOA, which limits the overvoltage imposing on the DC CB and absorbs the energy stored in the inductors and DC lines. Finally, the DC fault current gradually decreases to zero and the DC fault current is completely interrupted.

Generally, the MCB is an economical equipment for interrupting DC fault current because of its low on-state impedance and low conduction loss. However, there are some disadvantages such as long switching time and limited current breaking ability. The turn-off time of MCB usually takes dozens of milliseconds, which is difficult to meet the DC fault clearance speed for HVDC transmission

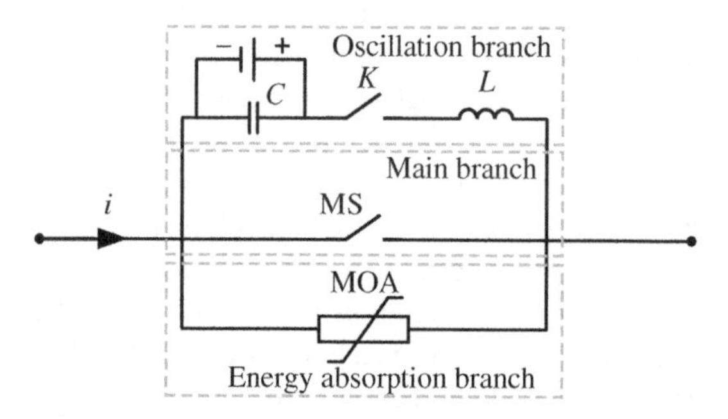

Figure 7.8 Basic topology of MCB.

system. In addition, the arc generated across MS during current breaking reduces the lifetime of the MS and increases the maintenance costs.

7.3.2 Semiconductor Circuit Breaker

The semiconductor circuit breakers (SCBs) adopt the power electronic devices as the main breaker (MB). With the appearance and development of power electronic devices, the CB with thyristors as switches appeared gradually in the late 1970s. From 80s to 90s, with the birth of fully controlled devices, SCB has developed rapidly. Compared with the MCB, its advantages of dynamic performance are obvious: (i) very fast switching speed, (ii) easy to achieve accurate and intelligent control, and (iii) no arc is generated when the switch is operated. According to the power electronic devices, SCB can be divided into two classes, including semi-controlled SCB and fully controlled SCB.

7.3.2.1 Semi-Controlled Semiconductor Circuit Breaker

Figure 7.9 depicts a typical circuit configuration of the semi-controlled SCB, which uses the thyristor as the steady-state flow branch. It consists of three parallel branches: main branch, oscillation branch, and energy absorption branch. The main branch is constituted with thyristors T_{main}. The oscillation branch contains a precharged capacitor C, an inductor L, and an auxiliary thyristor T_{aux}. The capacitor C is connected in parallel with an auxiliary power supply, and the C is precharged to a certain voltage. The energy-absorbing branch comprises series–parallel MOA.

In normal operation, T_{main} is turned on, and the system current flows through the main branch. When a short fault occurs on the DC side, the firing signal for the T_{main} is removed and the T_{aux} is triggered to generate the resonance current through high-frequency resonance of capacitor C and inductor L. The current of the main branch will drop to zero, and the T_{main} will be naturally turned off. Then,

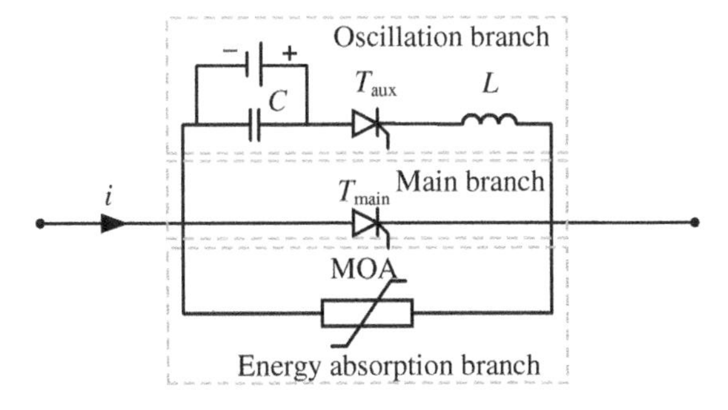

Figure 7.9 Basic topology of semi-controlled semiconductor circuit breaker.

the fault current reversely charges the capacitor *C*, and the capacitor voltage increases. When the amplitude reaches the action level of MOA, the current will be transferred to the energy-absorbing branch. The DC line energy is dissipated by MOA, and the DC line current is gradually reduced to zero, as a result, interruption of the semi-controlled SCB is completed.

The semi-controlled SCB is relatively mature, which can interrupt DC current faster than MCBs without generating arc. However, as the thyristor is a semi-controlled device, the semi-controlled SCB needs the oscillation branch to generate a current zero-crossing point to realize the DC fault interruption, which makes the structure more complicated. In addition, the relatively low working frequency of the thyristor limits the breaking speed of semi-controlled SCBs.

7.3.2.2 Fully Controlled Semiconductor Circuit Breaker

Figure 7.10 depicts a basic bidirectional configuration of the fully controlled SCB, which uses integrated gate-commutated thyristors (IGCTs) to constitute the steady-state flow branch. It consists of two parallel branches, including main branch and energy absorption branch. The main branch is mainly composed of anti-parallel connected IGCTs to provide a flow path for DC current in normal operation. The energy-absorbing branch is mainly composed of MOAs to absorb the energy stored in the inductive components of the system.

During normal operation, T_{main} is in on-state and conducts the current from source to the load. When a short-circuit fault is detected, T_{main} will be turned off and block the current flow. Then the load current commutates to MOA. Surge voltage across T_{main} is limited to the clamping voltage of MOA. When the energy is absorbed by MOA and the current becomes zero, interruption of the fully controlled SCB is finished.

Compared with MCBs, the fully controlled SCBs have a much faster switching speed to interrupt DC current without requiring a resonance current. However, the high on-state losses are the biggest barrier. In addition, the current and voltage rating of the individual devices is limited, which means a large number of the

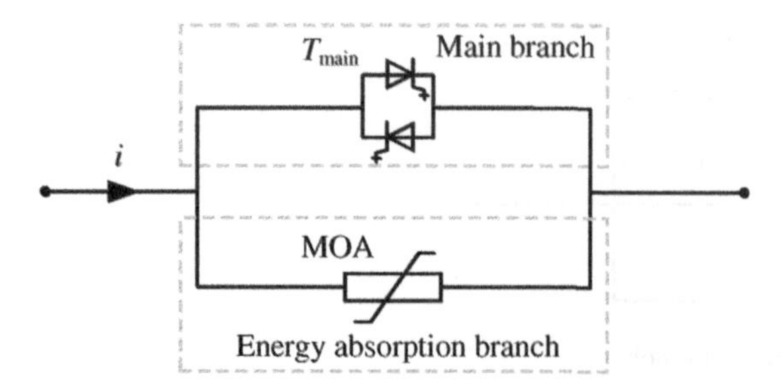

Figure 7.10 Fully controlled semiconductor circuit breaker.

semiconductor devices need to be series and parallel connected. Synchronous operation of these devices requires careful design of the driving circuit to ensure voltage and current sharing.

7.3.3 Hybrid Circuit Breaker

Integrating semiconductor devices with a mechanical breaker or disconnector in a combined configuration is called the hybrid circuit breaker. Generally, within an HCB, semiconductor switches only operate during the interruption process. All switches are controlled by electronic circuits. Recent developments in semiconductor switches and improvements in their characteristics, such as breakdown voltage, conduction losses, switching time, and reliability, bring about the possibility of using these devices as the main interrupters in CBs. In this section, two main structures are introduced, the conventional HCB and the proactive HCB.

7.3.3.1 Conventional Hybrid Circuit Breaker

The basic topology of the conventional HCB is shown in Figure 7.11. The main branch consists of MSs. IGCTs are used to form a transfer branch. The energy absorption branch is composed of MOA.

In normal operation, the system current flows through the mechanical switch MS. When a fault occurs, T_{main} is triggered and the mechanical switch MS is disconnected, and the current is forced by the arc voltage generated to transfer to T_{main}. Then T_{main} is turned off and blocks the current flow. The voltage across the breaker is clamped by the MOA. Finally, the whole interrupting process ends with the stored energy absorbed by the MOA and the fault is cleared.

The conventional HCB combines the merits of the low losses of the mechanical breaker and the fast switching speed of the semiconductor breaker. However, the switching speed strongly relies on the mechanical switch in the system. The arc

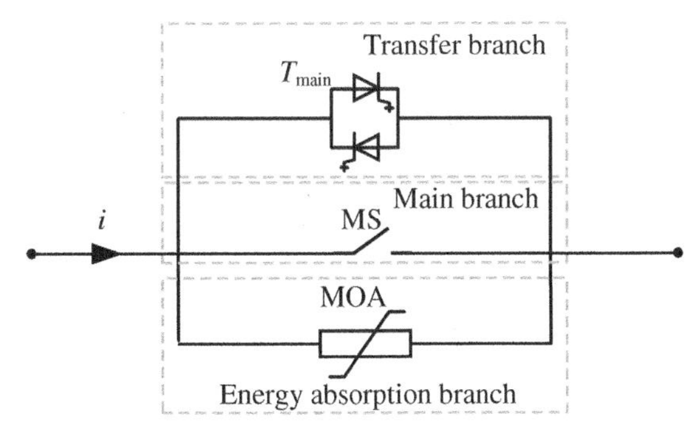

Figure 7.11 Typical topological configuration of the conventional hybrid circuit breaker.

voltage generated by the mechanical switch needs to be high enough to transfer the current from the main branch to the transfer branch. Otherwise, the mechanical switch may fail to open.

7.3.3.2 Proactive Hybrid Circuit Breaker

The typical structure of proactive HCBs can refer to the one launched by ABB in 2012. As shown in Figure 7.12, it consists of a bypass branch that is formed by a mechanical ultrafast disconnector (UFD) in series with a load commutation switch (LCS) and an MB branch consisting of a string of series-connected MB cells paralleled with individual MOAs. The LCS and MB adopt full-controlled semiconductors (such as IGBT module) to afford full fault current breaking capability.

In normal operation, the LCS is turned on and the UFD is closed. The current flows through the LCS and the UFD. When a DC-side fault is detected, the LCS is turned off and the MB is turned on, and the current is commutated to the MB. Once all the current is flowing through the MB branch, the UFD is opened. Then, the fault current can be interrupted by blocking the MB. Finally, the fault energies will be dissipated by the MOA. The typical HCB can interrupt currents from both directions by employing semiconductors reversely series connected in LCS and MB.

The proactive HCB has lower operating losses compared with the solid-state CB and provides faster operation compared with the MCB. In addition, it is capable of using the LCS to transfer the faulty current to the MB, which compensates for the time delay of the mechanical switch in the conventional HCB. However, its high costs can be a major barrier to its large-scale application. The MB requires a considerable number of series IGBTs to withstand high overvoltage during the current interruption, which leads to quite high costs of the HCB.

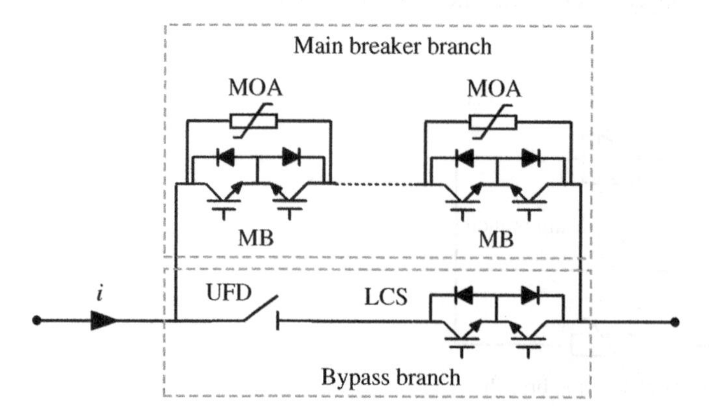

Figure 7.12 Structure of the proactive hybrid circuit breaker.

7.3.4 Multiterminal Circuit Breaker

The typical two-terminal HCB has shown good performances interrupting DC fault currents. However, its high costs can be a major obstacle to its large-scale application. The MB requires a considerable number of series IGBTs to withstand high overvoltage during the current interruption, which leads to quite high costs of a single HCB. In the DC grids, there are many nodes connecting with multiple lines. To isolate faults, the HCBs are required at both ends of a DC line. Since a DC gird usually has many DC lines, the costs of the HCBs will be relatively high.

Because the main costs of the HCB lie in its MB branch, several multiterminal HCBs have been presented [8–15] based on the concept of sharing the MB branch.

7.3.4.1 Assembly CB

The assembly CB uses the design of the HCB for reference, and it is more suitable for the HVDC grids with half-bridge MMCs and overhead lines. The basic structure of the assembly CB is illustrated in Figure 7.13.

The assembly CB consists of four main components, including the active short-circuit breaker (ASCB), the main mechanical disconnector, the MB, and the

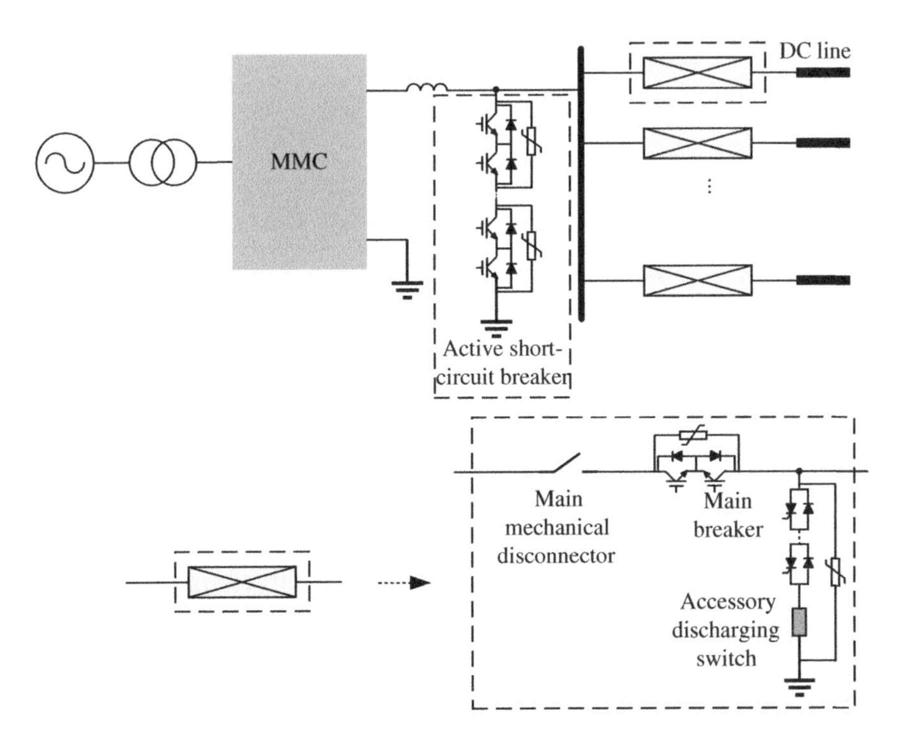

Figure 7.13 Structure of the assembly CB.

accessory discharging switch (ADS). The ASCB consists of a series of IGBT modules with arrester banks. The mechanical disconnector adopts the UFD that can be opened within several milliseconds under zero current condition. The MB consists of two IGBTs in reverse series, and it is required to break the current in either direction. The ADS consists of a semiconductor-based switch in series with a resistor.

The whole process of fault clearing is as follows. When the fault DC-side short-circuit fault is detected, the switches in ASCB and the ADS are turned on immediately. The fault current flowing from the MMC injects into the ASCB, the ADS, and the fault point at the same time. Then, the MB opens to isolate the faulted line from the MMC and the DC current decreases to zero. The energy stored in the DC line is released through the ADS and the fault point. When the DC current decreases to zero, the mechanical disconnector opens to realize the physical isolation. Finally, the ASCB opens and the remaining energy is released through the arrester banks. The current flowing through the ASCB decreases to zero and the DC fault is cleared completely.

7.3.4.2 Multiport CB

The multiport CB is benefiting from the HCB core concept for multiterminal HVDC application. The topology of the multiport CB with n ports is depicted in Figure 7.14.

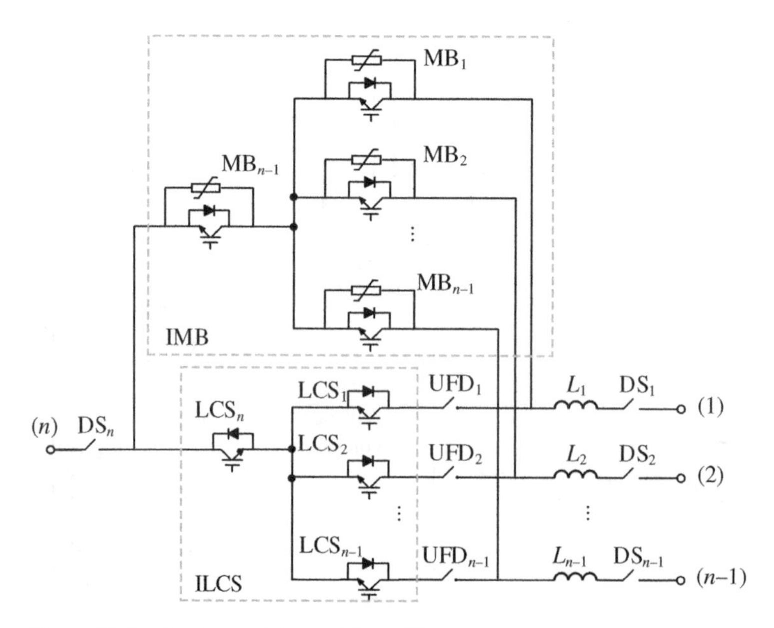

Figure 7.14 Structure of the multiport CB.

The multiport CB is composed of an integrated main breaker (IMB), an integrated load commutation switch (ILCS), n ultrafast disconnectors (UFD_1–UFD_{n-1}), n current-limiting inductors (L_1–L_{n-1}), and n disconnectors (DS_1–DS_{n-1}). The IMB unit consists of n main breaker (MB) subunits, incorporating series-connected IGBTs and surge arrestors (SAs). The ILCS consists of n LCS units. The ILCS is in the *on* state when all the LCS subunits are closed and is in the *off* state when the LCS subunits are opened. Port n is assumed to be connected to a DC bus and ports 1 to $n-1$ are assumed to be connected to $n-1$ adjacent transmission lines.

The whole process of fault clearing is as follows. Upon detection of a permanent DC fault, all adjacent lines must be isolated from the DC bus. The IMB will be closed after receiving DC fault trip command, and then the ILCS is opened in order to commutate the current into the IMB. Following the current commutation completion, UFD_1–UFD_{n-1} are opened. Finally, the IMB is opened and the currents will be redirected into the surge arrestors. The fault current will be reduced to zero by surge arrestor and the DC fault is cleared completely.

7.3.5 Superconducting Fault Current Limiter

The superconducting fault current limiter (SFCL) uses the quenching resistance of the yttrium barium copper oxide (YBCO) high-temperature superconducting tape to limit the short-circuit fault current. The YBCO high-temperature superconducting tape is composed of the YBCO layer and metal layers, which consist of stainless steel layer, Ag layer, Cu layer, and Hastelloy substrate layer, as shown in Figure 7.15. The equivalent resistance of the superconducting tape R_{sc} is mainly composed of the YBCO layer resistance R_{YBCO} and the metal layer resistance R_n in parallel.

- When the current flowing through the superconducting tape $i_j <$ superconducting critical current I_o, $R_{YBCO} \approx 0$, and $R_{YBCO} << R_n$. At this time, the equivalent resistance of the superconducting tape R_{sc} is almost zero.
- When i_j gradually increases and exceeds the critical current I_o, the equivalent resistance of the superconducting tape R_{sc} increases along with the increase of i_j.
- When i_j is significantly greater than the critical current I_o, the superconducting tape completely quenches, and R_{YBCO} increases and is much larger than R_n. The equivalent resistance of the superconducting tape R_{sc} is about R_n. At this time, R_{sc} plays a role in limiting the large current, as shown in Table 7.2.

7.3.6 SFCL-Based Circuit Breaker

Superconducting DC CBs mainly include the SFCL-based HCB, the SFCL-based self-oscillating CB, and the SFCL-based forced zero-crossing CB.

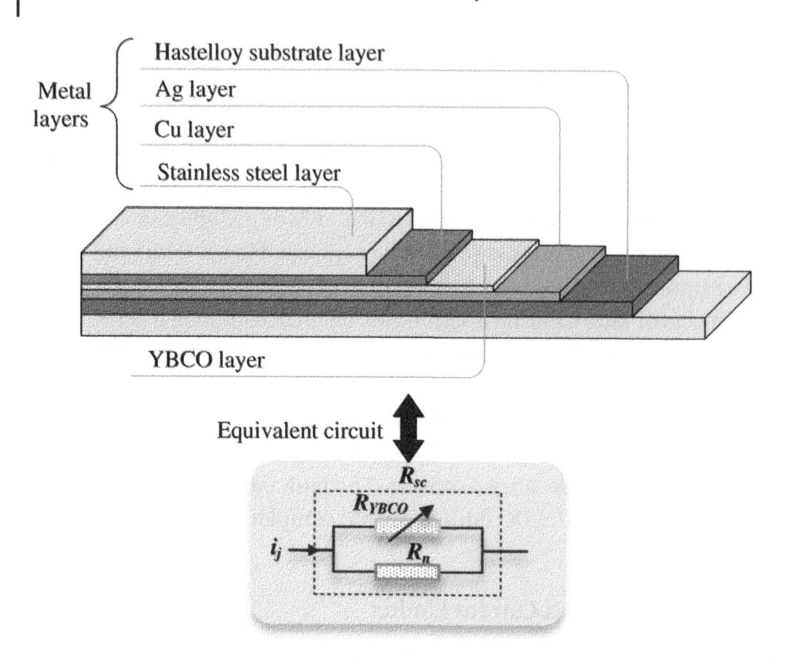

Figure 7.15 Schematic diagram and equivalent circuit of the superconducting tape.

Table 7.2 Equivalent impedance characteristics of the superconducting tape.

Current relationship	State of the superconducting tape	Resistance of the YBCO layer R_{YBCO}	Equivalent resistance of superconducting tape R_{sc}
$i_j < I_0$	Normal	$R_{YBCO} \approx 0$, $R_{YBCO} << R_n$	$R_{sc} \approx 0$
$i_j > I_0$	Quenching	$R_{YBCO} >> R_n$	$R_{sc} \approx R_n$

7.3.6.1 SFCL-Based Hybrid Circuit Breaker

The SFCL-based HCB [16] includes three parallel branches, as shown in Figure 7.16. The primary branch includes an SFCL, whose resistance is R_{sc}, in series with a mechanical switch M. The secondary branch consists of the semiconductor switch S. The third branch has an MOA.

During normal operation, the current flows through the SFCL and the closed mechanical switch M that has a low resistance and hence low losses. The semiconductor switch S is normally turned off.

When a short-circuit fault occurs, the current rises rapidly until the critical current of the SFCL is reached. The SFCL starts to quench and quickly develops resistance. The voltage appearing across the SFCL is used as a fault trigger signal

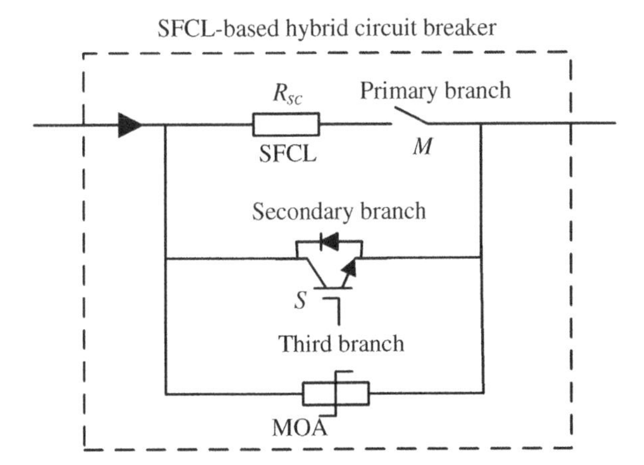

Figure 7.16 Schematic diagram of the SFCL-based hybrid circuit breaker.

of the semiconductor switch S, providing the secondary branch for the current to flow. When the current through the mechanical switch M has dropped to a sufficiently low level, M is released by the electromagnetic repulsion mechanism. The secondary branch continues to conduct the fault current until the M is completely opened. Then, the semiconductor switch S is turned off and blocks the current flow. The voltage across the SFCL-based HCB is clamped by the MOA. Finally, the whole interrupting process ends with the stored energy absorbed by the MOA and the fault is cleared.

The main advantage of this SFCL-based HCB is that it operates automatically and quickly, resulting in a lower potential DC fault current than other types of DC CBs. The disadvantage is that the switching speed strongly relies on the SFCL in the system. The voltage generated by the SFCL needs to be high enough to transfer the current from the primary branch to the secondary branch. Otherwise, the mechanical switch M may fail to open.

7.3.6.2 SFCL-Based Self-Oscillating Circuit Breaker

The SFCL-based self-oscillating CB [17, 18] combines the SFCL technology with self-oscillating DC switching technology to realize the interruption of the short-circuit current. Figure 7.17 shows the structure of the SFCL-based self-oscillating CB. The SFCL module and the breaking module are in series. The SFCL module includes an SFCL, whose resistance is R_{SC}. The breaking module consists of a mechanical switch M and a parallel branch including a precharged capacitor C, an inductor L, and a trigger switch K. The energy-consuming branch that has an MOA is connected in parallel at both ends of the SFCL module and the breaking module.

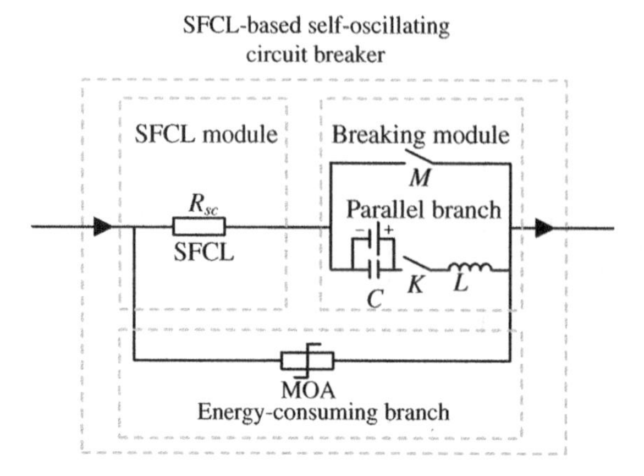

Figure 7.17 Schematic diagram of the SFCL-based self-oscillating circuit breaker.

During normal operation, the current flows through the SFCL and the closed mechanical switch *M* that has a low resistance and hence low losses. The trigger switch *K* is normally kept open.

When a short-circuit fault occurs, the short-circuit current is limited by the SFCL module to a very low level. When the current through the mechanical switch *M* has dropped to a sufficiently low level, *M* is quickly released by the electromagnetic repulsion mechanism. When the contacts of *M* separate, the arc across *M* is generated. Afterwards, the *K* is closed to generate a high-frequency oscillating current by the resonance of capacitor *C* and inductor *L*. The current flowing through *M* drops to zero rapidly and the arc is extinguished. Then the fault current is completely transferred to the energy-consuming branch, which absorbs the energy stored in the breaking module. Finally, the fault current gradually decreases to zero and the fault current is completely interrupted.

The main advantages of this SFCL-based self-oscillating CB are simple structure, high reliability, and low cost. The SFCL can effectively limit the fault current and provide sufficient interruption time for the mechanical switch *M*, which makes up for the disadvantages of long interruption time and small interruption current capacity of the self-oscillating breaker. The disadvantage is that the interruption time is closely related to the parameters of the breaking module, and the interruption speed is generally slow.

7.3.6.3 SFCL-Based Forced Zero-Crossing Circuit Breaker

The SFCL-based forced zero-crossing CB [19, 20] is composed of an SFCL, whose resistance is R_{sc}, and a DC vacuum CB in series, as show in Figure 7.18. The DC vacuum CB is composed of a vacuum interrupter MV and a commutation circuit in parallel. The commutation circuit comprises of a precharged capacitor *C*, an inductor *L*, and a mechanical switch *M* in series.

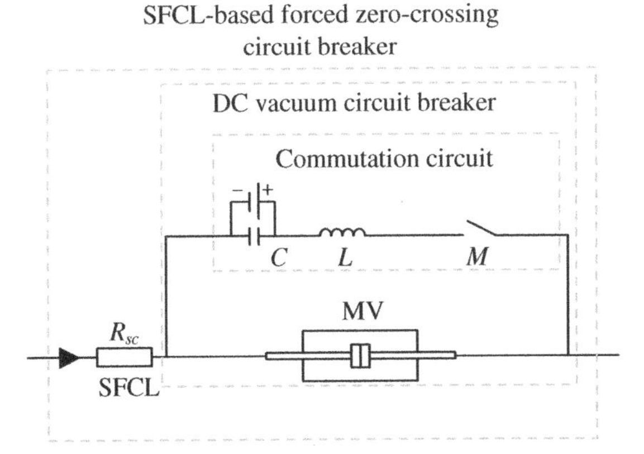

Figure 7.18 Schematic diagram of the SFCL-based forced zero-crossing circuit breaker.

During normal operation, the current flows through the SFCL and the closed vacuum interrupter MV that has a low resistance and hence low losses. The mechanical switch *M* is normally kept open.

When a short-circuit fault occurs, the SFCL greatly limits the rising rate and peak value of the fault current. Then, the vacuum interrupter MV is quickly released by the electromagnetic repulsion mechanism and the arc across MV is generated. To generate a zero current point, a reverse current is injected into MV by the commutation circuit after the *M* is closed, thereby generating an artificial zero-current condition to extinguish the arc of MV. Consequentially, the MV is completely opened and the fault is cleared.

The advantage of this SFCL-based forced zero-crossing CB is that it can significantly reduce the fault current and reduce the requirement on breaking capability. As the SFCL limits the fault current, there is no need to generate a high reverse current, and as a result, the capacitance of the precharged capacitor *C* can be small. The disadvantage is the need of additional precharged circuit and equipment and extra complex control.

7.4 Fault Blocking Converter Based Protection

Isolating the DC-side short-circuit faults is one of the challenges faced by the MMC in practical engineering. Since the application of the DC CB is restricted due to their immaturity and high costs, the MMCs with inherent capability to deal with DC-side fault have become attractive.

Several fault blocking SMs are presented, such as full-bridge (FB) SM [21, 22], clamp-circuit-double submodule (CC SM) [23], hybrid-double submodule (HD SM) [24], clamp-double submodule (CD SM) [25, 26], and switched capacitor submodule (SC SM) [27]. The hybrid MMC composed of the fault blocking SMs and half-bridge SMs is usually used when taking into account the cost and fault blocking capability. The fault blocking MMC relies on its own structure to absorb the additional energy generated during the short-circuit fault, which can quickly block the fault current and has the advantages of high reliability and strong practicability.

In this section, Section 7.4.1 introduces the hybrid MMC based on the FB SMs and HB SM. Section 7.4.2 introduces the fault blocking control of the hybrid MMC. Section 7.4.3 analyzes the ratio of FB SM in the hybrid MMC, and finally Section 7.4.4 introduces the alternative fault blocking SMs.

7.4.1 FB SM and HB SM Based Hybrid MMC

The hybrid MMCs can effectively interrupt the DC short-circuit fault current with the advantages of low cost and low power losses. Among them, the hybrid MMC composed of FB SMs and HB SMs has become the most attractive one owing to its merits like simplicity and more freedoms of control.

The topology of HB SM is shown in Figure 7.19a, which includes switches/diodes T_1/D_1, T_2/D_2, and a capacitor C. The operation modes of the HB SM have been introduced in Chapter 1.

Figure 7.19b shows the topology of FB SM, which includes four switch/diodes T_1/D_1, T_2/D_2, T_3/D_3, T_4/D_4, and a capacitor C. The five operation modes of the FB SM are listed in Table 7.3, which includes positive insert, negative insert, bypass I, bypass II, and block. Figure 7.20 shows the current paths of FB SM in various operation modes under $i_{arm} > 0$, and Figure 7.21 shows the current paths of FB SM under $i_{arm} < 0$, which are explained in detail as follows.

1) *Mode 1 (positive insert)*: T_1 and T_4 are turned on, T_2 and T_3 are turned off, and the FB SM is positively inserted into the arm circuit.
   - When the arm current $i_{arm} > 0$, the capacitor C is charged by i_{arm} and the capacitor voltage u_c is increased, as shown in Figure 7.20a.
   - When $i_{arm} < 0$, the C is discharged by i_{arm} and the u_c is reduced, as shown in Figure 7.21a.

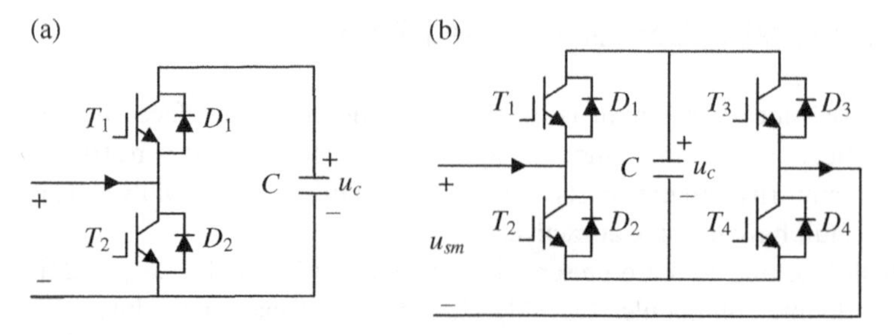

Figure 7.19 The topologies of (a) HB SM and (b) FB SM.

Table 7.3 The operation modes of the FB SM.

SM mode	SM state	T_1	T_2	T_3	T_4	u_{sm}	Arm current	Capacitor voltage u_c	Current paths
1	Positive insert	On	Off	Off	On	u_c	>0	Increased	Figure 7.20a
							<0	Reduced	Figure 7.21a
2	Negative insert	Off	On	On	Off	$-u_c$	>0	Reduced	Figure 7.20b
							<0	Increased	Figure 7.21b
3	Bypass I	On	Off	On	Off	0	>0	Unchanged	Figure 7.20c
							<0		Figure 7.21c
4	Bypass II	Off	On	Off	On	0	>0	Unchanged	Figure 7.20d
							<0		Figure 7.21d
5	Block	Off	Off	Off	Off	u_c	>0	Increased	Figure 7.20e
						$-u_c$	<0		Figure 7.21e

2) *Mode 2 (negative insert)*: T_2 and T_3 are turned on, T_1 and T_4 are turned off, and the FB SM is negatively inserted into the arm circuit.
   - When $i_{arm} > 0$, the C is discharged by i_{arm} and the capacitor voltage u_c is reduced, as shown in Figure 7.20b.
   - When $i_{arm} < 0$, the C is charged by i_{arm} and the u_c is increased, as shown in Figure 7.21b.
3) *Mode 3 (bypass I)*: T_1 and T_3 are turned on, T_2 and T_4 are turned off, and the FB SM is bypassed from the arm circuit. In this case, the C is bypassed and the u_c is unchanged, as shown in Figures 7.20c and 7.21c.
4) *Mode 4 (bypass II)*: T_2 and T_4 are turned on, T_1 and T_3 are turned off, and the FB SM is bypassed from the arm circuit. In this case, the C is bypassed and the u_c is unchanged, as shown in Figures 7.20d and 7.21d.
5) *Mode 5 (block)*: T_1, T_2, T_3, and T_4 are turned off.
   - When $i_{arm} > 0$, as shown in Figure 7.20e, the FB SM is positively inserted into the arm circuit, the C is charged, and the u_c is increased.
   - When $i_{arm} < 0$, as shown in Figure 7.21e, the FB SM is negatively inserted into the arm circuit, the C is charged, and the u_c is increased.

Figure 7.22 shows the topology of the hybrid MMC based on HB SMs and FB SMs. There are n SMs in each arm, including n_F FB SMs and $n - n_F$ HB SMs. The FB SM ratio in an arm is defined as

$$\eta_F = \frac{n_F}{n} \tag{7.2}$$

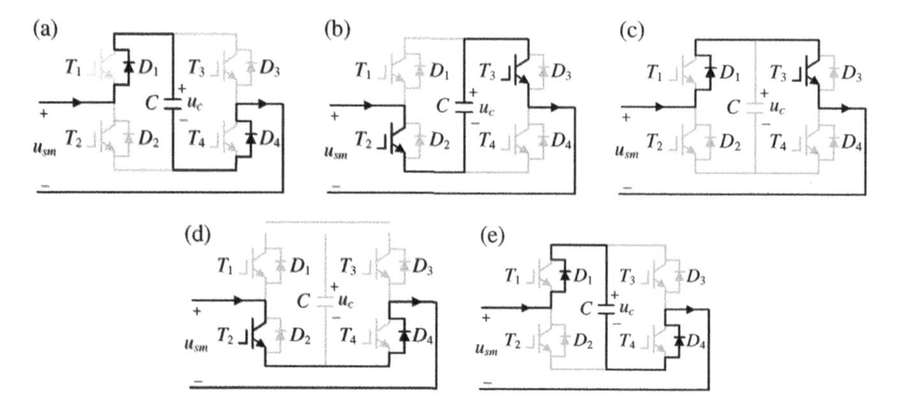

Figure 7.20 The current paths of FB SM in various operation modes under $i_{arm}>0$. (a) Positive insert. (b) Negative insert. (c) Bypass I. (d) Bypass II. (e) Block.

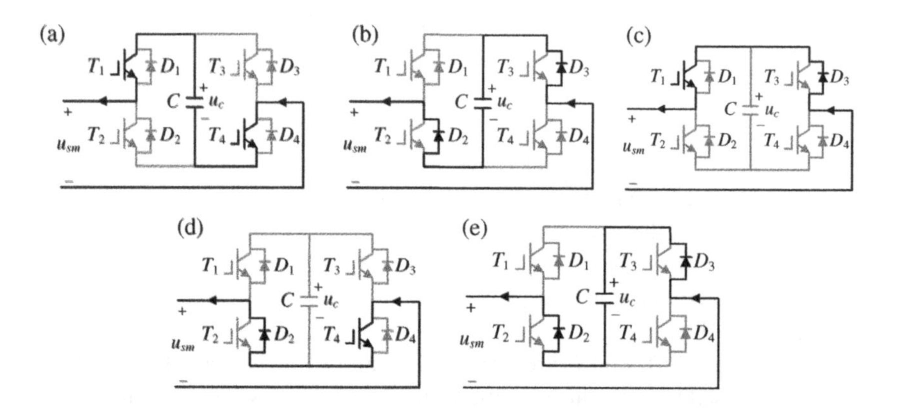

Figure 7.21 The current paths of FB SM in various operation modes under $i_{arm}<0$. (a) Positive insert. (b) Negative insert. (c) Bypass I. (d) Bypass II. (e) Block.

7.4.2 Fault Blocking Control

When a DC fault is detected, the switches in HB SMs and FB SMs will be immediately turned off to block the fault current. During the DC fault blocking period, since the output voltage direction of FB SMs is always opposite to the direction of the fault current, as shown in Figures 7.20e and 7.21e, the FB SMs will be charged by the DC fault current and form a voltage opposed to the fault current at DC side of the MMC. The fault energy stored in the DC line and arm inductors will be absorbed by the FB SMs, causing the capacitor voltages in the FB SMs to increase.

Figure 7.23 shows an example of the DC fault current loop after SMs blocking. The capacitor voltages of the FB SMs generate the reverse voltages to block DC fault

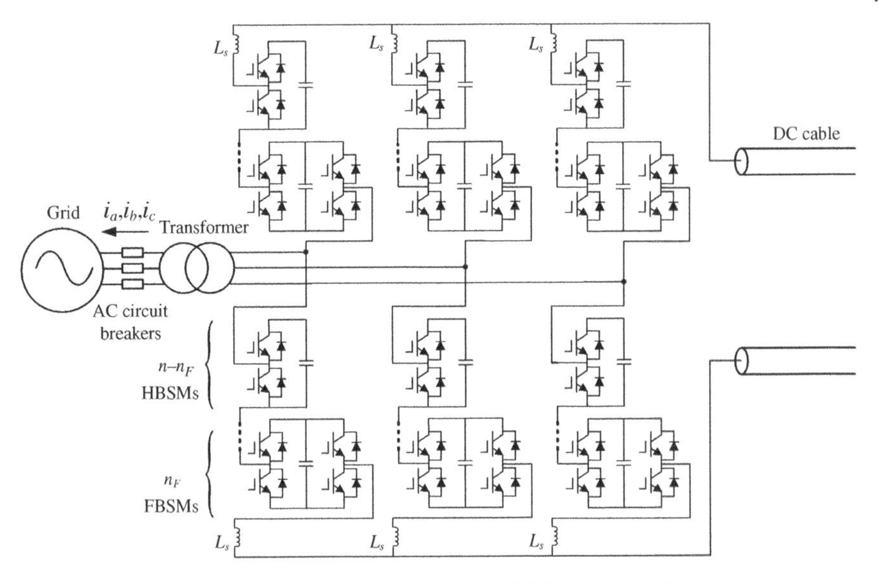

Figure 7.22 The topology of hybrid MMC based on FB SMs and HB SMs.

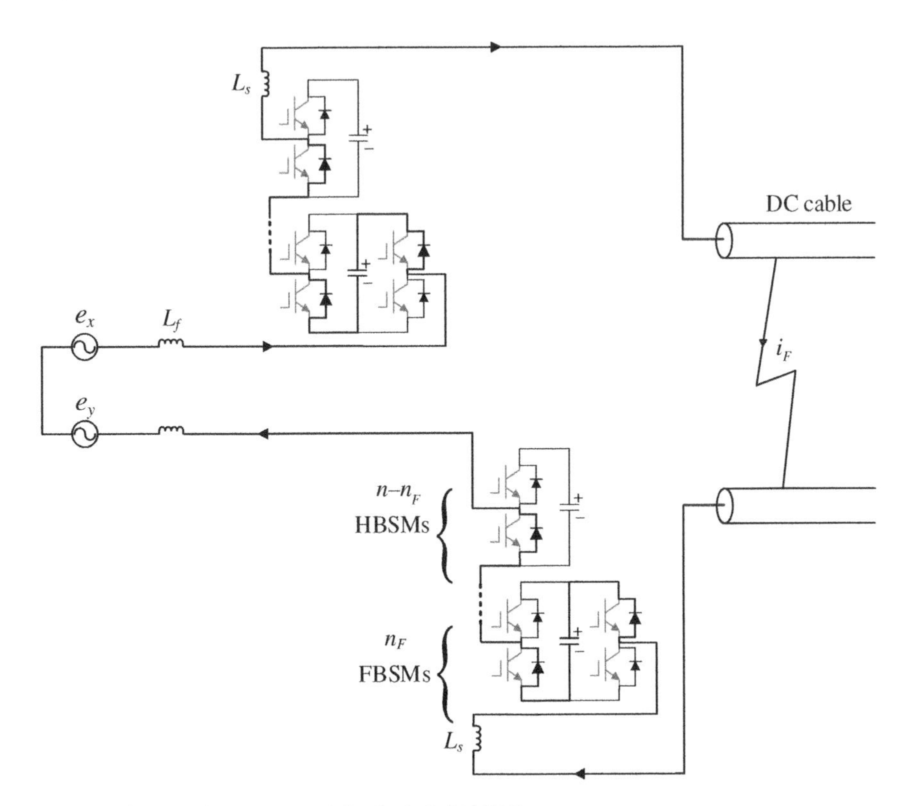

Figure 7.23 Fault current path in the hybrid MMC.

current. In Figure 7.23, x and y represent the phase legs of the upper and lower arms in the fault current loop, respectively, where $x \in [a, b, c]$, $y \in [a, b, c]$, and $x \neq y$.

7.4.3 FB SM Ratio

In Figure 7.23, the fault current flows through the lower diode of the HB SMs and through the SM capacitors of the FB SMs. The direction of the FB SM output voltage is opposed to the fault current direction. Figure 7.24 shows the equivalent circuit of the hybrid MMC. Assuming that the SM capacitor voltage is close to the rated capacitor voltage u_c, only when the sum of the capacitor voltage $2n_F u_c$ of the FB SM in the MMC is greater than the peak value of the AC-side line voltage e_{xy}, as

$$2n_F u_c > E_{llm} \tag{7.3}$$

the fault current is reduced, where the E_{llm} is the peak value of the AC-side line-to-line voltage.

According to the operating principle of the hybrid MMC, u_c, V_{dc}, and E_{llm} satisfy the following relationships

$$\begin{cases} V_{dc} = n u_c \\ E_{llm} = \dfrac{\sqrt{3}}{2} m V_{dc} \end{cases} \tag{7.4}$$

where m is the modulation index ($0 \leq m \leq 1$). Combining equations (7.2) and (7.3), the constraint for η_F can be obtained as

$$\frac{\sqrt{3}}{4} m \leq \eta_F \leq 1 \tag{7.5}$$

7.4.4 Alternative Fault Blocking SMs

The FB SM in the hybrid MMC mentioned above can also be replaced with other SMs with fault blocking capability. In recent years, several circuits with DC-side

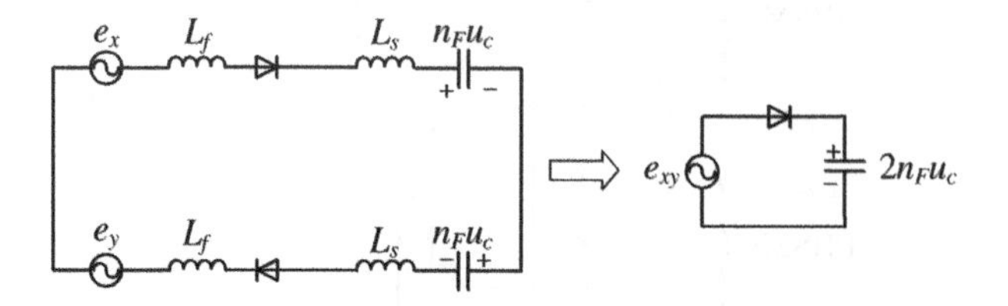

Figure 7.24 The fault equivalent circuit of the hybrid MMC.

short-circuit fault blocking capability have been reported. They can be divided into two types, including the circuit with asymmetrical blocking ability and the circuit with symmetrical blocking capability.

Figure 7.25 shows the SM circuit with asymmetrical DC-side short-circuit fault blocking capability and their operation modes, including hybrid-series-connected submodule (HS SM), HD SM, CD SM, diode-clamp submodule (DC SM), diode-clamp-double submodule (DCD SM), and SC SM. The SM circuit with asymmetrical DC-side short-circuit fault blocking capability has a weaker blocking capability under the condition of $i_{sm} < 0$ than the condition of $i_{sm} > 0$.

Figure 7.26 shows the SM circuit with symmetrical DC-side short-circuit fault blocking capability and their operation modes, including CC SM, cross-connected-double submodule (CCD SM), series-connected-double submodule (SCD SM), and unipolar-voltage FB SM (UFB SM). The SM circuits with symmetrical DC-side short-circuit fault blocking capability have the same blocking capability under any direction of current i_{sm}.

The circuits with asymmetrical DC-side short-circuit fault blocking capability require more series-connected SMs than the circuits with symmetrical DC-side short-circuit fault blocking capability to achieve the same blocking level under the condition of negative current direction, which will cause additional cost and power losses.

7.5 Bypass Thyristor MMC Based Protection

This section presents a BT-based MMC HVDC system with the current interruption capability and the corresponding protection scheme for the HVDC system under DC lines faults, where a thyristor is connected in anti-parallel with the bottom IGBT in each HB SM instead of the diode in the conventional MMC HVDC system. With this simple circuit configuration, the fault current can be effectively interrupted in quite a short time. Furthermore, the HVDC system can still be connected to the AC grid and continue operating and regulating the reactive power for the AC grid under DC line fault, which can short system restoring time after the fault is cleared, such as omitting capacitor precharging process and reducing restoring operation. In addition, the expensive AC CB and DC CB may be replaced with the cheap mechanical switch because of the current interruption capability of the MMC HVDC system, which can greatly save cost.

7.5.1 Bypass Thyristor MMC Configuration

Figure 7.27a shows the BT MMC based HVDC system, which consists of two MMCs (MMC 1 and 2). Each MMC is connected to an AC grid through a transformer and an AC CB. Figure 7.27b shows an MMC, which is composed of six

Circuits with asymmetrical blocking capability

HSSM

u_{c2}, D_5, T_5, D_6, T_6, D_3, T_3, D_4, T_4, u_{c1}, D_1, T_1, D_2, T_2, i_{sm}, u_{sm}

Normal operation	
Switches on	u_{sm}
T_1, T_3, T_5	0
T_2, T_4, T_5	0
T_1, T_4, T_5	u_{c1}
T_1, T_3, T_6	u_{c2}
T_2, T_4, T_6	u_{c2}
T_1, T_4, T_6	$u_{c1}+u_{c2}$
Blocking period	
i_{sm}	u_{sm}
>0	$u_{c1}+u_{c2}$
<0	$-u_{c1}$

HDSM

T_5, u_{c2}, T_6, D_5, D_6, D_3, D_4, T_4, u_{c1}, D_1, T_1, D_2, T_2, i_{sm}, u_{sm}

Normal operation	
Switches on	u_{sm}
T_2, T_4, T_5	0
T_1, T_4, T_5	u_{c1}
T_2, T_4, T_6	u_{c2}
T_1, T_4, T_6	$u_{c1}+u_{c2}$
Blocking period	
i_{sm}	u_{sm}
>0	$u_{c1}+u_{c2}$
<0	$-u_{c1}$

CDSM

D_3, T_3, u_{c2}, D_4, T_4, D_7, D_5, T_5, D_6, D_1, u_{c1}, D_2, T_1, T_2, i_{sm}, u_{sm}

Normal operation	
Switches on	u_{sm}
T_2, T_3, T_5	0
T_1, T_3, T_5	u_{c1}
T_2, T_4, T_5	u_{c2}
T_1, T_4, T_5	$u_{c1}+u_{c2}$
Blocking period	
i_{sm}	u_{sm}
>0	$u_{c1}+u_{c2}$
<0 ($u_{c1}<u_{c2}$)	$-u_{c1}$
<0 ($u_{c2}<u_{c1}$)	$-u_{c2}$

DCSM

T_1, u_{c1}, D_1, D_3, i_{sm}, u_{sm}, u_{c2}, T_2, D_2, T_4, D_4

Normal operation	
Switches on	u_{sm}
T_2, T_4	0
T_1, T_4	$u_{c1}+u_{c2}$
Blocking period	
i_{sm}	u_{sm}
>0	$u_{c1}+u_{c2}$
<0	$-u_{c1}$

DCDSM

D_3, T_3, u_{c3}, D_4, T_4, u_{c4}, D_5, D_7, D_6, T_5, D_8, T_6, u_{c1}, D_1, u_{c2}, D_2, T_1, T_2, i_{sm}, u_{sm}

Normal operation	
Switches on	u_{sm}
T_1, T_4, T_5, T_6	0
T_1, T_3, T_5, T_6	$u_{c3}+u_{c4}$
T_2, T_4, T_5, T_6	$u_{c1}+u_{c2}$
T_2, T_3, T_5, T_6	$u_{c1}+u_{c2}+u_{c3}+u_{c4}$
Blocking period	
i_{sm}	u_{sm}
>0	$u_{c1}+u_{c2}+u_{c3}+u_{c4}$
<0	$-(u_{c1}+u_{c4})$

SCSM

D_4, T_4, i_{sm}, u_{c2}, T_5, D_5, D_3, T_3, D_6, T_6, D_2, T_2, u_{sm}, u_{c1}, D_7, D_1, T_1

Normal operation	
Switches on	u_{sm}
T_1, T_3, T_5, T_6	0
T_1, T_2, T_4, T_5, T_6	$u_{c1}=u_{c2}$
T_1, T_2, T_3, T_4	$u_{c1}+u_{c2}$
Blocking period	
i_{sm}	u_{sm}
>0	$u_{c1}+u_{c2}$
<0	$-u_{c1}$

Figure 7.25 SM circuits with asymmetrical blocking capability.

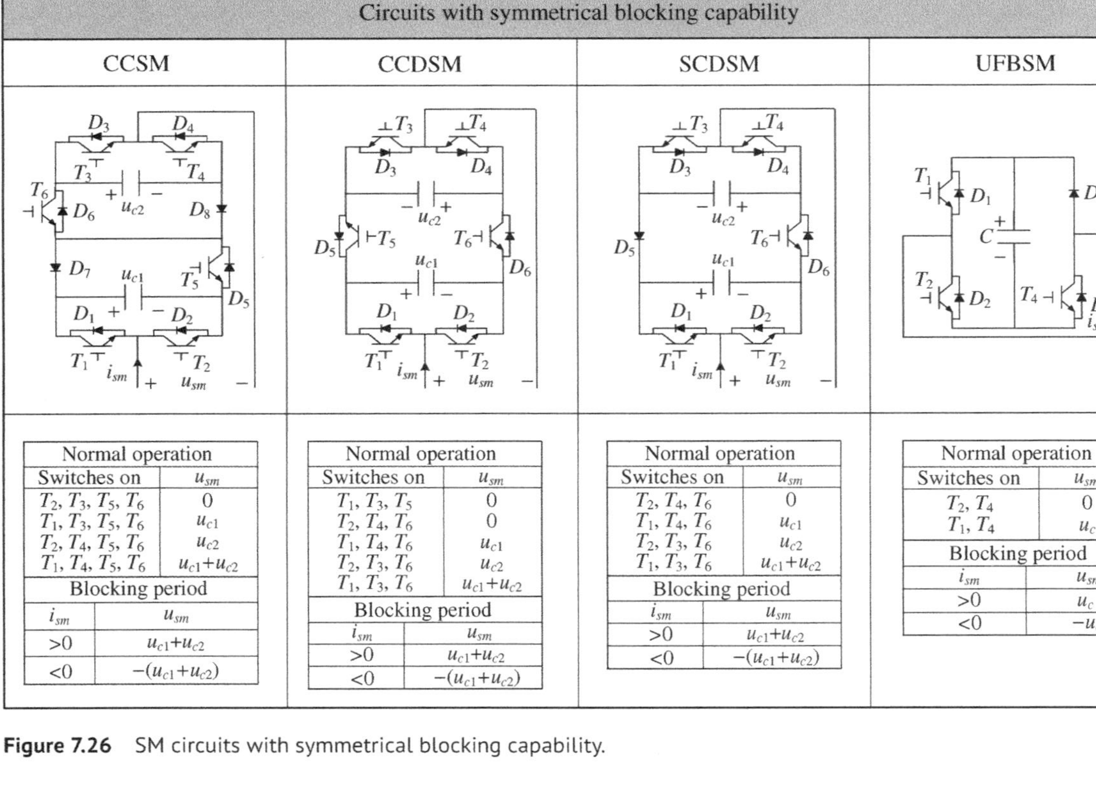

Figure 7.26 SM circuits with symmetrical blocking capability.

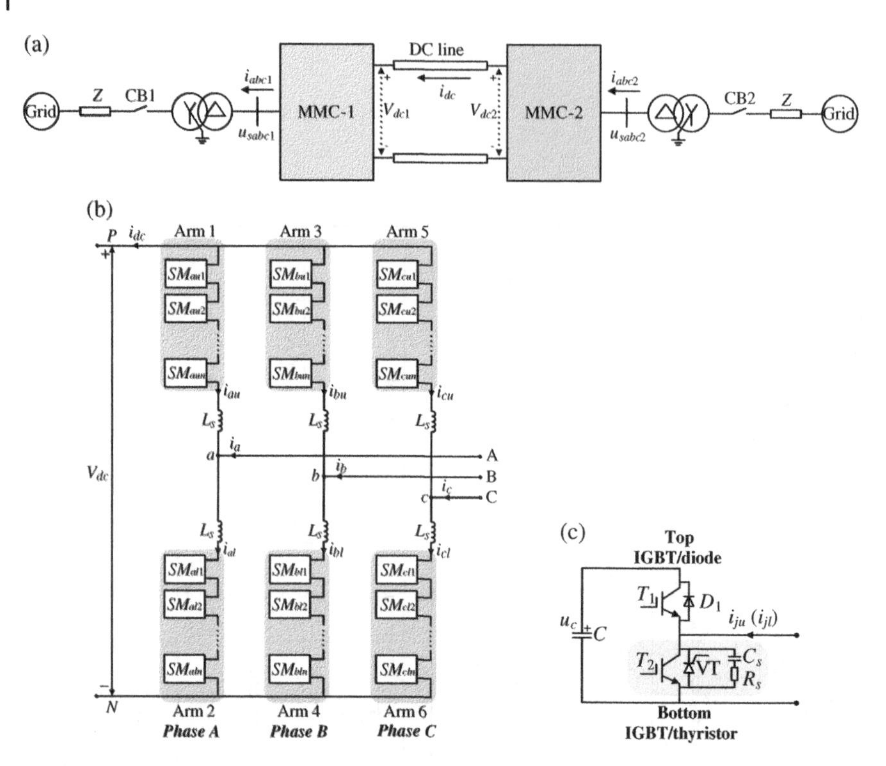

Figure 7.27 (a) Bypass thyristor MMC based HVDC system. (b) Bypass thyristor MMC. (c) SM configuration.

arms, and each arm contains n SMs connected in series and an arm inductor L_S. Figure 7.27c shows the SM configuration for the MMC with current interruption capability. The SM contains one top IGBT/Diode (T_1/D_1), one bottom IGBT/Thyristor (T_2/VT), and one DC capacitor C, where the thyristor VT is used to be connected in anti-parallel with the bottom IGBT T_2. The capacitor C_s and resistor R_s construct the snubber circuit of the thyristor VT.

7.5.2 SM Control

In normal operation, the SM has four operation modes, as shown in Table 7.4. Neglecting the snubber circuit, Figure 7.28 shows the four modes.

- In Mode 1, the arm current i_{ju} or i_{jl} ($j = a, b, c$) is positive. T_1 and T_2 are switched on and off, respectively. Here, the SM output voltage equals the capacitor voltage u_c, the C is charged, and the u_c increases.
- In Mode 2, the arm current is positive. T_1 and T_2 are switched off and on, respectively. Here, the SM output voltage is 0, the C is bypassed, and the u_c is unchanged.

Table 7.4 Operation of the SM in normal situation.

Mode	i_{ju} (i_{jl})	T_1	T_2	VT	Capacitor C	Capacitor voltage u_c
1	≥0	On	Off	Off	Charge	Increased
2	≥0	Off	On	Off	Bypass	Unchanged
3	<0	On	Off	Off	Discharge	Decreased
4	<0	Off	On	On	Bypass	Unchanged

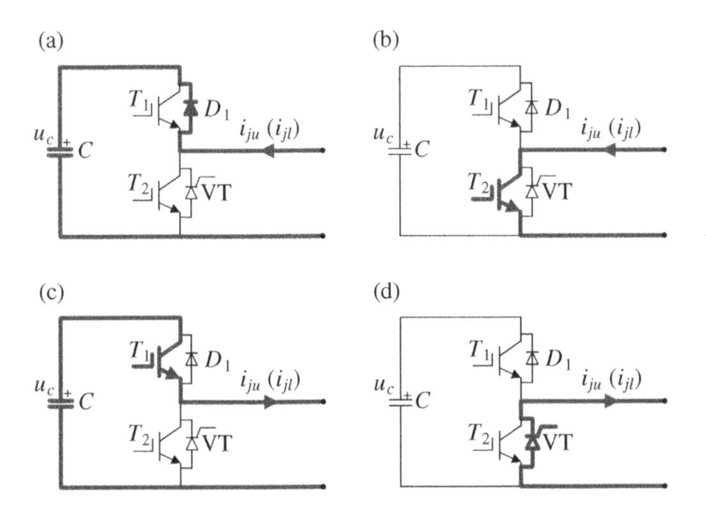

Figure 7.28 Four operation modes of the SM. (a) Mode 1. (b) Mode 2. (c) Mode 3. (d) Mode 4.

- In Mode 3, the arm current is negative. T_1 and T_2 are switched on and off, respectively. Here, the SM output voltage equals u_c, the C is discharged, and the u_c reduces.
- In Mode 4, the arm current is negative. T_1 and T_2 are switched off and on, respectively, and the VT is switched on. Here, the SM output voltage equals 0, the C is bypassed, and the u_c is unchanged.

7.5.3 Current Interruption Control

The current interruption operation of the BT MMC based HVDC system under DC line faults mainly contains three-phase rectifier period, one-phase current interruption moment, single-phase rectifier period, and three-phase current interruption moment.

7.5.3.1 Three-Phase Rectifier Period

In the BT MMC based HVDC system, all IGBTs should be blocked and the gate signals for all thyristors should be removed when the fault current is over the limit value. The gate-signal removed thyristors may still carry current until its current is lower than the holding current. Thus, the MMC may work in rectifier period with the still conducting thyristors. In this situation, the three-phase AC current will be fed into the DC link through the conducting arms in the MMC. Figure 7.29 gives an example for the three-phase rectifier mode of the MMC, where the upper arm of phase A, the upper arm of phase B and lower arm of phase C are conducting and feeding the three-phase AC current to the DC link. L_{eq} and R_{eq} are the equivalent inductance and resistance.

7.5.3.2 One-Phase Current Interruption Moment

The conducting thyristors in some phase will be first naturally blocked when the corresponding AC phase current is lower than the holding current, because the gate signals for all thyristors have been removed. At this moment, the capacitor voltages of the SMs in the blocked arm appear as series-connected voltage sources and have opposite polarity than the direction of the AC-side current. Figure 7.30 gives an example for the one-phase current interruption

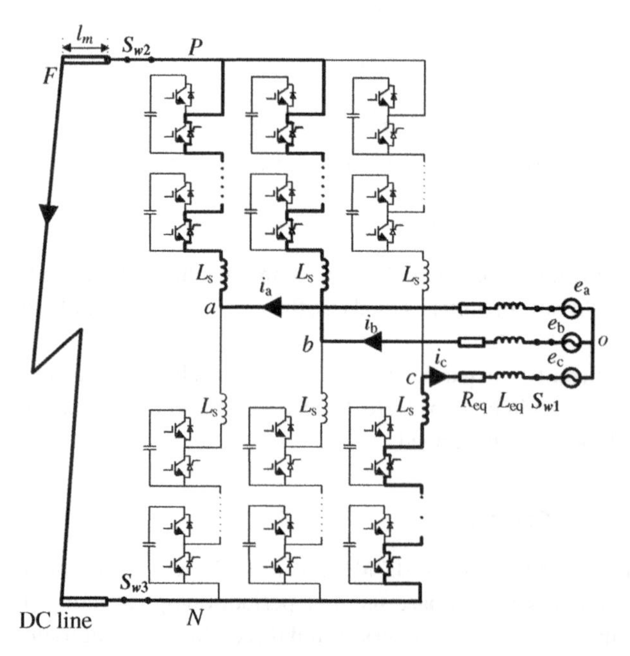

Figure 7.29 Three-phase current interruption period.

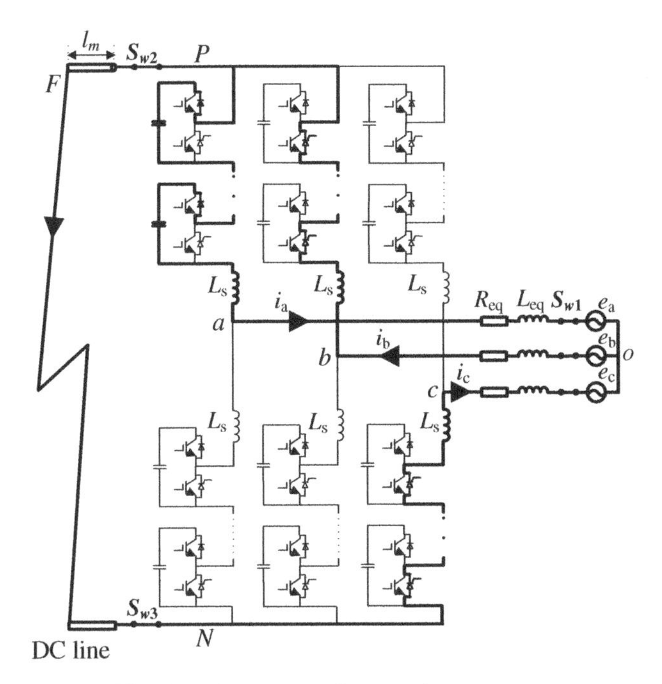

Figure 7.30 One-phase current interruption moment.

moment of the MMC, where the current in the phase A will be interrupted by the MMC.

The voltage imposed on the current interrupted arm is the line-to-line voltage, whose peak value E_{llm} is $\sqrt{3}mV_{dc}/2$. The voltage formed by the series-connected capacitor in the current interrupted arm is V_{dc} and obviously higher than the peak value E_{llm} of the line-to-line voltage, as

$$V_{dc} > E_{llm} = \frac{\sqrt{3}mV_{dc}}{2} \tag{7.6}$$

Consequently, the current can be interrupted by the MMC.

7.5.3.3 Single-Phase Rectifier Period

After the one-phase current interruption moment, the MMC will work in single-phase rectifier period with the still conducting thyristors in the other two arms, where the AC current will be fed into the DC link through the two conducting arms of the MMC. Figure 7.31 gives an example for the single-phase rectifier mode of the MMC, where the upper arm of phase B and the lower arm of phase C are conducting.

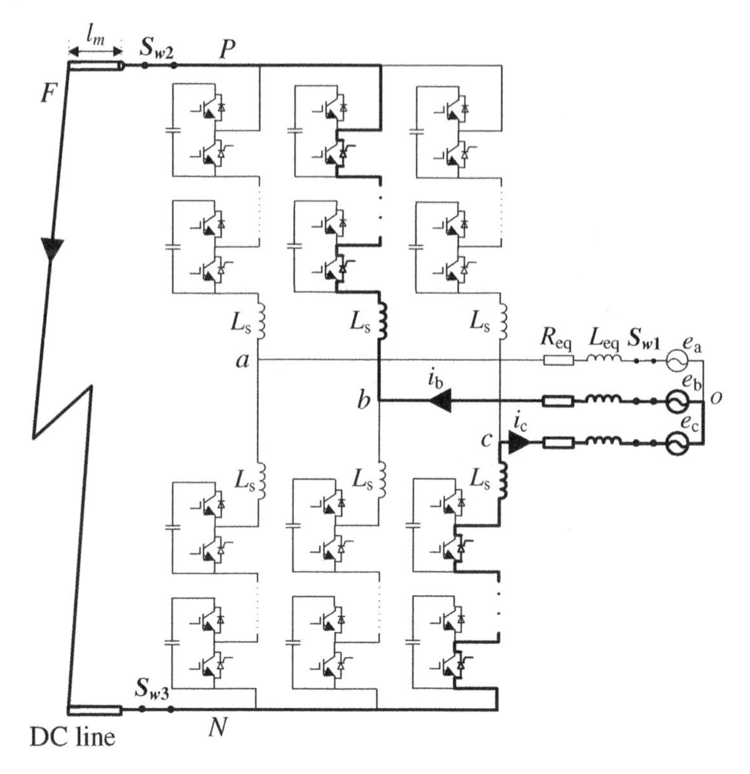

Figure 7.31 Single-phase rectifier period.

7.5.3.4 Three-Phase Current Interruption Moment

After the single-phase rectifier period, the conducting thyristors will be naturally blocked when the corresponding AC phase current is lower than the holding current because the gate signals for all thyristors have been removed. At this moment, all IGBTs and thyristors in MMCs are blocked. The SM capacitor voltages in each arm appear as series-connected voltage sources and have opposite polarity than the direction of the current from the AC side. Figure 7.32 gives an example for the three-phase current interruption moment of the MMC, where the currents in the upper arm of phase B and the lower arm of phase C are interrupted by the MMC, and accordingly the three-phase current in the MMC are all interrupted.

The voltage formed by the series-connected capacitors of the two arms (one upper arm and one lower arm) is $2V_{dc}$ and obviously higher than the peak value E_{llm} of the line-to-line AC voltage, as

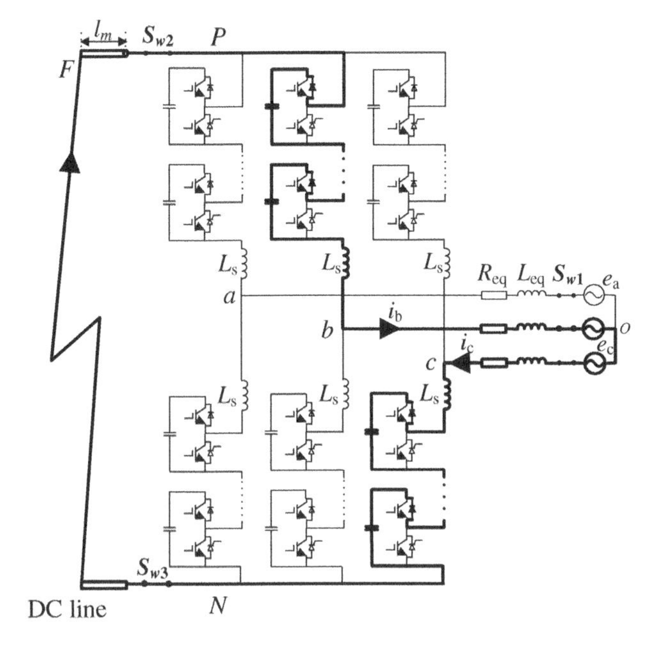

Figure 7.32 Three-phase current interruption moment.

$$2V_{dc} > E_{llm} = \frac{\sqrt{3}mV_{dc}}{2} \tag{7.7}$$

Consequently, the currents in the MMC can be interrupted.

7.5.4 Protection Operation

The protection scheme for the BT MMC based HVDC system is shown in Figure 7.33. First, the switches S_{w2} and S_{w3} at the DC terminal are closed. The IGBTs and thyristors are given the suitable signals to realize normal operation for the HVDC system. When the DC line fault is detected, all IGBTs are blocked and the gate signals for all thyristors are removed to interrupt the fault current. After the fault current is interrupted, the switches S_{w2} and S_{w3} at the DC terminal are opened to isolate the MMC from the faulty DC lines. Here, the MMC can still be connected to the AC grid under faults, where the switch S_{w1} shown in Figures 7.29–7.32 is not necessary to be opened. Before the DC line fault is cleared, all IGBTs and thyristors are deblocked again, where the MMC can keep its

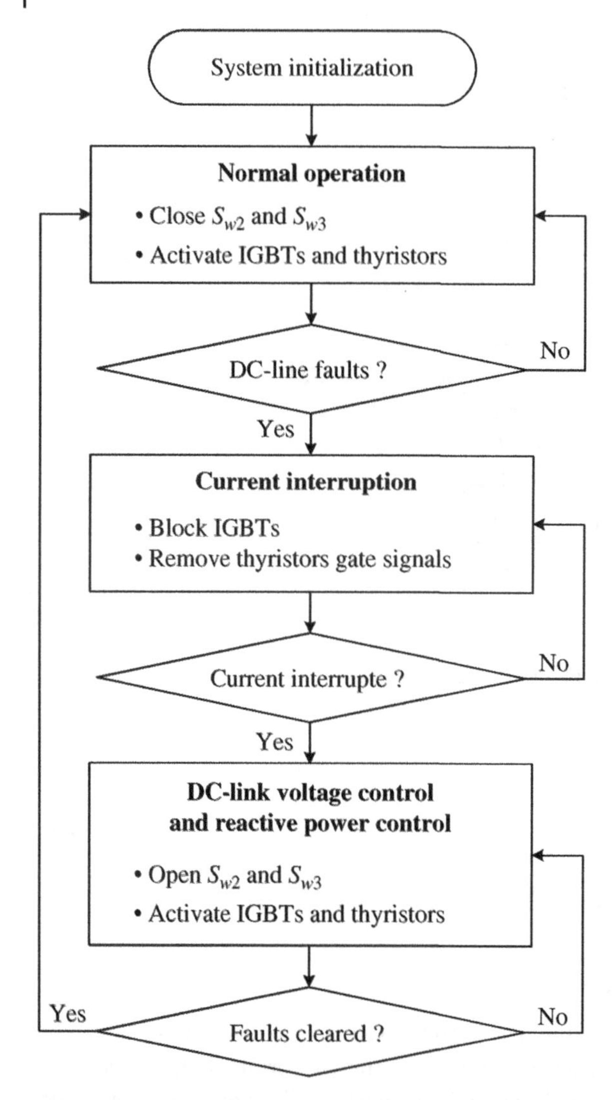

Figure 7.33 Three-phase current interruption scheme.

DC-link voltage constant and regulate the reactive power in the AC grid. Therefore, the HVDC system can improve the system performance under faults. In addition, the operation of the MMC during DC line fault can omit some processes when the HVDC restarts again, such as the capacitor precharging process. After the fault is cleared, the switches S_{w2} and S_{w3} are closed again and the HVDC system starts normal operation again.

Case Study 7.2 DC Fault Protection Based on Bypass Thyristor MMC

Objective: In this case study, an HVDC system based on bypass thyristor MMC, as shown in Figure 7.27a, is simulated with PSCAD/EMTDC, where the MMC-1 is used to keep the DC-link voltage V_{dc1} constant and the MMC-2 is used to send power to the MMC-1. The DC line short-circuit fault occurs at 20 km away from the MMC-2.
Parameters: The system parameters are shown in Table 7.5.

Simulation results and analysis
Figure 7.34 shows the performance of the MMC-2 in the HVDC system, where the DC line short-circuit fault occurs at 4.008 seconds. After the fault occurs, the DC-link voltage V_{dc2} drops and the DC-link current i_{dc} is increased, as shown in Figure 7.34a,b. Owing to all IGBTs are blocked and the gate signals for all thyristors are removed under faults, the fault current can be naturally interrupted by the thyristor when the arm current is lower than the holding current. The three-phase grid current i_{a2}, i_{b2}, and i_{c2} are interrupted in a short time and in approximate 15 ms, as shown in Figure 7.34c. In addition, the upper arm current i_{au} and the lower arm current i_{al} in phase A are interrupted in a short time, as shown in Figure 7.34d; the upper arm current i_{bu} and the lower arm current i_{bl} in phase B are interrupted in a short time, as shown in Figure 7.34e; the upper arm current i_{cu} and the lower arm current i_{cl} in phase C are interrupted in a short time, as shown in Figure 7.34f. After the fault current is interrupted, the mechanical switches S_{w2} and S_{w3} at the DC terminal of the MMC are opened to isolate the MMC from the faulty DC line. Afterwards, all IGBTs and thyristors are deblocked, and the MMC-2 starts to keep its DC-link voltage V_{dc2} and regulate the reactive power Q in the grid, as shown in Figure 7.34a,g.

Table 7.5 Parameters of the simulated system.

Parameters	Value
Rated power (MW)	200
Rated DC-link voltage V_{dc1} (kV)	150
DC cable length (km)	100
SM number per arm (n)	10
Arm inductance (mH)	8
SM capacitor C (mF)	1.5
Carrier frequency (Hz)	800
Transformer leakage reactance	15%
AC grid voltage (kV)	380
Impedance Z	0.32 + j4

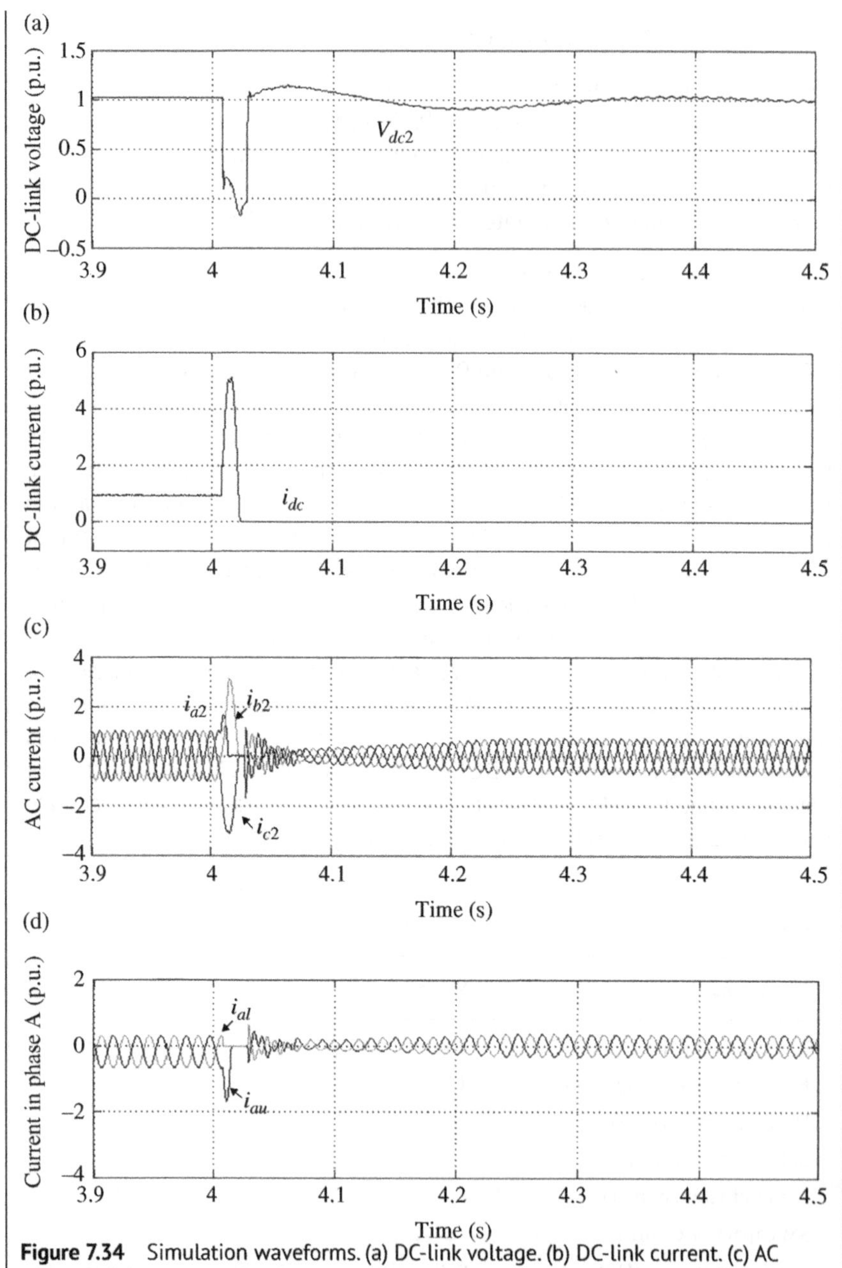

Figure 7.34 Simulation waveforms. (a) DC-link voltage. (b) DC-link current. (c) AC current. (d) Arm current in phase A. (e) Arm current in phase B. (f) Arm current in phase C. (g) Power.

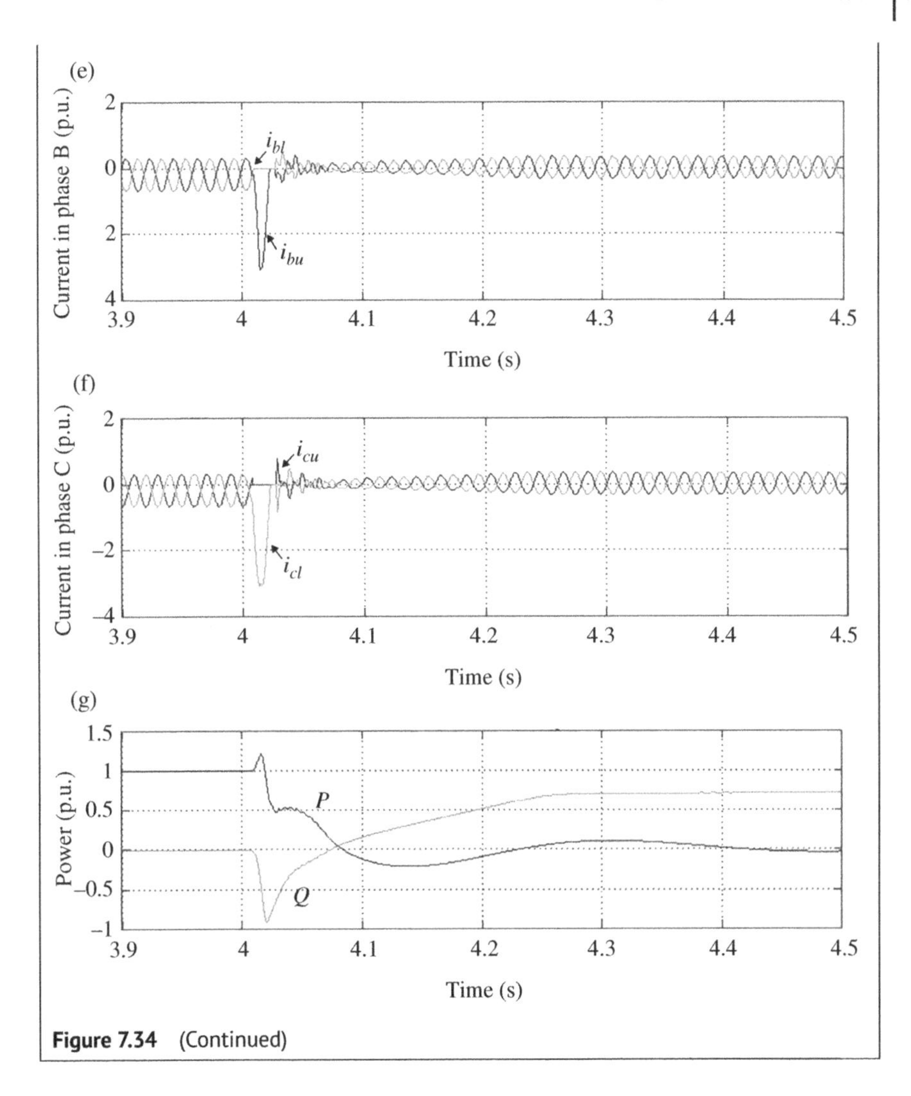

Figure 7.34 (Continued)

7.6 CTB-HMMC Based Protection

This section presents a crossing thyristor branches (CTB) based hybrid MMC (HMMC) to protect the HVDC system under DC line short-circuit faults. In the CTB-HMMC, the thyristor branches are employed to crossing-connect different arm inductors and each arm consists of unipolar full-bridge SMs (UFB SMs) and HB SMs, where the CTB effectively reduces the UFB SMs number in the arm of the HMMC with current interruption capability. The advantages of the CTB-HMMC are: (i) the number of UFB SMs in the arm is reduced and far less than that in conventional UFB SMs based HMMC, (ii) short current interruption time about several milliseconds, (iii) low semiconductor cost, and (iv) low power losses.

7.6.1 CTB-HMMC Configuration

The CTB-HMMC is shown in Figure 7.35a, where each phase contains upper and lower arms. Each arm consists of an inductor L_s and n SMs, which includes m UFB SMs and $(n-m)$ HB SMs. Figure 7.35b shows the HB SM, which contains two switches/diodes (T_1/D_1, T_2/D_2), a DC capacitor C, and a bypass switch K_h.

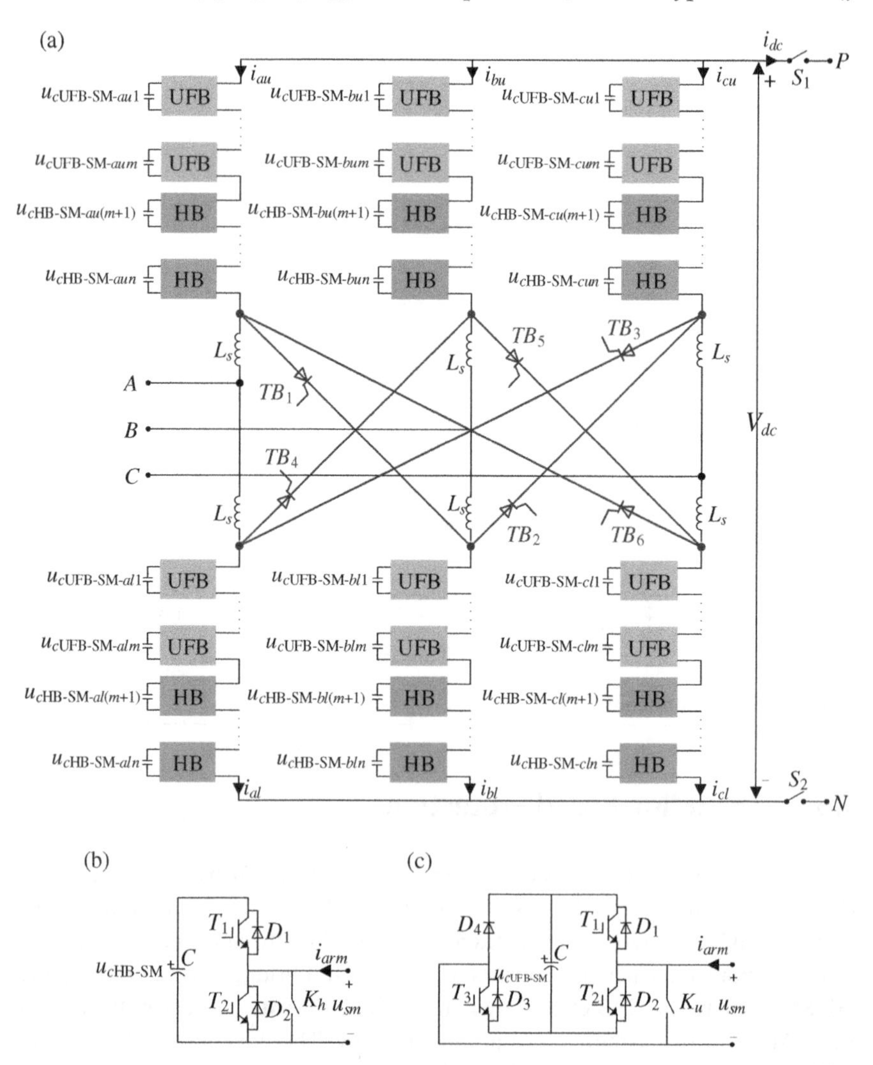

Figure 7.35 (a) CTB-HMMC. (b) HB SM. (c) UFB SM.

Figure 7.35c shows the UFB SM, which contains three switches/diodes (T_1/D_1, T_2/D_2, T_3/D_3), a diode D_4, a DC capacitor C, and a bypass switch K_u.

In the CTB-HMMC, six CTB TB_1–TB_6, which are composed of series-connected thyristors, are used to connect the upper arm inductors and the lower arm inductors in the different phases, as shown in Figure 7.35a. In the HVDC system, the six arm inductors are gathered in an indoor place, and therefore the connection of the CTB can be easily achieved. In addition, two fast MSs S_1 and S_2 are equipped at the DC side.

7.6.2 SM Operation Principle

The HB SM has three operation modes, as listed in Table 7.6, as follows.

- *Mode 1*: T_1 is switched on and T_2 is switched off. The HB SM is inserted into the arm and its terminal voltage u_{sm} is equal to the capacitor voltage $u_{cHB\text{-}SM}$.
- *Mode 2*: T_1 is switched off and T_2 is switched on. The HB SM is bypassed from the arm and $u_{sm} = 0$.
- *Mode 3*: T_1 and T_2 are both switched off and the HB SM is blocked. Here, the HB SM is inserted into the arm and $u_{sm} = u_{cHB\text{-}SM}$ if the arm current i_{arm} is positive; the HB SM is bypassed and $u_{sm} = 0$ if the i_{arm} is negative.

The UFB SM has four operation modes, as listed in Table 7.7, as follows.

- *Mode 1*: T_1, T_3 are both switched on and T_2 is switched off. The UFB SM is inserted into the arm and the terminal voltage u_{sm} of the UFB SM is equal to the capacitor voltage $u_{cUFB\text{-}SM}$.
- *Mode 2*: T_1 is switched off and T_2, T_3 are both switched on. The UFB SM is bypassed from the arm and $u_{sm} = 0$.
- *Mode 3*: T_1, T_2 are switched off and T_3 is switched on. Here, the UFB SM is inserted into the arm and $u_{sm} = u_{cUFB\text{-}SM}$ if the arm current i_{arm} is positive; the UFB SM is bypassed and $u_{sm} = 0$ if $i_{arm} < 0$.
- *Mode 4*: T_1, T_2, and T_3 are all switched off and the UFB SM is blocked. Here, the UFB SM is inserted into the arm and $u_{sm} = u_{cUFB\text{-}SM}$ if $i_{arm} > 0$; the UFB SM is reversely inserted into the arm and $u_{sm} = -u_{cUFB\text{-}SM}$ if $i_{arm} < 0$.

Table 7.6 Operation modes of the HB SM.

Mode	State	T_1	T_2	i_{arm}	u_{sm}
1	Insert	On	Off	>0 or <0	$u_{cHB\text{-}SM}$
2	Bypass	Off	On	>0 or <0	0
3	Insert	Off	Off	>0	$u_{cHB\text{-}SM}$
	Bypass			<0	0

Table 7.7 Operation modes of the UFB SM.

Mode	State	T_1	T_2	T_3	i_{arm}	u_{sm}
1	Insert	On	Off	On	>0 or <0	$u_{cUFB\text{-}SM}$
2	Bypass	Off	On	On	>0 or <0	0
3	Insert	Off	Off	On	>0	$u_{cUFB\text{-}SM}$
	Bypass				<0	0
4	Insert	Off	Off	Off	>0	$u_{cUFB\text{-}SM}$
					<0	$-u_{cUFB\text{-}SM}$

7.6.3 Operation Principle for DC Fault Protection

The operation principles and current interruption diagram of the CTB-HMMC under DC-side short-circuit fault are shown in Figures 7.36 and 7.37, respectively.

In normal operation, the CTB-HMMC is controlled according to its control objective such as active power, reactive power, and DC-link voltage control. Here, the CTB-HMMC works as the conventional MMC, where the HB-SMs work in their Mode 1 or Mode 2; the UFB SMs work in their Mode 1 or Mode 2; all CTB TB_1–TB_6 are blocked; the MSs S_1 and S_2 are closed.

If the DC-line short-circuit fault occurs in the HVDC system at t_0, the DC-side current i_{dc} of the CTB-HMMC would be increased, as shown in Figure 7.37. The

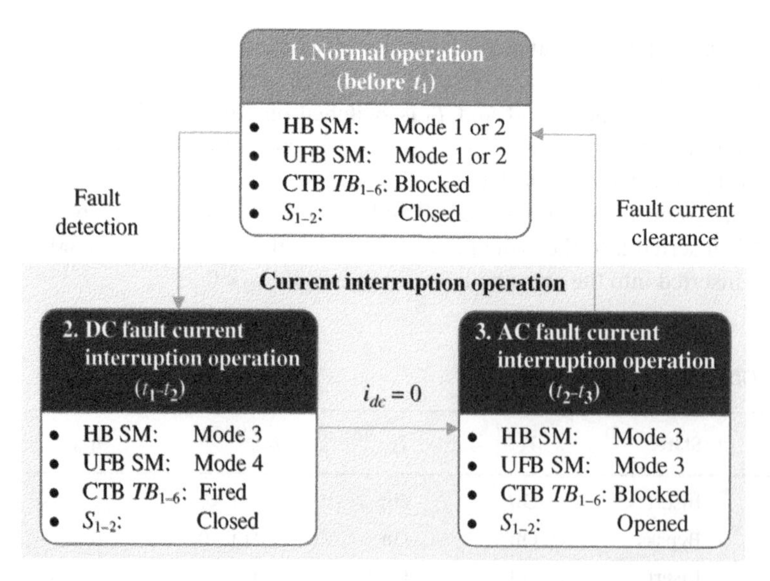

Figure 7.36 Operation for CTB-HMMCs under DC-line short-circuit faults.

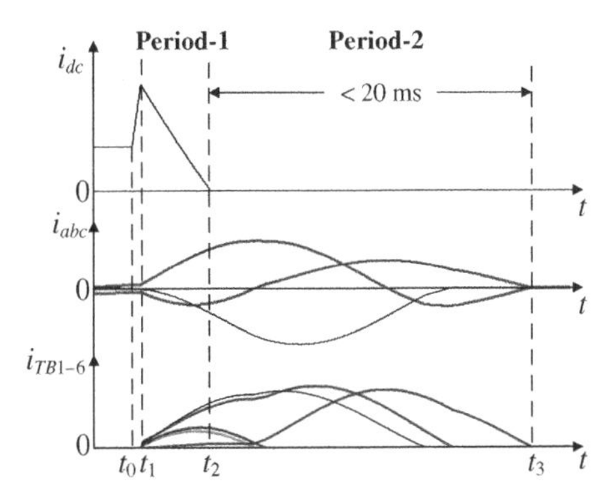

Figure 7.37 Diagram of current interruption period of the CTB-HMMC.

operation principle for the CTB-HMMC, as shown in Figure 7.36, can effectively interrupt fault current and protect HVDC system under DC-line short-circuit faults. The current interruption operation of the CTB-HMMC mainly contains DC-side current interruption operation in Period-1 (t_1, t_2) and AC-side current interruption operation in Period-2 (t_2, t_3), as shown in Figures 7.36 and 7.37, which are presented in the following sections, respectively.

7.6.4 DC-Side Current Interruption Operation

Once the fault is detected at t_1, the DC-side current interruption operation will be implemented. Here, the HB SM works in Mode 3, where all switches in the HB SMs are blocked; the UFB SM works in Mode 4, where all switches in the UFB SMs are blocked. In addition, the CTB TB_1–TB_6 are fired, and therefore the thyristor branches are connected in a circle as TB_1-TB_2-. . .-TB_6-TB_1 to construct the short-circuit path, as shown in Figure 7.38.

Figure 7.39 shows the equivalent circuit of the DC side of the CTB-HMMC during Period-1, where the fault distance is l_m and the fault resistor is R_{fr}. In Figure 7.39, the reverse voltage is constructed by the M blocked UFB SMs in each arm, whose positive pole is opposite to the direction of the fault current i_{dc}. As a result, the i_{dc} would be decreased by the reverse voltage injected by the blocked UFB SMs in the CTB-HMMC. At t_2, the DC-side current i_{dc} would be interrupted and reaches zero. Afterwards, the fast mechanical switch S_1 and S_2 at the DC terminal of the CTB-HMMC will be opened to isolate the CTB-HMMC and the faulty DC lines.

Suppose that all SM capacitor voltages are kept balanced by the capacitor voltage balancing control before t_1 as

$$u_{cHB-SM} = u_{cUFB-SM} = u_c = V_{dc} / n \tag{7.8}$$

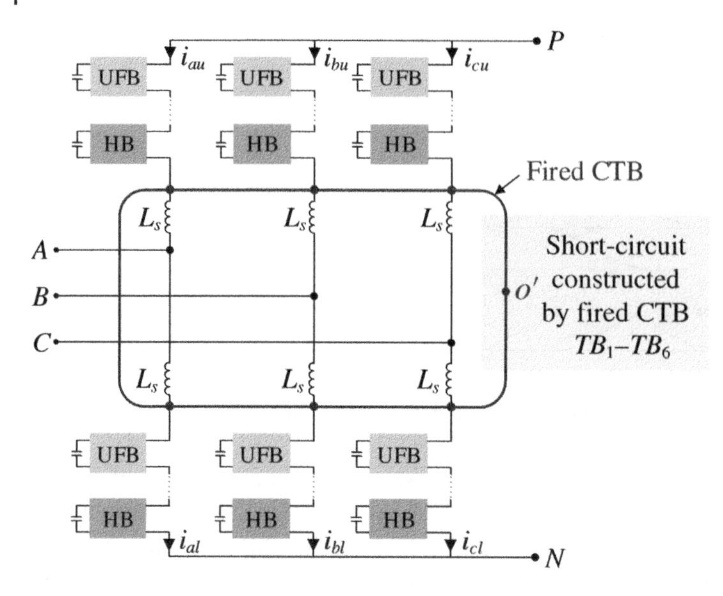

Figure 7.38 Short-circuit path constructed by fired CTB TB_1–TB_6 in Period-1.

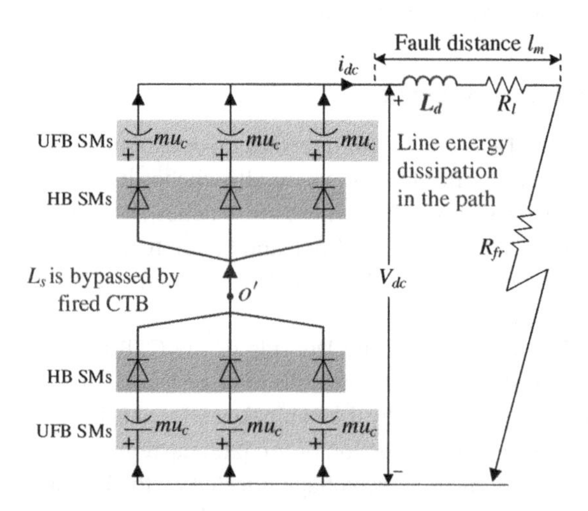

Figure 7.39 DC-side equivalent circuit under DC-side short-circuit faults.

the DC-side voltage V_{dc} of the CTB-HMMC during the Period-1 can be expressed as

$$\begin{cases} V_{dc}(t) = -\left[2mu_c + \dfrac{2m}{3C}\displaystyle\int_{t_1}^{t} i_{dc}(t)dt\right] \\ V_{dc}(t) = L_d \dfrac{di_{dc}(t)}{dt} + R_d i_{dc}(t) \end{cases} \tag{7.9}$$

where $R_d = R_l + R_{fr}$. L_d is line inductor and R_l is line resistor.

Based on equation (7.9), the DC-side current i_{dc} can be obtained as

$$i_{dc}(t) = e^{\alpha_c(t-t_1)}\left[C_1 \cos\beta_c(t-t_1) + C_2 \sin\beta_c(t-t_1)\right] \tag{7.10}$$

with

$$\begin{cases} \alpha_c = -\dfrac{R_d}{2L_d} \\ \beta_c = \dfrac{1}{2}\sqrt{\dfrac{8m}{3CL_d} - \left(\dfrac{R_d}{L_d}\right)^2} \\ C_1 = i_{dc}(t_1) \\ C_2 = -\dfrac{mu_c}{\beta_c L_d} + \dfrac{\alpha_c C_1}{\beta_c} \end{cases} \tag{7.11}$$

Figure 7.40 shows the current interruption time of the CTB-HMMC under various m/n and various fault distances l_m, which is derived from the simulation in case study. Along with the increase of m/n, the current interruption time is reduced, because higher reverse voltage is produced by more UFB SMs in the arm. In addition, along with the reduction of l_m, the current interruption time is reduced, because the inductance L_d and the energy stored in the faulty line is reduced along with the reduction of l_m.

7.6.5 Capacitor Voltage Increment

During the DC-side current interruption period, the UFB SMs would absorb the energy stored in the DC-side inductor L_d, as shown in Figure 7.39. Along with the reduction of the DC-side current i_{dc}, the capacitor voltages in the UFB SMs are gradually increased and the voltage increment $\Delta u_{cUFB\text{-}SM}$ of the UFB SMs can be expressed as

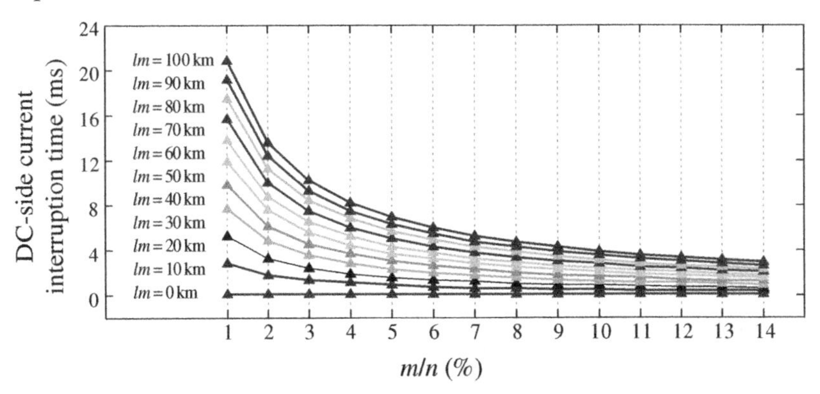

Figure 7.40 DC-side current interruption time under various m/n and l_m.

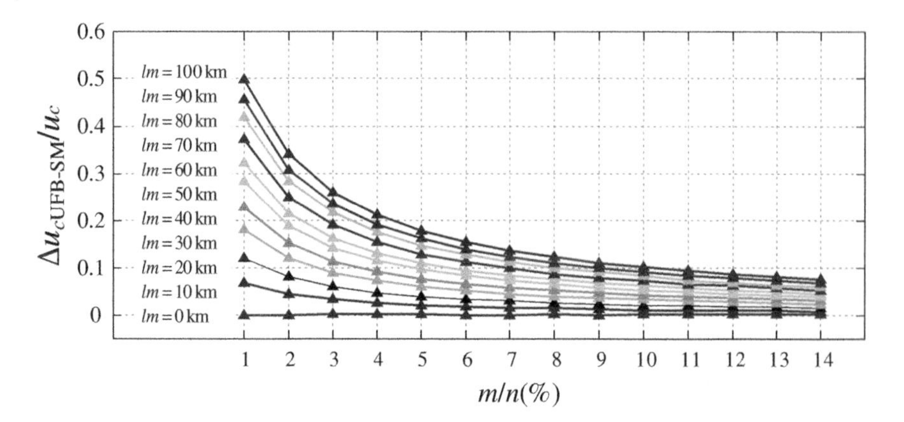

Figure 7.41 UFB SM voltage increment $\Delta u_{cUFB\text{-}SM}/u_c$ under various m/n and l_m.

$$\Delta u_{cUFB-SM} = \frac{1}{3C}\int_{t_1}^{t_2} i_{dc}(t)dt \tag{7.12}$$

Figure 7.41 shows the voltage increment $\Delta u_{cUFB\text{-}SM}/u_c$ of the UFB SM under various M/N and different fault distances l_m, which is derived from the simulation in case study. Along with the increase of l_m, the $\Delta u_{cUFB\text{-}SM}$ increases, because the inductance and the energy stored in the short-circuit line is increased along with the increase of l_m. In addition, along with the increase of the M/N, the $\Delta u_{cUFB\text{-}SM}$ reduces, because more capacitors in UFB SMs absorb the energy stored in the DC line.

One thing to mention is that since the arm inductors L_s are bypassed by the fired CTB, the energy stored in L_s does not need to be absorbed by the UFB SMs, which reduces the voltage increment $\Delta u_{cUFB\text{-}SM}$ of the UFB SMs. For example, the fault current can be eliminated immediately and UFB SMs would not be charged if the fault distance l_m is 0 km.

7.6.6 AC-Side Current Interruption Operation

According to DC-side current interruption operation during Period-1, once a DC fault is detected at t_1, the CTB TB_1–TB_6 would be fired, which would create a three-phase short-circuit path and result in AC-side short-circuit situation. Figure 7.42 shows the equivalent circuit of the AC side of the CTB-HMMC during Period-1.

Suppose the AC-side equivalent phase voltages e_a, e_b, and e_c are

$$\begin{cases} e_a(t) = E_m\cos(\omega t + \theta_0) \\ e_b(t) = E_m\cos(\omega t + \theta_0 - 2\pi/3) \\ e_c(t) = E_m\cos(\omega t + \theta_0 + 2\pi/3) \end{cases} \tag{7.13}$$

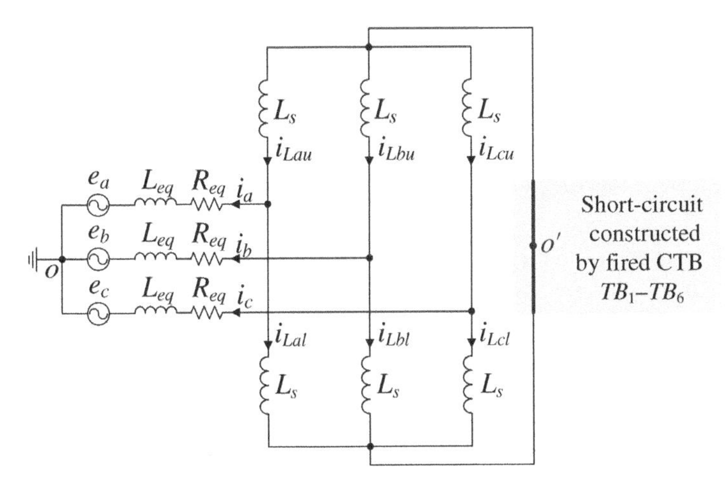

Figure 7.42 AC-side equivalent circuit during period-1.

where E_m and θ_0 are amplitude and initial phase angle, respectively. The voltage relationship in phase j ($j = a, b, c$) of Figure 7.42 can be expressed as

$$\begin{cases} u_{o'o} = L_s \dfrac{di_{Lju}}{dt} + L_{eq}\dfrac{di_j}{dt} + i_j R_{eq} + e_j \\ u_{o'o} = -L_s \dfrac{di_{Ljl}}{dt} + L_{eq}\dfrac{di_j}{dt} + i_j R_{eq} + e_j \end{cases} \tag{7.14}$$

where i_{Lju} and i_{Ljl} are the current flowing through upper and lower arm inductors of phase j, respectively. L_{eq} and R_{eq} are equivalent inductor and resistor at the AC side of the MMC, respectively.

Combining equations (7.13) and (7.14), the AC-side short-circuit current is

$$\begin{cases} i_a(t) = \left[i_a(t_1) + I_m\cos(\omega t_1 + \theta_0 - \varphi_s)\right]e^{-(t-t_1)/\tau_s} \\ \quad - I_m\cos(\omega t - \varphi_s) \\ i_b(t) = \left[i_b(t_1) + I_m\cos(\omega t_1 + \theta_0 - 2\pi/3 - \varphi_s)\right]e^{-(t-t_1)/\tau_s} \\ \quad - I_m\cos(\omega t - 2\pi/3 - \varphi_s) \\ i_c(t) = \left[i_c(t_1) + I_m\cos(\omega t_1 + \theta_0 + 2\pi/3 - \varphi_s)\right]e^{-(t-t_1)/\tau_s} \\ \quad - I_m\cos(\omega t + 2\pi/3 - \varphi_s) \end{cases} \tag{7.15}$$

with

$$\begin{cases} \tau_s = \left(L_{eq} + \dfrac{L_s}{2}\right) / R_{eq} \\ I_m = E_m / \sqrt{R_{eq}^{\ 2} + \omega^2 \left(L_{eq} + \dfrac{L_s}{2}\right)^2} \\ \varphi_s = \tan^{-1}\left[\omega\left(L_{eq} + \dfrac{L_s}{2}\right) / R_{eq}\right] \end{cases} \tag{7.16}$$

After the DC-side fault current is interrupted at t_2, the switches S_1 and S_2 are opened and the gate signals for CTB TB_1–TB_6 are removed. In the Period-2, the HB SMs still work in Mode 3, where T_1 and T_2 are both switched off; the UFB SMs are switched to Mode 3, where the T_1, T_2 are switched off and T_3 is switched on. It means that the HB SMs and the UFB SMs work in the same situation. Figure 7.43 shows the AC-side equivalent circuit in the Period-2. Since the electromotive force

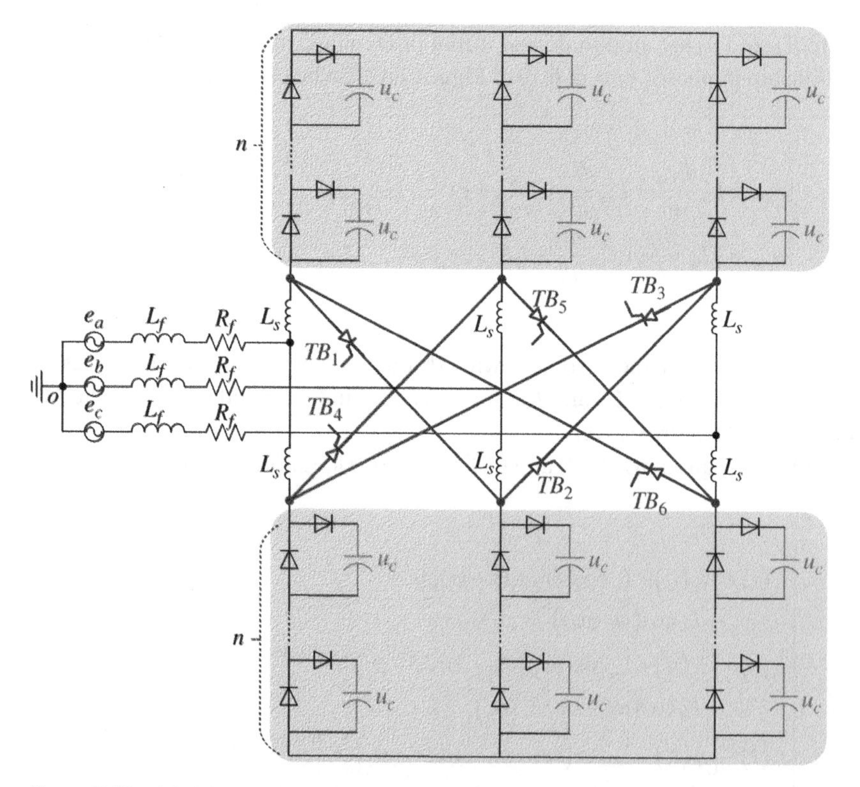

Figure 7.43 AC-side equivalent circuit in period-2.

constructed by N SMs in each arm is $nu_c = V_{dc}$, which is higher than the maximum line voltage of the AC side, the AC current would not flow through the SMs.

When the AC currents flowing through the thyristor branches are below the holding current of the thyristors, the thyristor branches without gate signals would be naturally turned off. As a result, the AC-side current in each branch would be interrupted. After the DC-line fault is cleared, the CTB-HMMCs would work in normal situation again.

7.6.7 MMC Comparison

In this section, the CTB-HMMC is compared with the possible MMC configurations with DC fault handling ability, including current blocking SM based methods and thyristor based methods, in terms of current interrupting time, semiconductor cost, and power losses.

7.6.7.1 Comparison with Current Blocking SM Based MMCs

Table 7.8 compares the possible current blocking SM based MMC configurations for the HVDC system, where all MMC configurations have the same output voltage levels in normal operation. In Table 7.8, CD SM based MMC [28], UFB SM based HMMC (50% UFB SMs and 50% HB SMs) [29], FB-SM based HMMC (50% FB-SMs and 50% HB SMs) [22], alternate arm converter (AAC) [30], FB-SM based MMC [31], and the CTB-HMMC are compared in terms of current interruption time, semiconductor cost, and power losses.

In Figure 7.44a, the current interruption time of the CD-MMC, UFB-HMMC, FB-HMMC, AAC, FB-MMC, and the CTB-HMMC are around several

Table 7.8 Comparisons of the current blocking SM based MMC configurations.

MMC configurations	Current interruption time (ms)	Semiconductor cost (p.u.)	Losses (%)
CD-MMC [28]	1.5–3.5	1.160	1.16
UFB-HMMC [29] (50% UFB SMs and 50% HB SMs)	1.5–3	1.078	1.16
FB-HMMC [22] (50% FB-SMs and 50% HB SMs)	1.5–3	1.194	1.16
AAC [30]	1–2	1.115	1.09
FB-MMC [31]	0.5–1	1.592	1.43
CTB-HMMC (14% UFB SMs and 86% HB SMs)	0–3	1	0.96

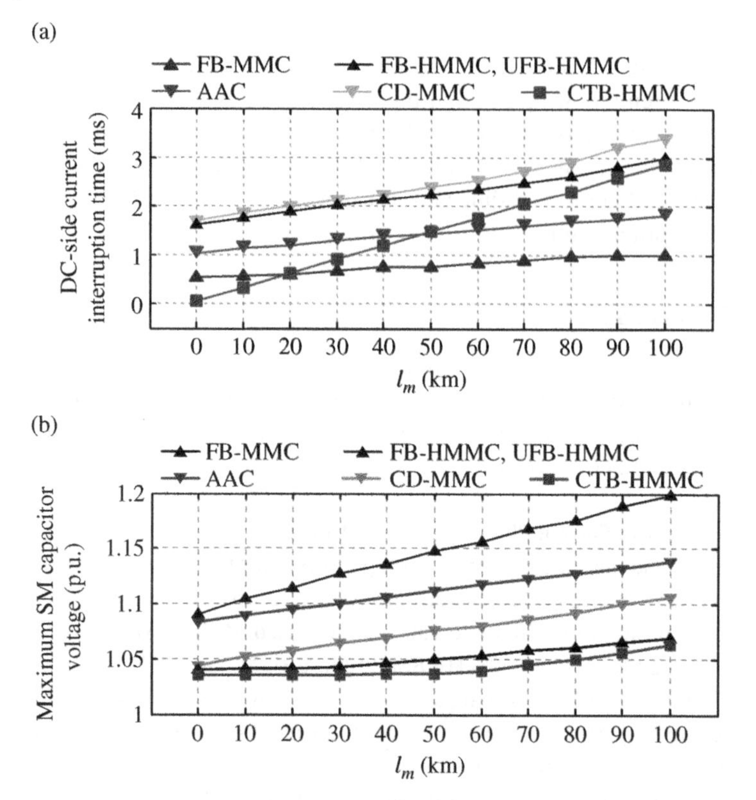

Figure 7.44 Comparison of current blocking SM based MMCs and the CTB-HMMC. (a) Current interruption time under various l_m. (b) Maximum SM capacitor voltage under various l_m.

milliseconds. In comparison with the CD-MMC, UFB-HMMC, FB-HMMC, AAC, and CTB-HMMC, the FB-MMC can interrupt the current in a shorter time but with more FB-SMs and more semiconductors. The current interruption time of the CTB-HMMC is sharply reduced along with the reduction of the fault distance l_m, which is shorter than that of the CD-MMC, UFB-HMMC, and FB-HMMC and shorter than that of the AAC when $l_m < 50$ km and even shorter than that of the FB-MMC when $l_m < 20$ km. In Figure 7.44b, the maximum SM capacitor voltage of the CTB-HMMC is lower than those in the CD-MMC, UFB-HMMC, FB-HMMC, AAC, and FB-MMC, because the arm inductors L_s are bypassed by the fired CTB and the energy stored in the arm inductors L_s is not required to be dissipated in the DC-side current interruption circuit. Besides, in the CTB-HMMC, the UFB SMs in the upper arm of the three phases and UFB SMs in the lower arm of the three phases can be charged in balance. However, in CD-MMC, UFB-HMMC,

FB-HMMC, AAC, and FB-MMC, the fault current only flows through the upper arm in the phase with the highest instantaneous AC voltage and through the lower arm in the phase with the lowest instantaneous AC voltage [32], which results in the unbalanced charge of SMs in different arms.

Suppose that the IGBT FZ1500R33HL3, freewheeling diode D1031SH45T, and thyristor T1930N32TOF are adopted to construct these MMCs for the HVDC system. Table 7.8 shows the total semiconductor cost of these MMCs based on the cost of IGBT, diode, and thyristor and the power losses of these MMCs based on the power losses calculation. Owing to that the CTB-HMMC only has 14% UFB SMs in the arm and the cost of the thyristor and diode in the CTB-HMMC is much cheaper than that of the IGBT, the total semiconductor cost of the CTB-HMMC is lower than those of the CD-MMC, UFB-HMMC, FB-HMMC, AAC, and FB-MMC. With less devices in current path, the power losses of the CTB-HMMC is lower than those of the CD-MMC, UFB-HMMC, FB-HMMC, AAC, and FB-MMC.

7.6.7.2 Comparison with Thyristor Based MMCs

Table 7.9 compares the CTB-HMMC and the possible thyristor based MMC configurations for the HVDC system, where all MMC configurations have the same output voltage levels in normal operation. In Table 7.9, HB-double-thyristor (DT) based MMC [33], crowbar circuit (CC) based MMC [32], BT based MMC [34], shadow rectifier bridge (SRB) based MMC [35], and the CTB-HMMC are compared in terms of current interruption time, semiconductor cost, and power losses.

Figure 7.45a shows the current interruption time of the DT-MMC, CC-MMC, SRB-MMC, BT-MMC, and CTB-HMMC under various fault distances l_m. The CTB-HMMC can interrupt the fault current in a much shorter time in comparison with the other thyristor based MMCs. Figure 7.45b shows the peak values of

Table 7.9 Comparisons of the thyristor based MMC configurations.

MMC configurations	Current interruption time (ms)	Semiconductor cost (p.u.)	Losses (%)
DT-MMC [33]	65–150	1.046	0.89
CC-MMC [32]	120–180	1.213	0.89
BT-MMC [34]	19–24	1.181	0.93
SRB-MMC [35]	15–18	1.155	1.12
CTB-HMMC (14% UFB SMs and 86% HB SMs)	0–3	1	0.96

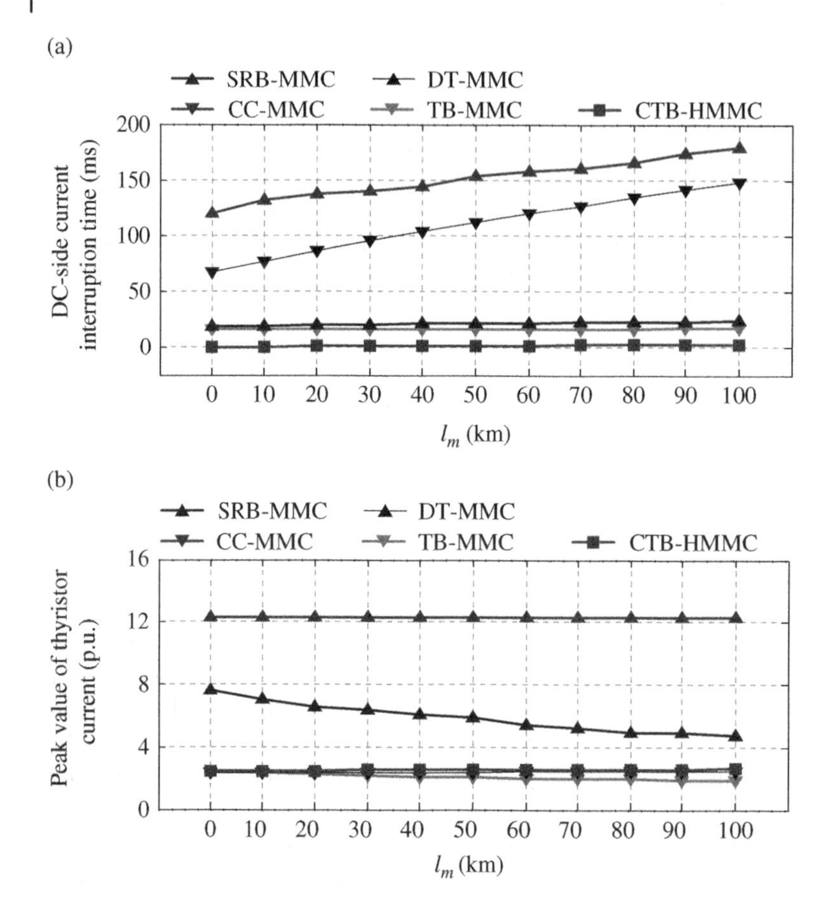

Figure 7.45 Comparison of thyristor based MMCs and CTB-HMMC. (a) Current interruption time under various l_m. (b) Peak value of thyristor current under various l_m.

thyristor current under various l_m. The peak value of thyristor current in the DT-MMC, BT-MMC, and CTB-HMMC are around 2.5 p.u., because the arm inductors are fully utilized to limit the AC-side currents. However, in the SRB-MMC and CC-MMC, the arm inductors are bypassed by the thyristors, which results in higher peak value of thyristor current. It means the SRB-MMC and CC-MMC require the thyristor with much higher current capability in comparison with the DT-MMC, BT-MMC, and the CTB-HMMC.

Table 7.9 shows the total semiconductor cost and the power losses of these MMCs. Since double-thyristors are required in the DT-MMC, thyristors with higher current withstanding ability are required in CC-MMC and SRB-MMC, and the cost of the CTB-HMMC is lower than the cost of the DT-MMC, CC-MMC, BT-MMC, and SRB-MMC.

Case Study 7.3 DC Fault Protection Based on CTB-HMMC

Objective: In this case study, an HVDC system based on CTB-HMMC, as shown in Figure 7.46, is simulated with PSCAD/EMTDC, where the MMC 1 transfers the power from AC to DC side and the MMC 2 keeps the DC link voltage constant.

Parameters: The system parameters are shown in Table 7.10.

Simulation results and analysis:
The DC-line short-circuit fault occurs at t_0 = 6.5 seconds and the fault point is 50 km far away from the CTB-HMMC 1 in the HVDC system. Here, the CTB-HMMC 1 transfers active power of 350 MW to the CTB-HMMC 2. Figure 7.47a–h shows the performance of the CTB-HMMC 1.

1) t_0–t_1: After the DC fault occurs at t_0 = 6.5 seconds, the DC-side voltage V_{dc} drops and the DC-side current i_{dc} increases, as shown in Figure 7.47a–d.

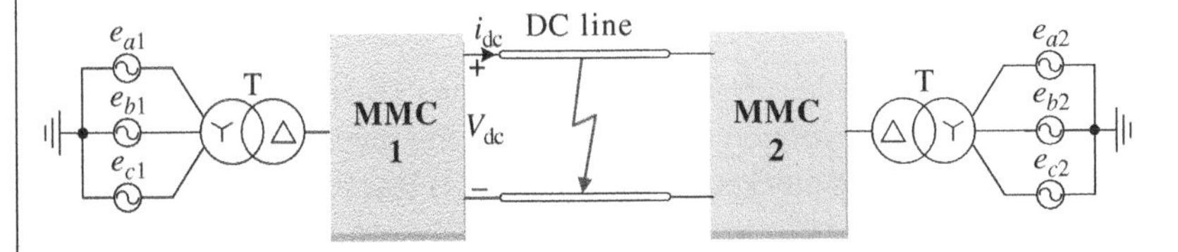

Figure 7.46 Diagram of the simulation system.

Table 7.10 Parameters of the simulated system.

Parameters	Values
Rated capacity P_N	500 MW
Rated DC-side voltage	320 kV
Rated AC-side voltage	166 kV
Arm inductor L_s	80 mH
Number of SMs per arm n	200
Number of UFB SMs per arm m	28
Number of HB SMs per arm n-m	172
SM capacitance C_{sm}	10 mF
Rated thyristor voltage	4 kV
Number of series thyristors per branch	200
DC line distance	100 km
Equivalent inductance of the DC lines	1.1 mH/km
Equivalent resistance of the DC lines	9.5 mΩ/km
DC-side short-circuit resistance R_{fr}	1 Ω

2) t_1–t_2: When the fault is detected at t_1, the CTB-HMMC 1 would work in the DC-side current interruption operation, where a reverse voltage produced by the UFB SMs is injected to the short-circuit path, as shown in Figure 7.47a, b. Consequently, the UFB SMs in the MMC would absorb the line energy in the short-circuit path and the capacitor voltages in the UFB SMs increase, while the capacitor voltages in the HB SMs are not changed because the capacitors in HB SMs are bypassed by the freewheeling diodes. Figure 7.47e,f shows the capacitor voltages of the HB SMs and UFB SMs in the upper and lower arms of phase A, respectively. The upper arm UFB SM capacitor voltage is only increased by 0.0069 p.u. and the lower arm UFB SM capacitor voltage is only increased by 0.08 p.u. In the meantime, along with the increase of capacitor voltages in the UFB SMs, the DC-side fault current would be rapidly reduced, as shown in Figure 7.47c,d, where the DC-side current i_{dc} can be interrupted in only 1.4 ms. Afterwards, the mechanical switches at the DC terminal are opened to isolate the CTB-HMMC 1 from the faulty line.
3) t_2–t_3: After the DC-side current i_{dc} reaches 0, the CTB-HMMC 1 works in the AC-side current interruption operation, where the thyristor branch without gate signal would be blocked when its current is less than the holding current of the thyristor, as shown in Figure 7.47g. As a result, the AC-side phase currents i_a, i_b, and i_c are interrupted in 17.8 ms, as shown in Figure 7.47h.

Suppose that the fault is cleared at 6.542 seconds, the mechanical switches at the DC terminal of the MMC would be closed again to reconnect the CTB-HMMC and the DC line. Afterwards, the CTB-HMMC based HVDC system recovers to its normal operation, where the DC-link voltage V_{dc} is rebuilt and the power transmission is restored.

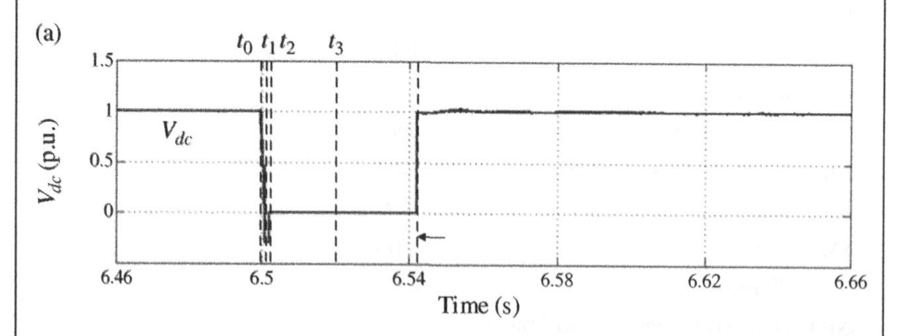

Figure 7.47 Simulation waveforms. (a) V_{dc}. (b) V_{dc} in short time scale. (c) i_{dc}. (d) i_{dc} in short time scale. (e) Capacitor voltages of HB SMs and UFB SMs in upper arm of phase A. (f) Capacitor voltages of HB SMs and UFB SMs in lower arm of phase A. (g) Thyristor branches TB_{1-6} currents i_{TB1-6}. (h) i_a, i_b, and i_c.

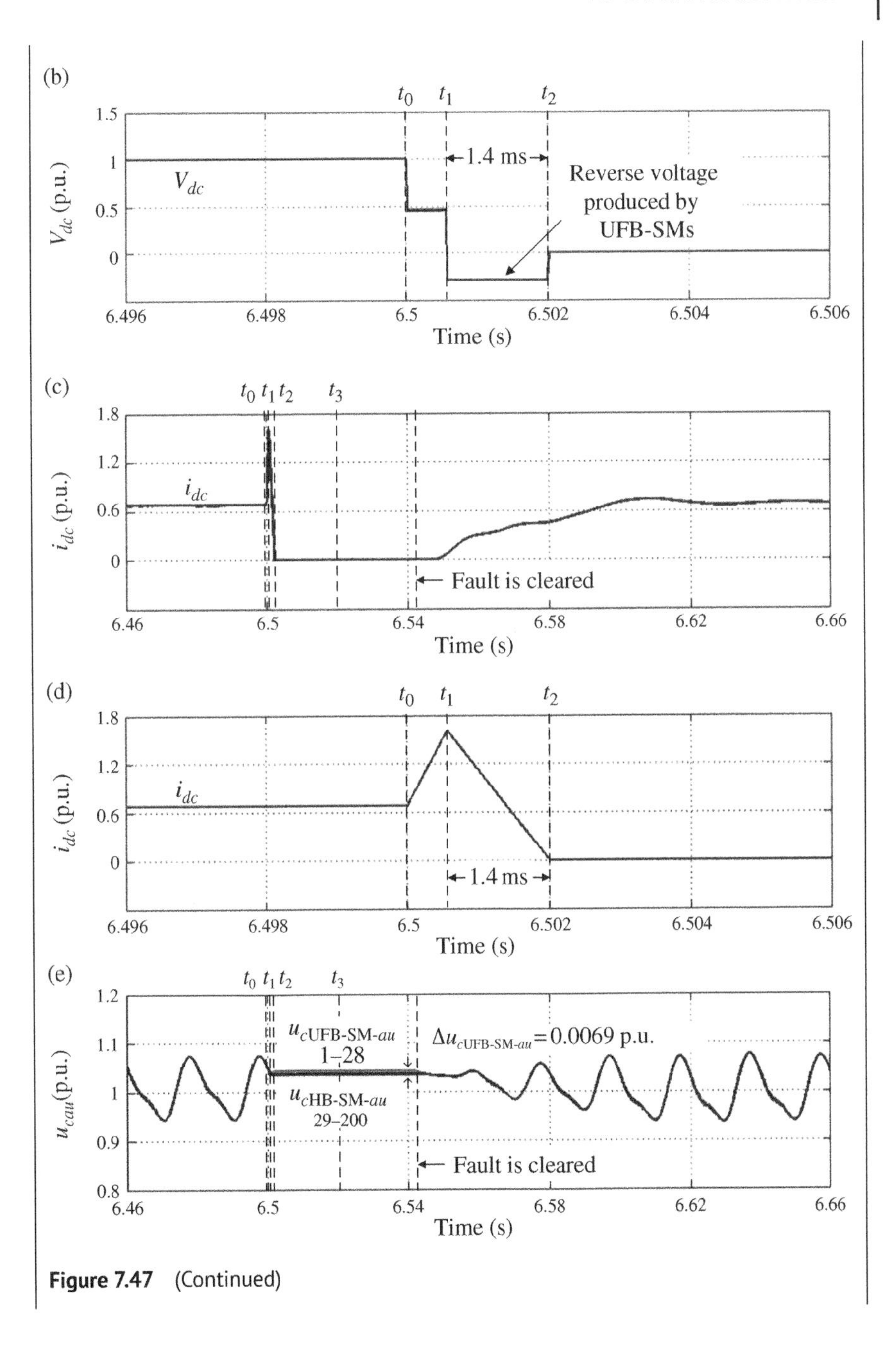

Figure 7.47 (Continued)

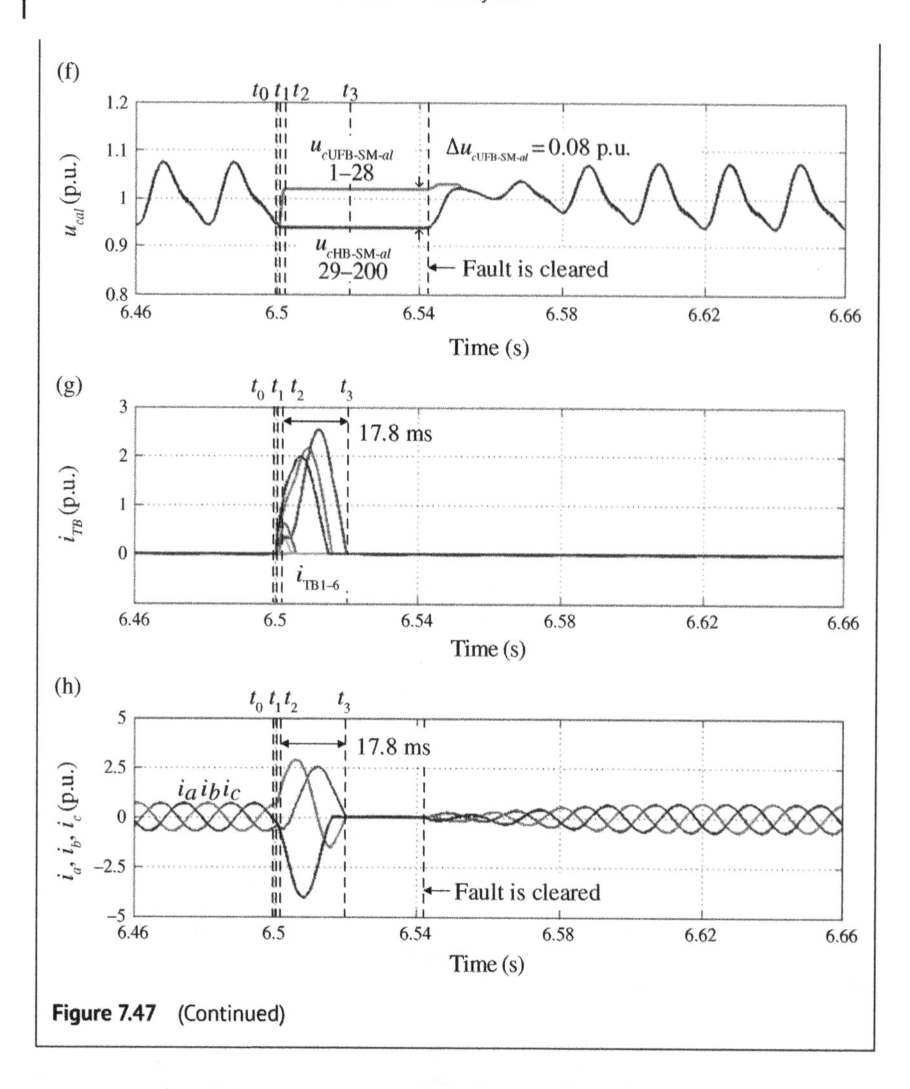

Figure 7.47 (Continued)

7.7 Summary

The DC line short-circuit fault poses challenges to the MMC based HVDC system because the DC voltage collapses and the DC current rising rate in DC grid is high after DC fault occurs. The surge DC fault current can cause damage to the semiconductor devices and other components of the MMC; therefore, it should be interrupted within several milliseconds. However, the HB SM based MMC lacks the ability to handle the DC fault current, which forms a rectifier bridge feeding the fault current to DC fault point after turning off all switches.

This chapter analyzes the performance of the HB SM based MMC under DC fault and presents the protection based on AC CBs. The DC protection method based on various kinds of DC CBs and fault blocking converters are presented, which can interrupt the DC fault current much faster compared with protection based on AC CBs. Furthermore, two advanced thyristor based protections are introduced. The BT MMC is able to interrupt the DC current with the help of natural turning off of the thyristor at current zero-crossing point. The CTB-HMMC bypasses the AC-side voltage and the arm inductors to reduce the UFB SM number required for fault blocking.

References

1 N. Flourentzou, V. G. Agelidis, and G. D. Demetriades, "VSC-based HVDC power transmission systems: an overview," *IEEE Trans. Power Electron.*, vol. 24, no. 3, pp. 592–602, Mar. 2009.

2 J. Yang, J. E. Fletcher, and J. O'Reilly, "Multiterminal DC wind farm collection grid internal fault analysis and protection design," *IEEE Trans. Power Delivery*, vol. 25, no. 4, pp. 2308–2318, Oct. 2010.

3 M. P. Bahrman and B. K. Johnson, "The ABCs of HVDC transmission technologies," *IEEE Power Energ. Mag.*, vol. 5, no. 2, pp. 32–44 Mar.–Apr. 2007.

4 J. Yang, J. E. Fletcher, and J. O'Reilly, "Short-circuit and ground fault analyses and location in VSC-based DC network cables," *IEEE Trans. Ind. Electron.*, vol. 59, no. 10, pp. 3827–3837, Oct. 2012.

5 L. Tang and B. Ooi, "Protection of VSC-multi-terminal HVDC against DC faults," in *2002 IEEE 33rd Annual IEEE Power Electronics Specialists Conference. Proceedings (Cat. No. 02CH37289)*, 2002, pp. 719–724, vol. 2.

6 G. Ding, G. Tang, Z. He, and M. Ding, "New technologies of voltage source converter (VSC) for HVDC transmission system based on VSC," in *2008 IEEE Power and Energy Society General Meeting – Conversion and Delivery of Electrical Energy in the 21st Century*, 2008, pp. 1–8.

7 G. Tang, Z. Xu, and Y. Zhou, "Impacts of three MMC-HVDC configurations on AC system stability under DC line faults," in *2015 IEEE Power & Energy Society General Meeting*, 2015, pp. 1–1.

8 W. Wen *et al.*, "Analysis and experiment of a micro-loss multi-port hybrid DCCB for MVDC distribution system," *IEEE Trans. Power Electron.*, vol. 34, no. 8, pp. 7933–7941, Aug. 2019.

9 G. Liu, F. Xu, Z. Xu, Z. Zhang, and G. Tang, "Assembly HVDC breaker for HVDC grids with modular multilevel converters," *IEEE Trans. Power Electron.*, vol. 32, no. 2, pp. 931–941, Feb. 2017.

10 F. Xu *et al.*, "Topology, control and fault analysis of a new type HVDC breaker for HVDC systems," in *2016 IEEE PES Asia-Pacific Power and Energy Engineering Conference (APPEEC)*, 2016, pp. 1959–1964.

11 E. Kontos, T. Schultz, L. Mackay, L. M. Ramirez-Elizondo, C. M. Franck, and P. Bauer, "Multiline breaker for HVdc applications," *IEEE Trans. Power Delivery*, vol. 33, no. 3, pp. 1469–1478, Jun. 2018.

12 W. Liu *et al.*, "A multi-port circuit breaker based multi-terminal DC system fault protection," *IEEE J. Emerging Sel. Top. Power Electron.*, vol. 7, no. 1, pp. 309–320, Mar. 2019.

13 A. Mokhberdoran, D. Van Hertem, N. Silva, H. Leite, and A. Carvalho, "Multiport hybrid HVDC circuit breaker," *IEEE Trans. Ind. Electron.*, vol. 65, no. 1, pp. 309–320, Jan. 2018.

14 C. Li, J. Liang, and S. Wang, "Interlink hybrid DC circuit breaker," *IEEE Trans. Ind. Electron.*, vol. 65, no. 11, pp. 8677–8686, Nov. 2018.

15 S. Wang, C. E. Ugalde-Loo, C. Li, J. Liang, and O. D. Adeuyi, "Bridge-type integrated hybrid DC circuit breakers," *IEEE J. Emerging Sel. Top. Power Electron.*, vol. 8, no. 2, pp. 1134–1151, Jun. 2020.

16 X. Pei, O. Cwikowski, A. C. Smith, and M. Barnes, "Design and experimental tests of a superconducting hybrid DC circuit breaker," *IEEE Trans. Appl. Supercond.*, vol. 28, no. 3, pp. 1–5 Apr. 2018.

17 B. Xiang, Z. Liu, Y. Geng, and S. Yanabu, "DC circuit breaker using superconductor for current limiting," *IEEE Trans. Appl. Supercond.*, vol. 25, no. 2, pp. 1–7, Apr. 2015.

18 S. Hwang, H. Choi, I. Jeong, and H. Choi, "Characteristics of DC circuit breaker applying transformer-type superconducting fault current limiter," *IEEE Trans. Appl. Supercond.*, vol. 28, no. 4, pp. 1–5, Jun. 2018.

19 K. Yang *et al.*, "Direct-current vacuum circuit breaker with superconducting fault-current limiter," *IEEE Trans. Appl. Supercond.*, vol. 28, no. 1, pp. 1–8, Jan. 2018.

20 K. Yang, H. Ge, X. Wang, Z. Liu, Y. Geng, and J. Wang, "A self-charging artificial current zero DC circuit breaker based on superconducting fault current limiter," *IEEE Trans. Appl. Supercond.*, vol. 29, no. 2, pp. 1–5, Mar. 2019.

21 D. Schmitt, Y. Wang, T. Weyh, and R. Marquardt, "DC-side fault current management in extended multiterminal-HVDC-grids," in *Proceedings of IEEE 9th International Multi-Conference on Systems, Signals, Devices,* 2012, pp. 1–5.

22 S. Cui and S. Sul, "A comprehensive DC short-circuit fault ride through strategy of hybrid modular multilevel converters (MMCs) for overhead line transmission," *IEEE Trans. Power Electron.*, vol. 31, no. 11, pp. 7780–7796, Nov. 2016.

23 R. Li, J. E. Fletcher, L. Xu, D. Holliday, and B. W. Williams, "A hybrid modular multilevel converter with novel three-level cells for DC fault blocking capability," *IEEE Trans. Power Delivery*, vol. 30, no. 4, pp. 1853–1862, Aug. 2015.

24 R. Zeng, L. Xu, and L. Yao, "An improved modular multilevel converter with DC fault blocking capability," in *2014 IEEE PES General Meeting | Conference & Exposition*, National Harbor, MD, USA, 2014, pp. 1–5.

25 R. Marquardt, "Modular multilevel converter topologies with DC-Short circuit current limitation," in *Proceedings of IEEE 8th International Conference on Power Electronics*, ECCE Asia, 2011, pp. 1425–1431.

26 D. Schmitt, Y. Wang, T. Weyh, and R. Marquardt, "DC-side fault current management in extended multiterminal-HVDC-grids," in *Proceedings of IEEE 9th International Multi-Conference on Systems, Signals, Devices*, 2012, pp. 1–5.

27 A. A. Elserougi, A. M. Massoud, and S. Ahmed, "A switched-capacitor submodule for modular multilevel HVDC converters with DC-fault blocking capability and a reduced number of sensors," *IEEE Trans. Power Delivery*, vol. 31, no. 1, pp. 313–322, Feb. 2016.

28 R. Marquardt, "Modular multilevel converter: an universal concept for HVDC-networks and extended DC-bus-applications," in *Proceedings of International Power Electronics Conference*, Jun. 2010, pp. 502–507.

29 J. Qin, M. Saeedifard, A. Rockhill, and R. Zhou, "Hybrid design of modular multilevel converters for HVDC systems based on various submodule circuits," *IEEE Trans. Power Delivery*, vol. 30, no. 1, pp. 385–394, Feb. 2015.

30 M. M. C. Merlin *et al.*, "The alternate arm converter: a new hybrid multilevel converter with DC-fault blocking capability," *IEEE Trans. Power Delivery*, vol. 29, no. 1, pp. 310–317, Feb. 2014.

31 C. Zhao, Y. Li, Z. Li, P. Wang, X. Ma, and Y. Luo, "Optimized design of full-bridge modular multilevel converter with low energy storage requirements for HVdc transmission system," *IEEE Trans. Power Electron.*, vol. 33, no. 1, pp. 97–109, Jan. 2018.

32 A. A. Elserougi, A. S. Abdel-Khalik, A. M. Massoud, and S. Ahmed, "A new protection scheme for HVDC converters against DC-side faults with current suppression capability," *IEEE Trans. Power Delivery*, vol. 29, no. 4, pp. 1569–1577, Aug. 2014.

33 X. Li, Q. Song, W. Liu, H. Rao, S. Xu and L. Li, "Protection of nonpermanent faults on DC overhead lines in MMC-based HVDC systems," *IEEE Trans. Power Delivery*, vol. 28, no. 1, pp. 483–490, Jan. 2013.

34 Q. Wang, F. Deng, C. Liu, Q. Heng, and Z. Chen, "Thyristor-based modular multilevel converter-HVDC systems with current interruption capability," *IET Power Electron.*, vol. 12, no. 12, pp. 3056–3067, Oct. 2019.

35 I. A. Gowaid, "A low-loss hybrid bypass for DC fault protection of modular multilevel converters," *IEEE Trans. Power Delivery*, vol. 32, no. 2, pp. 599–608, Apr. 2017.

Index

Modular Multilevel Converters: Control, Fault Detection, and Protection, First Edition.
Fujin Deng, Chengkai Liu, and Zhe Chen.

Published 2023 by John Wiley & Sons, Inc.

d

g

h

t

IEEE Press Series on Power and Energy Systems

Series Editor: Ganesh Kumar Venayagamoorthy, Clemson University, Clemson, South Carolina, USA.

The mission of the IEEE Press Series on Power and Energy Systems is to publish leading-edge books that cover a broad spectrum of current and forward-looking technologies in the fast-moving area of power and energy systems including smart grid, renewable energy systems, electric vehicles and related areas. Our target audience includes power and energy systems professionals from academia, industry and government who are interested in enhancing their knowledge and perspectives in their areas of interest.

1. *Electric Power Systems: Design and Analysis, Revised Printing*
   Mohamed E. El-Hawary
2. *Power System Stability*
   Edward W. Kimbark
3. *Analysis of Faulted Power Systems*
   Paul M. Anderson
4. *Inspection of Large Synchronous Machines: Checklists, Failure Identification, and Troubleshooting*
   Isidor Kerszenbaum
5. *Electric Power Applications of Fuzzy Systems*
   Mohamed E. El-Hawary
6. *Power System Protection*
   Paul M. Anderson
7. *Subsynchronous Resonance in Power Systems*
   Paul M. Anderson, B.L. Agrawal, J.E. Van Ness
8. *Understanding Power Quality Problems: Voltage Sags and Interruptions*
   Math H. Bollen
9. *Analysis of Electric Machinery*
   Paul C. Krause, Oleg Wasynczuk, and S.D. Sudhoff
10. *Power System Control and Stability, Revised Printing*
    Paul M. Anderson, A.A. Fouad
11. *Principles of Electric Machines with Power Electronic Applications, Second Edition*
    Mohamed E. El-Hawary

12. *Pulse Width Modulation for Power Converters: Principles and Practice*
D. Grahame Holmes and Thomas Lipo
13. *Analysis of Electric Machinery and Drive Systems, Second Edition*
Paul C. Krause, Oleg Wasynczuk, and S.D. Sudhoff
14. *Risk Assessment for Power Systems: Models, Methods, and Applications*
Wenyuan Li
15. *Optimization Principles: Practical Applications to the Operations of Markets of the Electric Power Industry*
Narayan S. Rau
16. *Electric Economics: Regulation and Deregulation*
Geoffrey Rothwell and Tomas Gomez
17. *Electric Power Systems: Analysis and Control*
Fabio Saccomanno
18. *Electrical Insulation for Rotating Machines: Design, Evaluation, Aging, Testing, and Repair*
Greg C. Stone, Edward A. Boulter, Ian Culbert, and Hussein Dhirani
19. *Signal Processing of Power Quality Disturbances*
Math H. J. Bollen and Irene Y. H. Gu
20. *Instantaneous Power Theory and Applications to Power Conditioning*
Hirofumi Akagi, Edson H.Watanabe and Mauricio Aredes
21. *Maintaining Mission Critical Systems in a 24/7 Environment*
Peter M. Curtis
22. *Elements of Tidal-Electric Engineering*
Robert H. Clark
23. *Handbook of Large Turbo-Generator Operation and Maintenance, Second Edition*
Geoff Klempner and Isidor Kerszenbaum
24. *Introduction to Electrical Power Systems*
Mohamed E. El-Hawary
25. *Modeling and Control of Fuel Cells: Distributed Generation Applications*
M. Hashem Nehrir and Caisheng Wang
26. *Power Distribution System Reliability: Practical Methods and Applications*
Ali A. Chowdhury and Don O. Koval
27. *Introduction to FACTS Controllers: Theory, Modeling, and Applications*
Kalyan K. Sen, Mey Ling Sen

28. *Economic Market Design and Planning for Electric Power Systems*
James Momoh and Lamine Mili
29. *Operation and Control of Electric Energy Processing Systems*
James Momoh and Lamine Mili
30. *Restructured Electric Power Systems: Analysis of Electricity Markets with Equilibrium Models*
Xiao-Ping Zhang
31. *An Introduction to Wavelet Modulated Inverters*
S.A. Saleh and M.A. Rahman
32. *Control of Electric Machine Drive Systems*
Seung-Ki Sul
33. *Probabilistic Transmission System Planning*
Wenyuan Li
34. *Electricity Power Generation: The Changing Dimensions*
Digambar M. Tagare
35. *Electric Distribution Systems*
Abdelhay A. Sallam and Om P. Malik
36. *Practical Lighting Design with LEDs*
Ron Lenk, Carol Lenk
37. *High Voltage and Electrical Insulation Engineering*
Ravindra Arora and Wolfgang Mosch
38. *Maintaining Mission Critical Systems in a 24/7 Environment, Second Edition*
Peter Curtis
39. *Power Conversion and Control of Wind Energy Systems*
Bin Wu, Yongqiang Lang, Navid Zargari, Samir Kouro
40. *Integration of Distributed Generation in the Power System*
Math H. Bollen, Fainan Hassan
41. *Doubly Fed Induction Machine: Modeling and Control for Wind Energy Generation Applications*
Gonzalo Abad, Jesús López, Miguel Rodrigues, Luis Marroyo, and Grzegorz Iwanski
42. *High Voltage Protection for Telecommunications*
Steven W. Blume
43. *Smart Grid: Fundamentals of Design and Analysis*
James Momoh

44. *Electromechanical Motion Devices, Second Edition*
Paul Krause Oleg, Wasynczuk, Steven Pekarek

45. *Electrical Energy Conversion and Transport: An Interactive Computer-Based Approach, Second Edition*
George G. Karady, Keith E. Holbert

46. *ARC Flash Hazard and Analysis and Mitigation*
J.C. Das

47. *Handbook of Electrical Power System Dynamics: Modeling, Stability, and Control*
Mircea Eremia, Mohammad Shahidehpour

48. *Analysis of Electric Machinery and Drive Systems, Third Edition*
Paul C. Krause, Oleg Wasynczuk, S.D. Sudhoff, Steven D. Pekarek

49. *Extruded Cables for High-Voltage Direct-Current Transmission: Advances in Research and Development*
Giovanni Mazzanti, Massimo Marzinotto

50. *Power Magnetic Devices: A Multi-Objective Design Approach*
S.D. Sudhoff

51. *Risk Assessment of Power Systems: Models, Methods, and Applications, Second Edition*
Wenyuan Li

52. *Practical Power System Operation*
Ebrahim Vaahedi

53. *The Selection Process of Biomass Materials for the Production of Bio-Fuels and Co-Firing*
Najib Altawell

54. *Electrical Insulation for Rotating Machines: Design, Evaluation, Aging, Testing, and Repair, Second Edition*
Greg C. Stone, Ian Culbert, Edward A. Boulter, and Hussein Dhirani

55. *Principles of Electrical Safety*
Peter E. Sutherland

56. *Advanced Power Electronics Converters: PWM Converters Processing AC Voltages*
Euzeli Cipriano dos Santos Jr., Edison Roberto Cabral da Silva

57. *Optimization of Power System Operation, Second Edition*
Jizhong Zhu

58. *Power System Harmonics and Passive Filter Designs*
J.C. Das

59. *Digital Control of High-Frequency Switched-Mode Power Converters*
Luca Corradini, Dragan Maksimoviæ, Paolo Mattavelli, Regan Zane

60. *Industrial Power Distribution, Second Edition*
    Ralph E. Fehr, III
61. *HVDC Grids: For Offshore and Supergrid of the Future*
    Dirk Van Hertem, Oriol Gomis-Bellmunt, Jun Liang
62. *Advanced Solutions in Power Systems: HVDC, FACTS, and Artificial Intelligence*
    Mircea Eremia, Chen-Ching Liu, Abdel-Aty Edris
63. *Operation and Maintenance of Large Turbo-Generators*
    Geoff Klempner, Isidor Kerszenbaum
64. *Electrical Energy Conversion and Transport: An Interactive Computer-Based Approach*
    George G. Karady, Keith E. Holbert
65. *Modeling and High-Performance Control of Electric Machines*
    John Chiasson
66. *Rating of Electric Power Cables in Unfavorable Thermal Environment*
    George J. Anders
67. *Electric Power System Basics for the Nonelectrical Professional*
    Steven W. Blume
68. *Modern Heuristic Optimization Techniques: Theory and Applications to Power Systems*
    Kwang Y. Lee, Mohamed A. El-Sharkawi
69. *Real-Time Stability Assessment in Modern Power System Control Centers*
    Savu C. Savulescu
70. *Optimization of Power System Operation*
    Jizhong Zhu
71. *Insulators for Icing and Polluted Environments*
    Masoud Farzaneh, William A. Chisholm
72. *PID and Predictive Control of Electric Devices and Power Converters Using MATLAB®/Simulink®*
    Liuping Wang, Shan Chai, Dae Yoo, Lu Gan, Ki Ng
73. *Power Grid Operation in a Market Environment: Economic Efficiency and Risk Mitigation*
    Hong Chen
74. *Electric Power System Basics for Nonelectrical Professional, Second Edition*
    Steven W. Blume
75. *Energy Production Systems Engineering Thomas*
    Howard Blair

76. *Model Predictive Control of Wind Energy Conversion Systems*
Venkata Yaramasu, Bin Wu

77. *Understanding Symmetrical Components for Power System Modeling*
J.C. Das

78. *High-Power Converters and AC Drives, Second Edition*
Bin Wu, Mehdi Narimani

79. *Current Signature Analysis for Condition Monitoring of Cage Induction Motors: Industrial Application and Case Histories*
William T. Thomson, Ian Culbert

80. *Introduction to Electric Power and Drive Systems*
Paul Krause, Oleg Wasynczuk, Timothy O'Connell, Maher Hasan

81. *Instantaneous Power Theory and Applications to Power Conditioning, Second Edition*
Hirofumi, Edson Hirokazu Watanabe, Mauricio Aredes

82. *Practical Lighting Design with LEDs, Second Edition*
Ron Lenk, Carol Lenk

83. *Introduction to AC Machine Design*
Thomas A. Lipo

84. *Advances in Electric Power and Energy Systems: Load and Price Forecasting*
Mohamed E. El-Hawary

85. *Electricity Markets: Theories and Applications*
Jeremy Lin, Jernando H. Magnago

86. *Multiphysics Simulation by Design for Electrical Machines, Power Electronics and Drives*
Marius Rosu, Ping Zhou, Dingsheng Lin, Dan M. Ionel, Mircea Popescu, Frede Blaabjerg, Vandana Rallabandi, David Staton

87. *Modular Multilevel Converters: Analysis, Control, and Applications*
Sixing Du, Apparao Dekka, Bin Wu, Navid Zargari

88. *Electrical Railway Transportation Systems*
Morris Brenna, Federica Foiadelli, Dario Zaninelli

89. *Energy Processing and Smart Grid*
James A. Momoh

90. *Handbook of Large Turbo-Generator Operation and Maintenance, 3rd Edition*
Geoff Klempner, Isidor Kerszenbaum

91. *Advanced Control of Doubly Fed Induction Generator for Wind Power Systems*
Dehong Xu, Dr. Frede Blaabjerg, Wenjie Chen, Nan Zhu

92. *Electric Distribution Systems, 2nd Edition*
Abdelhay A. Sallam, Om P. Malik

93. *Power Electronics in Renewable Energy Systems and Smart Grid: Technology and Applications*
Bimal K. Bose

94. *Distributed Fiber Optic Sensing and Dynamic Rating of Power Cables*
Sudhakar Cherukupalli, and George J Anders

95. *Power System and Control and Stability, Third Edition*
Vijay Vittal, James D. McCalley, Paul M. Anderson, and A. A. Fouad

96. *Electromechanical Motion Devices: Rotating Magnetic Field-Based Analysis and Online Animations, Third Edition*
Paul Krause, Oleg Wasynczuk, Steven D. Pekarek, and Timothy O'Connell

97. *Applications of Modern Heuristic Optimization Methods in Power and Energy Systems*
Kwang Y. Lee and Zita A. Vale

98. *Handbook of Large Hydro Generators: Operation and Maintenance*
Glenn Mottershead, Stefano Bomben, Isidor Kerszenbaum, and Geoff Klempner

99. *Advances in Electric Power and Energy: Static State Estimation*
Mohamed E. El-hawary

100. *Arc Flash Hazard Analysis and Mitigation, Second Edition*
J.C. Das

101. *Maintaining Mission Critical Systems in a 24/7 Environment, Third Edition*
Peter M. Curtis

102. *Real-Time Electromagnetic Transient Simulation of AC-DC Networks*
Venkata Dinavahi and Ning Lin

103. *Probabilistic Power System Expansion Planning with Renewable Energy Resources and Energy Storage Systems*
Jaeseok Choi and Kwang Y. Lee

104. *Power Magnetic Devices: A Multi-Objective Design Approach, Second Edition*
Scott D. Sudhoff

105. *Optimal Coordination of Power Protective Devices with Illustrative Examples*
Ali R. Al-Roomi

106. *Resilient Control Architectures and Power Systems*
Craig Rieger, Ronald Boring, Brian Johnson, and Timothy McJunkin

107. *Alternative Liquid Dielectrics for High Voltage Transformer Insulation Systems: Performance Analysis and Applications*
Edited by U. Mohan Rao, I. Fofana, and R. Sarathi

108. *Introduction to the Analysis of Electromechanical Systems*
Paul C. Krause, Oleg Wasynczuk, and Timothy O'Connell

109. *Power Flow Control Solutions for a Modern Grid using SMART Power Flow Controllers*
Kalyan K. Sen and Mey Ling Sen

110. *Power System Protection: Fundamentals and Applications*
John Ciufo and Aaron Cooperberg

111. *Soft-Switching Technology for Three-phase Power Electronics Converters*
Dehong Xu, Rui Li, Ning He, Jinyi Deng, and Yuying Wu

112. *Power System Protection, Second Edition*
Paul M. Anderson, Charles Henville, Rasheek Rifaat, Brian Johnson, and Sakis Meliopoulos

113. *High Voltage and Electrical Insulation Engineering, Second Edition*
Ravindra Arora and Wolfgang Mosch.

114. *Modeling and Control of Modern Electrical Energy Systems*
Masoud Karimi-Ghartemani

115. *Control of Power Electronic Converters with Microgrid Applications*
Armdam Ghosh and Firuz Zare

116. *Coordinated Operation and Planning of Modern Heat and Electricity Incorporated Networks*
Mohammadreza Daneshvar, Behnam Mohammadi-Ivatloo, and Kazem Zare

117. *Smart Energy for Transportation and Health in a Smart City*
Chun Sing Lai, Loi Lei Lai, and Qi Hong Lai

118. *Wireless Power Transfer: Principles and Applications*
Zhen Zhang and Hongliang Pang

119. *Intelligent Data Mining and Analysis in Power and Energy Systems: Models and Applications for Smarter Efficient Power Systems*
Zita Vale, Tiago Pinto, Michael Negnevitsky, and Ganesh Kumar Venayagamoorthy

120. *Introduction to Modern Analysis of Electric Machines and Drives*
Paul C. Krause and Thomas C. Krause

121. *Electromagnetic Analysis and Condition Monitoring of Synchronous Generators*
Hossein Ehya and Jawad Faiz.

122. *Transportation Electrification: Breakthroughs in Electrified Vehicles, Aircraft, Rolling Stock, and Watercraft*
Ahmed A. Mohamed, Ahmad Arshan Khan, Ahmed T. Elsayed, Mohamed A. Elshaer

123. *Modular Multilevel Converters: Control, Fault Detection, and Protection*
Fuji Deng, Chengkai Liu, and Zhe Chen

Printed and bound by CPI Group (UK) Ltd, Croydon, CR0 4YY

16/04/2025

14658583-0003